International Workshop on

APPLIED DIFFERENTIAL EQUATIONS

International Workshop on

APPLIED DIFFERENTIAL EQUATIONS

4-7 June 1985
Beijing, China

Edited by

Xiao Shutie
Pu Fuquan

World Scientific

Published by

World Scientific Publishing Co Pte Ltd.
P. O. Box 128, Farrer Road, Singapore 9128.
242, Cherry Street, Philadelphia PA 19106-1906, USA

Library of Congress Cataloging-in-Publication Data

International Workshop on Applied Differential Equations.

1. Differential equations--Congresses. I. Xiao, Shutie. II. Pu, Fuquan. III. Title.
QA370.I58 1985 515.3'5 86-23337
ISBN 9971-50-143-0

Printed in Singapore by Kim Hup Lee Printing Co. Pte. Ltd.

PREFACE

The international workshop on Applied Differential Equations was held at and sponsored by Tsinghua University, Beijing, People's Republic of China from June 4 to June 7, 1985. It was the first time to organize such an international conference in mathematical field at Tsinghua University since 1976. There are in all 36 papers collected in this proceedings in which a greater part of them are printed in full and the others abstracts.

Since it was a workshop on allied differential equations, so the topics were diversified in various fields relating to both ODE and PDE, the problems were either theoretical or computational including Navier-Stokes equations in fluid mechanics, porous media equations in filtration problems, boundary and eigenvalue problems, nonlinear evolution equations, nonlinear hyperbolic equations, control theory and rational mechanics.

The academic committee of this Workshop consisted of Professors L. Collatz, W.N. Everitt, Xiao Shutie, Dong Jingzhu (Tung Chin-Chu), Li Daqian, Jiang Li-Shang and Li Yishen. In virtue of their efforts, the program of the conference was smoothly prepared and successfully proceeded.

The invited one hour speakers are: Professors J. Abrachet (West Germany), M. Chavent (France), L. Collatz (West Germany), Feng Kang (PRC), A. Friedman (USA), K.H. Hoffman (West Germany), Li Yishen (PRC), J. Mawhin (Belgium), A.B. Mingarelli (Canada), C.D. Pagani (Italy), L. Raphael (USA), R. Rautmann (West Germany), D.G. Schaffer (USA), Zhou Yulin (PRC).

We thank all the participants from six foreign countries and various areas of China for their assistance in attending the conference. Thanks are also due to the graduate students who did a lot in organizing the conference.

There are two points we would like to mention. We include the paper of D.G. Schaeffer and M. Shearer though Professor Schaeffer could not attend the workshop. The theses of A. Friedman and K.H. Hoffman are only abstracts. Their full paper can be found elsewhere.

April 7, 1986

Editors: Xiao Shutie
Pu Fuquan

CONTENTS

EINE EINHEITLICHE HERLEITUNG VON EINSCHLIESSUNGSSÄTZEN FÜR EIGENWERTE

Friedrich Goerisch und Julius Albrecht

Symmetrisable eigenvalue problems of the form $M\phi = \lambda N\phi$ are considered. A simple method for obtaining inclusion theorems for eigenvalues is presented. It consists in combining a fundamental inclusion theorem with procedures for calculating upper bounds for the norm of the solutions of certain linear equations, $Nw = h$ or $M\tilde{w} = \tilde{h}$. Thus, numerous inclusion theorems are derived in a uniform manner.

Inhalt

Einleitung: Die Methode zur Herleitung von Einschließungssätzen

Für eine Klasse allgemeiner Eigenwertaufgaben $M\phi = \lambda N\phi$ wird eine einfache Methode zur Aufstellung von Einschließungssätzen für Eigenwerte geschildert. Sie besteht darin, Verfahren, welche die Berechnung oberer Schranken für die Norm der Lösung gewisser linearer Gleichungen $Nw = h$ oder $M\tilde{w} = \tilde{h}$ ermöglichen, mit einem grundlegenden Einschließungssatz zu kombinieren.

§1 Zunächst werden einige Voraussetzungen zusammengestellt, die für die gesamte Arbeit gelten.

Voraussetzung V1

1. $(H, \langle\cdot|\cdot\rangle)$ sei ein reeller Hilbertraum.
2. $D(M)$, $D(N)$ und $D(S)$ seien reelle Vektorräume mit $D(M) \subset D(N) \subset D(S)$.
3. $M: D(M) \to H$, $N: D(N) \to H$ und $S: D(S) \to H$ seien lineare Abbildungen.
4. Für alle $f,g \in D(N)$ gelte $\langle Sf|Ng\rangle = \langle Sg|Nf\rangle$; für alle $f \in D(N)$ mit $f \neq 0$ gelte $\langle Sf|Nf\rangle > 0$.
5. Für alle $f,g \in D(N^{-1}M)$ [1] gelte $\langle Sf|Mg\rangle = \langle Sg|Mf\rangle$.

Ähnliche Voraussetzungen wurden von Schäfke und Schneider [48], [49] bei theoretischen Untersuchungen über Eigenwertaufgaben mit Systemen gewöhnlicher Differentialgleichungen verwendet. - Bei vielen Aufgaben ist $D(N) \subset H$, und dann kann meistens S als die identische Abbildung I gewählt werden. Hier wird $S = I$ jedoch nicht vorausgesetzt, so daß auch Eigenwertaufgaben mit Differentialgleichungen, bei denen der Eigenwert in den Randbedingungen auftritt, erfaßt werden können; dies soll an Hand einer Stekloffschen Eigenwertaufgabe erläutert werden:

Beispiel 1 [2]

Eigenwertaufgabe: $\Delta\phi = 0$ in Ω, $\frac{\partial\phi}{\partial n} = \lambda\phi$ auf $\partial\Omega$

Einordnung: $H := L_2(\partial\Omega)$, $\langle f|g\rangle := \int_{\partial\Omega} fg\,ds$ für $f,g \in H$;

$D(M) := D(N) := D(S) := \{f \in C^\infty(\bar{\Omega}): \Delta f = 0 \text{ in } \Omega\}$,

$Mf := \frac{\partial f}{\partial n}$, $Nf := Sf := f_{|\partial\Omega}$.

---

1) Ist P eine lineare Abbildung, so bezeichnet D(P) ihren Definitionsbereich.

2) Im folgenden bezeichnet Ω ein beschränktes, glattes (vgl. [5], S. 190, Def. 3-1) Gebiet im $\mathbb{R}^2$, $\partial\Omega$ den Rand von Ω und $\frac{\partial}{\partial n}$ die Ableitung in Richtung der äußeren Normalen von $\partial\Omega$.

Definition

Auf D(N) werden durch

$\{f|g\} := \langle Sf|Ng\rangle$ für $f,g \in D(N)$, $\|f\| := \sqrt{\langle Sf|Nf\rangle}$ für $f \in D(N)$

ein Skalarprodukt und eine Norm definiert.

Voraussetzung V2

Es gebe eine Folge $(\lambda_i)_{i\in\mathbb{N}}$ von Eigenwerten und eine Folge $(\phi_i)_{i\in\mathbb{N}}$ von Eigenelementen der Aufgabe $M\phi = \lambda N\phi$ mit der Eigenschaft, daß

$M\phi_i = \lambda_i N\phi_i$, $\{\phi_i|\phi_k\} = \delta_{ik}$ für $i,k \in \mathbb{N}$ (δ_{ik} Kronecker-Symbol)

und $\{f|f\} = \sum_{i=1}^{\infty} \{f|\phi_i\}^2$ für alle $f \in D(N^{-1}M)$ gilt.

Die Grundlage für alle weiteren Überlegungen bildet der folgende Einschließungssatz:

Satz 1

Es seien $\varrho,\tau \in \mathbb{R}$; U sei ein r-dimensionaler Unterraum von $D(N^{-1}M)$, und für alle $u \in U$ gelte $\|N^{-1}Mu - \varrho u\| \leq \tau\|u\|$.

Dann liegen mindestens r Eigenwerte[1)] der Aufgabe $M\phi = \lambda N\phi$ im Intervall $[\varrho-\tau,\varrho+\tau]$.

Beweis: Annahme: Das Intervall $[\varrho-\tau,\varrho+\tau]$ enthält höchstens r-1 Eigenwerte von $M\phi = \lambda N\phi$. E sei der Raum, der von denjenigen Eigenelementen aufgespannt wird, die zu den in $[\varrho-\tau,\varrho+\tau]$ gelegenen Eigenwerten gehören; da seine Dimension höchstens r-1 ist, gibt es Elemente $u \in U$ mit

$u \neq 0$, $\{u|\phi_i\} = 0$ für alle $i \in \mathbb{N}$ mit $\lambda_i \in [\varrho-\tau,\varrho+\tau]$.

Ein solches u sei fest gewählt; nach der Besselschen Ungleichung gilt

$$\sum_{i=1}^{\infty} (\lambda_i-\varrho)^2\{u|\phi_i\}^2 \leq \{N^{-1}Mu-\varrho u|N^{-1}Mu-\varrho u\},$$

und aus $\|N^{-1}Mu-\varrho u\| \leq \tau\|u\|$ folgt mit Hilfe der Voraussetzung V2

$$\sum_{\substack{i\in\mathbb{N}\\ \lambda_i\notin[\varrho-\tau,\varrho+\tau]}} (\lambda_i-\varrho)^2\{u|\phi_i\}^2 \leq \tau^2 \sum_{\substack{i\in\mathbb{N}\\ \lambda_i\notin[\varrho-\tau,\varrho+\tau]}} \{u|\phi_i\}^2 .$$

---

1) Eigenwerte werden stets entsprechend ihrer Vielfachheit gezählt.

Da $(\lambda_i-\varrho)^2 > \tau^2$ für alle $i \in \mathbb{N}$ mit $\lambda_i \notin [\varrho-\tau,\varrho+\tau]$ gilt, ergibt sich $\{u|\phi_i\}^2 = 0$ für alle $i \in \mathbb{N}$. Dies steht im Widerspruch zu $u \neq 0$.

Für bestimmte Zwecke ist eine andere Fassung des grundlegenden Einschließungssatzes etwas günstiger:

Satz 2

M sei injektiv. Es seien $\varrho,\tau \in \mathbb{R}$ mit $0 \leq \tau < |\varrho|$; V sei ein r-dimensionaler Unterraum von $D(M^{-1}N)$, und für alle $v \in V$ gelte

$$\|v - \frac{\varrho^2-\tau^2}{\varrho} M^{-1}Nv\| \leq \frac{\tau}{|\varrho|} \|v\|. \qquad (1)$$

Dann liegen mindestens r Eigenwerte der Aufgabe $M\phi = \lambda N\phi$ im Intervall $[\varrho-\tau,\varrho+\tau]$.

Beweis (durch Zurückführung auf Satz 1): $U := \{M^{-1}Nv: v \in V\}$ ist ein r-dimensionaler Unterraum von $D(N^{-1}M)$. Gilt $u \in U$ und setzt man $v := N^{-1}Mu$, so folgt wegen $v \in V$ zunächst

$$\{v - \frac{\varrho^2-\tau^2}{\varrho} u \mid v - \frac{\varrho^2-\tau^2}{\varrho} u\} \leq \frac{\tau^2}{\varrho^2}\{v|v\}$$ und hieraus dann

$$\{v-\varrho u|v-\varrho u\} \leq \tau^2\{u|u\}, \text{ also } \|N^{-1}Mu-\varrho u\| \leq \tau \|u\|.$$

Bemerkung: Sind die Voraussetzungen von Satz 2 erfüllt und ist ϱ positiv, so ergibt sich bei Verwendung der Abkürzungen

$$\lambda_* := \frac{\varrho^2-\tau^2}{\varrho}, \quad \varepsilon_* := \frac{\tau}{\varrho}$$

für das Einschließungsintervall die Darstellung

$$[\varrho-\tau,\varrho+\tau] = [\frac{\lambda_*}{1+\varepsilon_*}, \frac{\lambda_*}{1-\varepsilon_*}] .$$

§2 Die Einschließungssätze für Eigenwerte können folgendermaßen eingeteilt werden:

I Sätze, bei denen Paare (u,v) mit $u \in D(M)$, $v \in D(N)$, $Mu = Nv$ zur Einschließung benötigt werden,

II Sätze, bei deren Anwendung man zwar Elemente u aus $D(N^{-1}M)$ benutzen muß, $v := N^{-1}Mu$ aber nicht auszurechnen braucht, und

III Sätze, welche eine Bestimmung von Einschließungsintervallen ohne Verwendung von Elementen aus $D(N^{-1}M)$ ermöglichen.

Die Methode zur Herleitung von Einschließungssätzen besteht darin, eine der beiden Fassungen des grundlegenden Satzes aus §1 mit Abschätzungen für die Norm der Lösung linearer Gleichungen zu kombinieren. (Einen Überblick vermittelt ein Diagramm am Schluß der Arbeit.)

Soll ein unter II fallender Satz hergeleitet werden, so verwendet man Satz 1 zum Beweis, faßt die dort auftretende Größe $N^{-1}Mu - \varrho u$ als Lösung w der linearen Gleichung

$Nw = h$ (mit $h := Mu - \varrho Nu$)

auf und zieht zum Nachweis, daß die Voraussetzungen von Satz 1 erfüllt sind, ein Verfahren heran, das obere Schranken für $\|w\|$ liefert.

Wenn ein unter III fallender Satz hergeleitet werden soll, benutzt man Satz 2 zum Beweis, sieht die in (1) auftretende Größe $v - (\varrho^2-\tau^2)\varrho^{-1}M^{-1}Nv$ als Lösung $\tilde{w}$ der linearen Gleichung

$M\tilde{w} = \tilde{h}$ (mit $\tilde{h} := Mv - (\varrho^2-\tau^2)\varrho^{-1}Nv$)

an - dies ist natürlich nur unter der Zusatzvoraussetzung $v \in D(M)$ möglich - und macht beim Nachweis, daß (1) erfüllt ist, von Abschätzungen für $\|\tilde{w}\|$ Gebrauch.

Die unter I zusammengefaßten Sätze, die klassischen Einschließungssätze, werden nicht gesondert bewiesen, da sie sich unmittelbar als Spezialfälle der zu II gehörenden Sätze ergeben.

In der geschilderten Weise können zwar viele, aber nicht sämtliche Einschließungssätze hergeleitet werden. Resultate, die man nach der Methode der Zwischenaufgaben ("intermediate problems" [56]) oder nach der Methode der Orthogonalinvarianten [16] erhält, ordnen sich hier nicht ein.

Die Möglichkeit, Einschließungssätze in einheitlicher Weise herzuleiten, wurde zuerst von Wielandt [57] erkannt; er wies nach, daß der Satz von Kryloff-Bogoliubov, der Satz von Temple und Collatz sowie der Collatzsche Quotienten-Einschließungssatz für Matrizen aus einem grundlegenden Satz folgen. Dieser Gedanke wurde in verschiedenen Richtungen weiterentwickelt (z. B. [7], [15], [46]). In der vorliegenden Arbeit wird aufgezeigt, daß man sowohl die klassischen als auch die unter II und III einzuordnenden Einschließungssätze in einheitlicher Weise gewinnen kann, wenn man Normabschätzungen für die Lösung linearer Gleichungen als weiteren wesentlichen Bestandteil der Beweismethode hinzunimmt.

An Hand von zwei Beispielen soll nun erläutert werden, daß es nicht immer möglich und sinnvoll ist, Paare (u,v) mit der Eigenschaft

$$u \in D(M),\ v \in D(N),\ Mu = Nv \tag{2}$$

zur Eigenwerteinschließung zu verwenden, und daß es deshalb wichtig ist, auch andere als die klassischen Einschließungssätze zur Verfügung zu haben.

Beispiel 2

Eigenwertaufgabe: $\Delta^2\phi = \lambda(-\Delta\phi + 2d\,\frac{\partial^2\phi}{\partial x\partial y})$ in Ω (mit $d \in \mathbb{R}$, $|d| < 1$),

$$\phi = 0,\ \frac{\partial\phi}{\partial n} = 0 \qquad \text{auf } \partial\Omega$$

Einordnung: $H := L_2(\Omega)$, $\langle f|g\rangle := \int_\Omega fg\,dxdy$ für $f,g \in H$,

$D(M) := \{f \in C^\infty(\overline{\Omega}):\ f = 0,\ \frac{\partial f}{\partial n} = 0 \text{ auf } \partial\Omega\}$,

$D(N) := \{f \in C^\infty(\overline{\Omega}):\ f = 0 \text{ auf } \partial\Omega\}$, $\quad D(S) := H$,

$Mf := \Delta^2 f$, $\ Nf := -\Delta f + 2d\,\frac{\partial^2 f}{\partial x\partial y}$, $\ Sf := f$.

Die Bedingung (2) lautet hier

$$\Delta^2 u = -\Delta v + 2d\,\frac{\partial^2 v}{\partial x\partial y} \text{ in } \Omega,\quad u = 0,\ \frac{\partial u}{\partial n} = 0,\ v = 0 \text{ auf } \partial\Omega.$$

Wählt man ein $u \in D(M)$, so muß man v als exakte Lösung einer Randwertaufgabe bestimmen; Entsprechendes gilt, wenn man mit der Wahl eines $v \in D(N)$ beginnt.

Beispiel 3

Eigenwertaufgabe: $-\Delta\phi = \lambda\phi$ in Ω, $\phi = 0$ auf $\partial\Omega$

Einordnung: $H := L_2(\Omega) \times L_2(\partial\Omega)$; mit $\omega \in \mathbb{R}$, $\omega > 0$ sei

$$\left\langle \begin{pmatrix} f_1 \\ f_2 \end{pmatrix} \middle| \begin{pmatrix} g_1 \\ g_2 \end{pmatrix} \right\rangle := \int_\Omega f_1 g_1\,dxdy + \omega\int_{\partial\Omega} f_2 g_2\,ds \text{ für } \begin{pmatrix} f_1 \\ f_2 \end{pmatrix}, \begin{pmatrix} g_1 \\ g_2 \end{pmatrix} \in H,$$

$D(M) := D(N) := D(S) = C^\infty(\overline{\Omega})$,

$$Mf := \begin{pmatrix} -\Delta f \\ f_{|\partial\Omega} \end{pmatrix},\ Nf := Sf := \begin{pmatrix} f \\ 0 \end{pmatrix}.$$

Paare (u,v), die der Bedingung (2) genügen, für die also

$-\Delta u = v$ in Ω, $u = 0$ auf $\partial\Omega$

gilt, wird man im allgemeinen leicht finden, schwierig kann es jedoch sein, die Paare (u,v) außerdem noch so zu wählen, daß sie sowohl eine für die Rechnung hinreichend einfache Gestalt haben als auch genaue Einschließungen liefern.

I Die klassischen Einschließungssätze für Aufgaben $M\phi = \lambda N\phi$

Zunächst werden Sätze zusammengestellt, bei denen Paare (u,v) mit der Eigenschaft $u \in D(M)$, $v \in D(N)$, $Mu = Nv$ zur Eigenwerteinschließung verwendet werden. Diese Sätze gehen - im Gegensatz zu den weiter unten besprochenen - im wesentlichen auf Veröffentlichungen aus der Zeit vor 1950 zurück und werden deshalb als klassische Einschließungssätze bezeichnet; sie sind Spezialfälle der in §§5 und 6 bewiesenen Sätze.

§3 Eine von Lehmann [37], [38], [39] entdeckte Verallgemeinerung des Satzes von Kryloff-Bogoliubov ermöglicht die Bestimmung von Einschließungsintervallen mit vorgeschriebenem Mittelpunkt ϱ:

Satz 3

Voraussetzungen und Bezeichnungen:

1. $u_1,\dots,u_n$ seien linear unabhängige Elemente von $D(N^{-1}M)$; für $i=1,\dots,n$ sei $v_i := N^{-1}Mu_i$.
2. Es seien $\varrho \in \mathbb{R}$,
   $A := (\{v_i|v_k\} - 2\varrho\{v_i|u_k\} + \varrho^2\{u_i|u_k\})_{i,k=1,\dots,n}$ und
   $B := (\{u_i|u_k\})_{i,k=1,\dots,n}$.
3. Für $i=1,\dots,n$ sei μ_i der i-tkleinste Eigenwert der Matrix-Eigenwertaufgabe $Ax = \mu Bx$.

Behauptung: Für $r=1,\dots,n$ liegen im Intervall $[\varrho-\sqrt{\mu_r},\varrho+\sqrt{\mu_r}]$ mindestens r Eigenwerte der Eigenwertaufgabe $M\phi = \lambda N\phi$.

Beweis: Die Behauptung ergibt sich aus Satz 14 mit Hilfe von (10).

Zur Berechnung von Einschließungsintervallen mit vorgeschriebenem rechten Endpunkt L dient die folgende ebenfalls auf Lehmann [37], [38], [39] zurückgehende Verallgemeinerung des Satzes von Temple und Collatz:

Satz 4

Voraussetzungen und Bezeichnungen:

1. $u_1,\dots,u_n$ seien linear unabhängige Elemente von $D(N^{-1}M)$; für $i=1,\dots,n$ sei $v_i := N^{-1}Mu_i$.
2. Es seien $L \in \mathbb{R}$,
   $A := (\{v_i|u_k\}-L\{u_i|u_k\})_{i,k=1,\dots,n}$ und
   $B := (\{v_i|v_k\}-2L\{v_i|u_k\} + L^2\{u_i|u_k\})_{i,k=1,\dots,n}$;
   B sei positiv definit.[1)]
3. Für $i=1,\dots,n$ sei μ_i der i-tkleinste Eigenwert der Matrix-Eigenwertaufgabe $Ax = \mu Bx$.

Behauptung: Für alle $r \in \mathbb{N}$ mit $r \leq n$ und $\mu_r < 0$ enthält das Intervall $[L + \frac{1}{\mu_r}, L]$ mindestens r Eigenwerte der Eigenwertaufgabe $M\phi = \lambda N\phi$.

Beweis: Die Behauptung ergibt sich aus Satz 13 (für $t:=0$, $N_0:=N$).

Satz 4, der sich numerisch sehr gut bewährt hat (s. z. B. [3]), wird in [37], [38], [39] für den Fall $D(S) = H$, $S = I$ bewiesen; für $n = 1$, $D(S) = H$, $S = I$ reduziert er sich auf den Satz von Temple [54] und Collatz ([11], S. 174). Ergebnisse für den Fall, daß $n = 1$, aber S nicht notwendig die identische Abbildung ist, finden sich in zahlreichen Arbeiten ([4], [9], [13], [29]). Rektorys ([44], S. 493) hat den Satz von Temple und Collatz auf Eigenwertaufgaben in Bilinearformgestalt übertragen (vgl. auch [45]).

Zur Anwendung des nun folgenden Quotienten-Einschließungssatzes, der auf einem Grundgedanken von Collatz [10] beruht,

---

1) Ist L nicht Eigenwert der Aufgabe $M\phi = \lambda N\phi$, so ist B stets positiv definit.

braucht man - anders als bei den Sätzen 3 und 4 - keine Skalarprodukte auszurechnen; statt dessen muß man jedoch im allgemeinen Infima und Suprema reellwertiger Funktionen bestimmen.

Satz 5

Voraussetzungen und Bezeichnungen:

1. Auf $D(N)$ sei eine Halbordnung definiert, und zwar so, daß $(D(N), \leq)$ ein halbgeordneter linearer Raum ist und aus $f,g \in D(N)$, $f \geq 0$, $g \geq 0$ stets $\{f|g\} \geq 0$ folgt.
2. Es sei $u \in D(N^{-1}M)$ mit $u \neq 0$ und $u \geq 0$; es gebe $\underline{\alpha}, \overline{\alpha} \in \mathbb{R}$ mit $\underline{\alpha}u \leq N^{-1}Mu \leq \overline{\alpha}u$.
3. $\alpha_{min} := \sup\{\alpha \in \mathbb{R}: \alpha u \leq N^{-1}Mu\}$; $\alpha_{max} := \inf\{\alpha \in \mathbb{R}: N^{-1}Mu \leq \alpha u\}$.

Behauptung: Im Intervall $[\alpha_{min}, \alpha_{max}]$ liegt mindestens ein Eigenwert der Eigenwertaufgabe $M\phi = \lambda N\phi$.

Beweis: Die Behauptung ergibt sich aus Satz 10, wenn man dort $t := \frac{1}{2}(\alpha_{max}+\alpha_{min})$, $N_o := N$ setzt und berücksichtigt, daß

$\sup\{\beta \in \mathbb{R}: \beta u \leq N^{-1}Mu - tu\} = \alpha_{min} - t \leq 0$,

$\inf\{\beta \in \mathbb{R}: N^{-1}Mu - tu \leq \beta u\} = \alpha_{max} - t \geq 0$ gilt.

Satz 5 enthält als Spezialfall den in [1] für Eigenwertaufgaben mit gewöhnlichen Differentialgleichungen aufgestellten Einschließungssatz; der Kürze halber wird er hier nur für solche Aufgaben formuliert, bei denen alle Randbedingungen wesentlich sind:

Satz 6

Voraussetzungen und Bezeichnungen:

1. Es seien $m,n \in \mathbb{N}_o$ mit $m > n$; $x_o, x_1 \in \mathbb{R}$ mit $x_o < x_1$; $p_i \in C^i[x_o,x_1]$ für $i=0,\dots,m$; $q_i \in C^i[x_o,x_1]$ für $i=0,\dots,n$; $J_1, J_2 \subset \mathbb{N}_o$ mit $J_1 \cap J_2 = \emptyset$, $J_1 \cup J_2 = \{0,1,\dots,n\}$, $n \in J_1$. Es gelte $p_m(x) > 0$ für $x \in [x_o,x_1]$; $q_i(x) > 0$ für $x \in [x_o,x_1]$, $i \in J_1$; $q_i(x) = 0$ für $x \in [x_o,x_1]$, $i \in J_2$.
2. Es seien

$D(M) := \{f \in C^{2m}[x_o,x_1]: f^{(k)}(x_o) = f^{(k)}(x_1) = 0 \text{ für } k=0,\dots,m-1\}$,

$D(N) := \{f \in C^{2n}[x_o,x_1]: f^{(k)}(x_o) = f^{(k)}(x_1) = 0 \text{ für } k=0,\dots,n-1\}$,

$Mf := \sum_{i=0}^{m} (-1)^i (p_i f^{(i)})^{(i)}$ und $Nf := \sum_{i=0}^{n} (-1)^i (q_i f^{(i)})^{(i)}$.

3. Es seien $u \in D(M)$ und $v \in D(N)$ mit $Mu = Nv$.
Für alle $x \in [x_o, x_1]$ mit Ausnahme von höchstens endlich vielen und für alle $i \in J_1$ gelte $u^{(i)}(x) \neq 0$. Es gebe $\underline{\alpha}, \overline{\alpha} \in \mathbb{R}$ mit der Eigenschaft, daß $\underline{\alpha} \leq \frac{v^{(i)}(x)}{u^{(i)}(x)} \leq \overline{\alpha}$ für alle $i \in J_1$ und alle $x \in [x_o, x_1]$ mit $u^{(i)}(x) \neq 0$ gilt.

4. $$\alpha_{min} := \min_{i \in J_1} \inf_{\substack{x \in [x_o, x_1] \\ u^{(i)}(x) \neq 0}} \frac{v^{(i)}(x)}{u^{(i)}(x)} ; \qquad (3.1)$$

$$\alpha_{max} := \max_{i \in J_1} \inf_{\substack{x \in [x_o, x_1] \\ u^{(i)}(x) \neq 0}} \frac{v^{(i)}(x)}{u^{(i)}(x)} . \qquad (3.2)$$

Behauptung: Im Intervall $[\alpha_{min}, \alpha_{max}]$ liegt mindestens ein Eigenwert der Eigenwertaufgabe $M\phi = \lambda N\phi$.

Beweis: Es seien $H := L_2[x_o, x_1]$, $\langle f|g \rangle := \int_{x_o}^{x_1} fg \, dx$ für $f, g \in H$, $D(S) := D(N)$, $S := I$. Dann sind die Voraussetzungen V1 und V2 aus §1 erfüllt (bezüglich V2 s. [11], S. 141). Für $f, g \in D(N)$ sei $f \leq g$ gleichbedeutend damit, daß $f^{(i)}(x)u^{(i)}(x) \leq g^{(i)}(x)u^{(i)}(x)$ für alle $i \in J_1$ und alle $x \in [x_o, x_1]$ gilt. Aus Satz 5 ergibt sich nun die Behauptung.

Die Frage, für welches i das Minimum in (3.1) und das Maximum in (3.2) angenommen wird, ist von Held [27],[28] untersucht worden.
Im Spezialfall $J_1 = \{n\}$ - die Aufgabe $M\phi = \lambda N\phi$ gehört dann zur "Eingliedklasse" - reduziert sich Satz 6 auf einen von Collatz ([11], S. 132) bewiesenen Einschließungssatz.

Angemerkt sei noch, daß Quotienten-Einschließungssätze auch unter ganz anderen Voraussetzungen als den hier betrachteten bewiesen werden können; so wird beispielsweise in [12] unter anderem die Existenz einer nichtnegativen Eigenfunktion vorausgesetzt.

II Einschließung ohne Berechnung von $N^{-1}Mu$

Als nächstes werden Einschließungssätze, zu deren Anwendung man zwar Elemente u aus $D(N^{-1}M)$ benötigt, $N^{-1}Mu$ aber nicht auszurechnen braucht, aus dem grundlegenden Satz hergeleitet. Hierbei spielen Abschätzungen für die Norm der Lösung w einer linearen Gleichung

$$Nw = h \tag{4}$$

eine entscheidende Rolle. Die Norm von w kann man im allgemeinen dann leicht abschätzen, wenn

a) N ein Operator von monotoner Art ist (§4),

b) ein leicht invertierbarer Operator N_o bekannt ist, der in einem gewissen Sinn kleiner als N ist (§5), oder

c) die Lösung von (4) durch komplementäre Extremalprinzipien charakterisiert werden kann (§6).

§4 Wenn N (bezüglich geeigneter Halbordnungen in D(N) und H, s.u.) ein Operator von monotoner Art ist, erweist sich die folgende Abschätzung als nützlich:

Satz 7

Voraussetzungen und Bezeichnungen:

1. Auf H sei eine Halbordnung erklärt, und zwar so, daß $(H, \leq)$ ein halbgeordneter linearer Raum ist und aus $f, g \in H$, $f \geq 0$, $g \geq 0$ stets $\langle f|g\rangle \geq 0$ folgt.
2. Für alle $f, g \in D(N)$ mit $Nf \leq Ng$ gelte $Sf \leq Sg$.
3. Es seien $w, z \in D(N)$ und $h \in H$; es gelte $Nw = h$ und $-Nz \leq h \leq Nz$.

Behauptung: $\|w\| \leq \|z\|$

Beweis: Aus $-Nz \leq Nw \leq Nz$ folgt zunächst $-Sz \leq Sw \leq Sz$ und weiter $0 \leq \langle Nz + Nw \mid Sz - Sw\rangle = \{z|z\} - \{w|w\}$.

Bemerkung: Ist die Voraussetzung 1 von Satz 7 erfüllt, so kann man auf D(N) eine Halbordnung $\leq$ durch die Festsetzung definieren, daß $f \leq g$ gleichbedeutend mit $Sf \leq Sg$ ist. Die Voraussetzung 2 besagt dann gerade, daß $N : (D(N), \leq) \to (H, \leq)$ ein Operator von monotoner Art ist.

Kombiniert man die Abschätzung aus Satz 7 mit dem

grundlegenden Satz 1, so erhält man den folgenden Quotienten-Einschließungssatz:

Satz 8

Voraussetzungen und Bezeichnungen:

1. Die Voraussetzungen 1 und 2 von Satz 7 seien erfüllt.
2. Es sei $u \in D(N^{-1}M)$ mit $u \neq 0$ und $Nu \geq 0$; es gebe $\underline{\beta}, \overline{\beta} \in \mathbb{R}$ mit $\underline{\beta}Nu \leq Mu \leq \overline{\beta}Nu$.
3. $\beta_{min} := \sup\{\beta \in \mathbb{R}: \beta Nu \leq Mu\}$; $\beta_{max} := \inf\{\beta \in \mathbb{R}: Mu \leq \beta Nu\}$.

Behauptung: Im Intervall $[\beta_{min}, \beta_{max}]$ liegt mindestens ein Eigenwert der Aufgabe $M\phi = \lambda N\phi$.

Beweis: Da $Nu \geq 0$, $Nu \neq 0$ gilt, ist die Menge $\{\beta \in \mathbb{R}: \beta Nu \leq Mu\}$ durch $\overline{\beta}$ nach oben und die Menge $\{\beta \in \mathbb{R}: Mu \leq \beta Nu\}$ durch $\underline{\beta}$ nach unten beschränkt. Es sei $\varepsilon \in \mathbb{R}$ mit $\varepsilon > 0$; mit den Bezeichnungen $\varrho := \frac{1}{2}(\beta_{max} + \beta_{min})$ und $\tau := \frac{1}{2}(\beta_{max} - \beta_{min})$ ergibt sich

$$(\varrho-\tau-\varepsilon)Nu = (\beta_{min} - \varepsilon)Nu \leq Mu \leq (\beta_{max} + \varepsilon)Nu = (\varrho+\tau+\varepsilon)Nu,$$

also $-(\tau+\varepsilon)Nu \leq Mu - \varrho Nu \leq (\tau+\varepsilon)Nu$. Setzt man in Satz 7 $w := N^{-1}Mu - \varrho u$, $z := (\tau+\varepsilon)u$ und $h := Mu - \varrho Nu$, so erhält man $\|N^{-1}Mu - \varrho u\| \leq (\tau+\varepsilon)\|u\|$. Da dies für beliebiges positives ε gilt, ergibt sich $\|N^{-1}Mu - \varrho u\| \leq \tau\|u\|$. Die Behauptung folgt nun aus Satz 1.

Der Ausgangspunkt für die Untersuchungen [18],[19], die auf Satz 8 führten, war ein Problem, das Collatz ([11], S. 304) im Zusammenhang mit den von ihm entdeckten Quotienten-Einschließungssätzen aufgeworfen hat. - Einige mit Hilfe von Satz 8 gewonnene numerische Resultate für Eigenwertaufgaben mit gewöhnlichen Differentialgleichungen finden sich in [2], [47].

Ein auf Bartsch [6] zurückgehender Quotienten-Einschließungssatz (vgl. [26]) läßt sich in ähnlicher Weise wie Satz 8 aus dem grundlegenden Einschließungssatz herleiten.

§5 Die Norm der Lösung w einer linearen Gleichung $Nw = h$ kann man auch dann in einfacher Weise abschätzen, wenn man einen leicht invertierbaren Operator N_0 kennt, der im Sinne einer gewissen Halbordnung (s.u.) kleiner als N ist.

Satz 9

Voraussetzungen:

1. $D(N_o)$ sei ein reeller Vektorraum mit $D(N) \subset D(N_o) \subset D(S)$;
   $N_o: D(N_o) \to H$ sei eine lineare Abbildung.
2. Für alle $f,g \in D(N_o)$ gelte $\langle Sf|N_o g\rangle = \langle Sg|N_o f\rangle$;
   für alle $f \in D(N_o)$ mit $f \neq 0$ gelte $\langle Sf|N_o f\rangle > 0$.
3. Für alle $f \in D(N)$ gelte $\langle Sf|N_o f\rangle \leq \langle Sf|Nf\rangle$.
4. Auf $D(N_o)$ sei eine Halbordnung definiert, und zwar so, daß $(D(N_o), \leq)$ ein halbgeordneter linearer Raum ist und aus $f,g \in D(N_o)$, $f \geq 0$, $g \geq 0$ stets $\langle Sf|N_o g\rangle \geq 0$ folgt.
5. Es seien $\alpha \in \mathbb{R}$, $w,z \in D(N)$ und $h \in H$; es gelte $|\alpha| \leq 1$, $Nw = h$, $h + \alpha Nz \in D(N_o^{-1})$ und $-z \leq N_o^{-1}(h + \alpha Nz) - \alpha z \leq z$.

Behauptung: $\|w\| \leq \|z\|$

Beweis: Aus $(1-\alpha)z + N_o^{-1}(h+\alpha Nz) \geq 0$ und $(1+\alpha)z - N_o^{-1}(h+\alpha Nz) \geq 0$ folgt

$$\langle S((1-\alpha)z + N_o^{-1}(h+\alpha Nz)) \mid N_o((1+\alpha)z - N_o^{-1}(h+\alpha Nz))\rangle \geq 0 . \qquad (5)$$

Da $\alpha^2 \leq 1$ gilt, erhält man

$$(1-\alpha^2)\langle Sz|(N-N_o)z\rangle \geq 0. \qquad (6)$$

Für alle $\begin{pmatrix} f_1 \\ f_2 \end{pmatrix}, \begin{pmatrix} g_1 \\ g_2 \end{pmatrix} \in D(N_o) \times D(N)$ sei

$$b(\begin{pmatrix} f_1 \\ f_2 \end{pmatrix}, \begin{pmatrix} g_1 \\ g_2 \end{pmatrix}) := \langle Sf_1|N_o g_1\rangle + \langle Sf_2|(N-N_o)g_2\rangle;$$

ferner sei $\hat{z} := \begin{pmatrix} N_o^{-1}(h+\alpha Nz) - \alpha z \\ -\alpha z \end{pmatrix}$.

Mit Hilfe von (5) und (6) ergibt sich

$$0 \leq b(\begin{pmatrix} z \\ z \end{pmatrix} + \hat{z}, \begin{pmatrix} z \\ z \end{pmatrix} - \hat{z}) = \{z|z\} - b(\hat{z},\hat{z}).$$

Da $b(\cdot,\cdot)$ eine positiv semidefinite Bilinearform ist, gilt

$$0 \leq b(\hat{z} - \begin{pmatrix} w \\ w \end{pmatrix}, \hat{z} - \begin{pmatrix} w \\ w \end{pmatrix}) = b(\hat{z},\hat{z}) - 2\langle Sw|h\rangle + \langle Sw|Nw\rangle$$

$$= b(\hat{z},\hat{z}) - \{w|w\}.$$

Insgesamt erhält man also $0 \leq \{z|z\} - \{w|w\}$.

Bemerkung: Auf der Menge aller linearen Abbildungen $P: D(P) \to H$ mit der Eigenschaft, daß $D(P) \subset D(S)$ und $\langle Sf|Pg\rangle = \langle Sg|Pf\rangle$ für alle $f,g \in D(P)$ gilt, kann man eine Halbordnung $\leq$ durch die Festlegung

$$\left.\begin{array}{l} P_1 \leq P_2 \quad \Leftrightarrow \\ D(P_2) \subset D(P_1),\ \langle Sf|P_1 f\rangle \leq \langle Sf|P_2 f\rangle \text{ für alle } f \in D(P_2) \end{array}\right\} \quad (7)$$

definieren. Ist speziell $D(S) = H$, $S = I$, so stimmt diese Halbordnung mit der für symmetrische Operatoren üblichen überein. Der Operator N_o ist also im Sinne der soeben erklärten Halbordnung kleiner als N, wenn die Voraussetzungen von Satz 9 erfüllt sind.

Kombiniert man die Abschätzung aus Satz 9 mit dem grundlegenden Satz 1, so erhält man

Satz 10

Voraussetzungen und Bezeichnungen:

1. Die Voraussetzungen 1 bis 4 von Satz 9 seien erfüllt.
2. Es seien $t \in \mathbb{R}$ und $u \in D(N^{-1}M)$; es gelte $u \neq 0$, $u \geq 0$ und $Mu - tNu \in D(N_o^{-1})$. Es gebe $\underline{\beta}, \overline{\beta} \in \mathbb{R}$ mit $\underline{\beta}u \leq N_o^{-1}(Mu-tNu) \leq \overline{\beta}u$.
3. $\beta_{min} := \sup\{\beta \in \mathbb{R}: \beta \leq 0,\ \beta u \leq N_o^{-1}(Mu-tNu)\}$;
   $\beta_{max} := \inf\{\beta \in \mathbb{R}: \beta \geq 0,\ N_o^{-1}(Mu-tNu) \leq \beta u\}$.

Behauptung: Im Intervall $[\beta_{min}+t,\ \beta_{max}+t]$ liegt mindestens ein Eigenwert der Eigenwertaufgabe $M\phi = \lambda N\phi$.

Beweis: Es sei $\varepsilon \in \mathbb{R}$ mit $\varepsilon > 0$; zur Abkürzung wird $\sigma := \frac{1}{2}(\beta_{max} - \beta_{min}) + \varepsilon$ gesetzt. Die in Voraussetzung 5 von Satz 9 auftretenden Größen werden folgendermaßen definiert:
$\alpha := \frac{1}{2\sigma}(\beta_{max} + \beta_{min})$, $w := N^{-1}Mu - (t+\alpha\sigma)u$, $z := \sigma u$,
$h := Mu - (t+\alpha\sigma)Nu$.
Dann gilt $\sigma(\alpha-1) = \beta_{min} - \varepsilon < 0$ und $\sigma(\alpha+1) = \beta_{max} + \varepsilon > 0$; hieraus erhält man $\alpha^2 - 1 < 0$. Da $u \geq 0$ ist, ergibt sich
$\sigma(\alpha-1)u = (\beta_{min} - \varepsilon)u \leq N_o^{-1}(Mu-tNu) \leq (\beta_{max} + \varepsilon)u = \sigma(\alpha+1)u$,
also $(\alpha-1)z \leq N_o^{-1}(h+\alpha Nz) \leq (\alpha+1)z$. Nach Satz 9 folgt
$\|N^{-1}Mu - (t+\frac{1}{2}(\beta_{max} + \beta_{min}))u\| \leq \frac{1}{2}(\beta_{max} - \beta_{min} + 2\varepsilon)\|u\|$.

Wenn man hierin ε gegen 0 gehen läßt, erhält man mit Hilfe von Satz 1 die Behauptung.

Der für die Anwendung wichtigste Spezialfall (s. [18])
$D(S) = H,\ S = I,\ N_o = \tilde{\kappa} I$ (mit $\tilde{\kappa} \in \mathbb{R}$)
wurde an Hand von verschiedenen Eigenwertaufgaben mit gewöhnlichen Differentialgleichungen numerisch erprobt [47]. - Setzt man $N_o := N$, so reduziert sich Satz 10 auf den klassischen Quotienten-Einschließungssatz (Satz 5).
Die Kenntnis eines leicht invertierbaren Operators N_o, der im Sinne der Halbordnung (7) kleiner als N ist, ermöglicht auch die Aufstellung von komplementären Extremalprinzipien und kann auf diese Weise zur Herleitung eines weiteren Einschließungssatzes (Satz 13) verwendet werden.

§6 Die folgende Abschätzung für die Norm der Lösung w von (4) hängt eng mit komplementären Extremalprinzipien zusammen:

Satz 11

Voraussetzungen:

1. X sei ein reeller Vektorraum.
2. $b(\cdot,\cdot)$ sei eine symmetrische Bilinearform auf X; für alle $f \in X$ gelte $b(f,f) \geq 0$.
3. $T : D(N) \to X$ sei eine lineare Abbildung; für alle $f,g \in D(N)$ gelte $b(Tf,Tg) = \{f|g\}$.
4. Es seien $w \in D(N)$ und $h \in H$ mit $Nw = h$.
5. Es sei $\tilde{z} \in X$; für alle $g \in D(N)$ gelte $b(Tg,\tilde{z}) = \langle Sg|h\rangle$.

Behauptung: $\|w\|^2 \leq b(\tilde{z},\tilde{z})$

Beweis: $0 \leq b(\tilde{z}-Tw,\tilde{z}-Tw) = b(\tilde{z},\tilde{z}) - 2\langle Sw|h\rangle + \{w|w\}$
$= b(\tilde{z},\tilde{z}) - \{w|w\}$.

Bemerkung: Sind die Voraussetzungen 1 bis 4 von Satz 11 erfüllt, so läßt sich w durch komplementäre Extremalprinzipien charakterisieren:

$$\max_{f \in D(N)} (2\langle Sf|h\rangle - \{f|f\}) = \|w\|^2 = \min_{\tilde{z} \in \tilde{X}} b(\tilde{z},\tilde{z}) \tag{8}$$

mit $\tilde{X} := \{\tilde{z} \in X:\ b(Tg,\tilde{z}) = \langle Sg|h\rangle \text{ für alle } g \in D(N)\}$;

das Maximum wird dabei für $f = w$ angenommen. Die linke Gleichung in (8) ergibt sich daraus, daß

$0 \leq \{w-f|w-f\} = \{w|w\} - 2\langle Sf|h\rangle + \{f|f\}$ für alle $f \in D(N)$

gilt, die rechte folgt aus Satz 11 wegen $Tw \in \tilde{X}$. - Viele bekannte (vgl. [55]) komplementäre Extremalprinzipien lassen sich in die Form (8) bringen und ermöglichen es, $\|w\|$ geeignet abzuschätzen.

Aus den Sätzen 1 und 11 erhält man

Satz 12

Voraussetzungen und Bezeichnungen:

1. Die Voraussetzungen 1 bis 3 von Satz 11 seien erfüllt.
2. Es seien $u_1,\ldots,u_n \in D(N^{-1}M)$ und $v_1,\ldots,v_n \in X$; $u_1,\ldots,u_n$ seien linear unabhängig; für alle $g \in D(N)$ und $i=1,\ldots,n$ gelte $b(Tg,v_i) = \langle Sg|Mu_i\rangle$.
3. Es seien $L \in \mathbb{R}$,
   $A := (\langle Su_i|Mu_k\rangle - L\langle Su_i|Nu_k\rangle)_{i,k=1,\ldots,n}$ und
   $B := (b(v_i,v_k) - 2L\langle Su_i|Mu_k\rangle + L^2\langle Su_i|Nu_k\rangle)_{i,k=1,\ldots,n}$;
   B sei positiv definit[1].
4. Für $i=1,\ldots,n$ sei μ_i der i-tkleinste Eigenwert der Matrix-Eigenwertaufgabe $Ax = \mu Bx$.

Behauptung: Für alle $r \in \mathbb{N}$ mit $r \leq n$ und $\mu_r < 0$ enthält das Intervall $[L + \frac{1}{\mu_r}, L]$ mindestens r Eigenwerte der Eigenwertaufgabe $M\phi = \lambda N\phi$.

Beweis: Es sei $r \in \mathbb{N}$ mit $r \leq n$ und $\mu_r < 0$. $\tilde{U}$ sei der Unterraum des $\mathbb{R}^n$, der von denjenigen Eigenvektoren der Aufgabe $Ax = \mu Bx$ aufgespannt wird, die zu einem Eigenwert $\mu_i \leq \mu_r$ gehören. Dann gilt $\dim \tilde{U} \geq r$ und $c'Ac \leq \mu_r c'Bc$ für alle $c \in \tilde{U}$.

$U := \{\sum_{i=1}^{n} c_i u_i : (c_1,\ldots,c_n)' \in \tilde{U}\}$ ist ein Unterraum von $D(N^{-1}M)$ mit $\dim U \geq r$.

Es sei nun $u \in U$. Mit den eindeutig bestimmten Koeffizienten c_i

---

[1] Ist L nicht Eigenwert der Aufgabe $M\phi = \lambda N\phi$, so ist B stets positiv definit.

der Darstellung $u = \sum_{i=1}^{n} c_i u_i$ wird $v := \sum_{i=1}^{n} c_i v_i$ definiert. Dann gilt

$\langle Su|Mu\rangle - L\langle Su|Nu\rangle \leq \mu_r(b(v,v) - 2L\langle Su|Mu\rangle + L^2\langle Su|Nu\rangle)$.

Unter Berücksichtigung der Beziehungen $\langle Su|Nu\rangle = b(Tu,Tu)$ und $\langle Su|Mu\rangle = b(Tu,v)$ ergibt sich hieraus

$$b(v - (\frac{1}{2\mu_r}+L)Tu,\ v - (\frac{1}{2\mu_r}+L)Tu) \leq (\frac{1}{2\mu_r})^2\{u|u\}. \tag{9}$$

Setzt man in Satz 11 $w := N^{-1}Mu - (\frac{1}{2\mu_r}+L)u$, $h := Mu - (\frac{1}{2\mu_r}+L)Nu$ und $\tilde{z} := v - (\frac{1}{2\mu_r}+L)Tu$, so erhält man

$$\|N^{-1}Mu - (\frac{1}{2\mu_r}+L)u\|^2 \leq b(v - (\frac{1}{2\mu_r}+L)Tu,\ v - (\frac{1}{2\mu_r}+L)Tu).$$

Mit Hilfe von (9) folgt hieraus

$$\|N^{-1}Mu - (\frac{1}{2\mu_r}+L)u\| \leq |\frac{1}{2\mu_r}|\,\|u\|.$$

Satz 1 liefert nun unmittelbar die Behauptung.

Der Spezialfall

$$\left.\begin{array}{ll} X := D(N),\ Tf := f & \text{für } f \in D(N), \\ b(f,g) := \langle Sf|Ng\rangle & \text{für } f,g \in D(N) \end{array}\right\} \tag{10}$$

von Satz 12 stimmt mit dem Satz von Temple-Collatz-Lehmann (Satz 4) überein. Bei geeigneter Wahl von X, b und T ist Satz 12 jedoch leichter zu handhaben als Satz 4 und besitzt einen größeren Anwendungsbereich.

Zur Behandlung von Beispiel 2 aus §2 kann man

$X := C^\infty(\bar{\Omega})$, $\quad Tf := f \quad$ für $f \in D(N)$,

$b(f,g) := \int_\Omega (\frac{\partial f}{\partial x}\frac{\partial g}{\partial x} + \frac{\partial f}{\partial y}\frac{\partial g}{\partial y} + 2d\frac{\partial f}{\partial x}\frac{\partial g}{\partial y})\,dxdy$ für $f,g \in X$

wählen; die in Satz 12 auftretenden Funktionen v_i brauchen dann lediglich den Bedingungen

$$v_i \in C^\infty(\bar{\Omega}),\quad -\Delta v_i + 2d\frac{\partial^2 v_i}{\partial x \partial y} = \Delta^2 u_i \quad \text{in } \Omega$$

zu genügen, also keine Randbedingungen zu erfüllen.

Es empfiehlt sich, für die Funktionen v_i einen Ansatz, der noch freie Parameter enthält, zu machen; die Gleichungen für die optimale Wahl dieser Parameter stimmen dann mit den Trefftzschen Gleichungen überein (s. [19], [20]). Numerisch erprobt wurde Satz 12 an Hand von Beispiel 2 aus §2 [36]. - Vor kurzem [25] wurde ein zu Satz 12 analoger Satz, bei dem nicht N, sondern M als positiv definit vorausgesetzt wird, bewiesen; er hat sich bei zahlreichen Aufgaben, z. B. bei verschiedenen Stekloffschen Eigenwertaufgaben [23] und bei einem System partieller Differentialgleichungen [21], gut bewährt.

Satz 12 kann durch geeignete Wahl von X, b und T auch an die Situation, daß man einen leicht invertierbaren Operator N_o kennt, der im Sinne der Halbordnung (7) kleiner als N ist, angepaßt werden; man erhält dann

Satz 13

Voraussetzungen und Bezeichnungen:

1. Die Voraussetzungen 1 bis 3 von Satz 9 seien erfüllt.
2. Es seien $u_1,\dots,u_n \in D(N^{-1}M) \cap D(N_o^{-1}M) \cap D(N_o^{-1}N)$.
3. Es seien $L,t \in \mathbb{R}$,
   $A := (\langle Su_i|Mu_k\rangle - L\langle Su_i|Nu_k\rangle)_{i,k=1,\dots,n}$ und
   $B := (\langle SN_o^{-1}(Mu_i-tNu_i) \mid Mu_k-tNu_k\rangle - 2(L-t)\langle Su_i|Mu_k\rangle$
   $+ (L^2-t^2)\langle Su_i|Nu_k\rangle)_{i,k=1,\dots,n}$;
   B sei positiv definit.
4. Für i=1,...,n sei μ_i der i-tkleinste Eigenwert der Matrix-Eigenwertaufgabe $Ax = \mu Bx$.

Behauptung: Für alle $r \in \mathbb{N}$ mit $r \leq n$ und $\mu_r < 0$ enthält das Intervall $[L + \frac{1}{\mu_r}, L]$ mindestens r Eigenwerte der Eigenwertaufgabe $M\phi = \lambda N\phi$.

Beweis: Setzt man $X := D(N_o) \times D(N)$,

$$b\left(\begin{pmatrix} f_1 \\ f_2 \end{pmatrix}, \begin{pmatrix} g_1 \\ g_2 \end{pmatrix}\right) := \langle Sf_1|N_o g_1\rangle + \langle Sf_2|(N-N_o)g_2\rangle \text{ für alle}$$

$$\begin{pmatrix} f_1 \\ f_2 \end{pmatrix}, \begin{pmatrix} g_1 \\ g_2 \end{pmatrix} \in X, \quad Tf := \begin{pmatrix} f \\ f \end{pmatrix} \text{ für alle } f \in D(N) \text{ und}$$

$$v_i := \begin{pmatrix} N_o^{-1}(Mu_i-tNu_i) + tu_i \\ tu_i \end{pmatrix} \text{ für } i=1,\dots,n,$$ so erhält man die Behauptung unmittelbar aus Satz 12.

Geeignete Operatoren N_o kann man sich beispielsweise dann verschaffen, wenn

α) der Spezialfall $D(S) = H$, $S = I$ vorliegt,

β) eine Orthonormalbasis $(\psi_i)_{i\in\mathbb{N}}$ des Hilbertraums $(H,\langle\cdot|\cdot\rangle)$ bekannt ist, die aus Eigenelementen der Eigenwertaufgabe $N\psi = \kappa\psi$ besteht, und

γ) die Folge $(\kappa_i)_{i\in\mathbb{N}}$ der zu den Eigenelementen ψ_i gehörenden Eigenwerte κ_i der Aufgabe $N\psi = \kappa\psi$ monoton nicht fallend ist.

Man kann dann jeden der Operatoren $N_{o,j}: H \to H$,

$$N_{o,j}f := \sum_{i=1}^{j} \kappa_i \langle f|\psi_i\rangle\psi_i + \kappa_{j+1}\Big(f - \sum_{i=1}^{j} \langle f|\psi_i\rangle\psi_i\Big) \quad (j\in\mathbb{N}) \tag{11}$$

als Abbildung N_o in Satz 13 verwenden; $N_{o,j}^{-1}$ läßt sich unmittelbar angeben:

$$N_{o,j}^{-1}f = \sum_{i=1}^{j} \frac{1}{\kappa_i}\langle f|\psi_i\rangle\psi_i + \frac{1}{\kappa_{j+1}}\Big(f - \sum_{i=1}^{j} \langle f|\psi_i\rangle\psi_i\Big) \quad (j\in\mathbb{N}).$$

Operatoren der Form (11) wurden von Bazley und Fox [8] im Rahmen der Methode der Zwischenaufgaben ("intermediate problems") eingeführt. Der Vorschlag, solche Operatoren im Zusammenhang mit Einschließungssätzen zu benutzen, stammt von Schellhaas [50], [51], auf den auch zwei Spezialfälle von Satz 13 zurückgehen. Ferner hat Schellhaas analoge Resultate für Eigenwertaufgaben $M\phi = \lambda N\phi$, bei denen nicht N, sondern M positiv definit ist, angegeben und zur Berechnung von Beulwerten von Platten verwendet ([50], [51]).

Während man in Satz 12 den rechten Endpunkt L der Einschließungsintervalle frei wählen kann, ermöglicht der nun folgende Satz die Bestimmung von Einschließungsintervallen mit vorgeschriebenem Mittelpunkt ϱ; ansonsten sind die beiden Sätze sehr ähnlich, da sie beide auf der Normabschätzung aus Satz 11 beruhen.

Satz 14

Voraussetzungen und Bezeichnungen:

1. Die Voraussetzungen 1 bis 3 von Satz 11 und die Voraussetzung 2 von Satz 12 seien erfüllt.

2. Es seien $\varrho \in \mathbb{R}$,

   $A := (b(v_i,v_k) - 2\varrho\langle Su_i|Mu_k\rangle + \varrho^2\langle Su_i|Nu_k\rangle)_{i,k=1,...,n}$ und

   $B := (\{u_i|u_k\})_{i,k=1,...,n}$.

3. Für $i=1,...,n$ sei μ_i der i-tkleinste Eigenwert der Matrix-Eigenwertaufgabe $Ax = \mu Bx$.

Behauptung: Für $r=1,...,n$ liegen im Intervall $[\varrho - \sqrt{\mu_r}, \varrho + \sqrt{\mu_r}]$ mindestens r Eigenwerte der Eigenwertaufgabe $M\phi = \lambda N\phi$.

Beweis: Es sei $r \in \mathbb{N}$ mit $r \leq n$. $\tilde{U}$ sei der Unterraum des $\mathbb{R}^n$, der von denjenigen Eigenvektoren der Aufgabe $Ax = \mu Bx$ aufgespannt wird, die zu einem Eigenwert $\mu_i \leq \mu_r$ gehören. Dann gilt $\dim \tilde{U} \geq r$ und $c'Ac \leq \mu_r c'Bc$ für alle $c \in \tilde{U}$.

$U := \{ \sum_{i=1}^{n} c_i u_i : (c_1,...,c_n)' \in \tilde{U}\}$ ist ein Unterraum von $D(N^{-1}M)$ mit $\dim U \geq r$.

Es sei nun $u \in U$. Mit den eindeutig bestimmten Koeffizienten c_i der Darstellung $u = \sum_{i=1}^{n} c_i u_i$ wird $v := \sum_{i=1}^{n} c_i v_i$ definiert. Dann gilt

$b(v,v) - 2\varrho\langle Su|Mu\rangle + \varrho^2\langle Su|Nu\rangle \leq \mu_r\{u|u\}$.

Unter Berücksichtigung der Beziehungen $\langle Su|Nu\rangle = b(Tu,Tu)$ und $\langle Su|Mu\rangle = b(Tu,v)$ ergibt sich hieraus $b(v-\varrho Tu,v-\varrho Tu) \leq \mu_r\{u|u\}$. Setzt man in Satz 11 $w := N^{-1}Mu - \varrho u$, $h := Mu - \varrho Nu$ und $\tilde{z} := v - \varrho Tu$, so erhält man $\|N^{-1}Mu - \varrho u\|^2 \leq b(v-\varrho Tu,v-\varrho Tu)$; aus Satz 1 folgt dann die Behauptung.

Der Satz von Kryloff-Bogoliubov-Lehmann (Satz 3) ergibt sich unmittelbar aus Satz 14, wenn man X, b und T wie in (10) wählt. Die Sätze 12, 13 und 14 enthalten ferner einige Einschließungssätze, die eng mit den von Falk [15] für normale Matrizenpaare aufgestellten zusammenhängen, als Spezialfälle.

III Einschließung ohne Verwendung von Elementen aus $D(N^{-1}M)$

Im folgenden werden Einschließungssätze, zu deren Anwendung man keine Elemente von $D(N^{-1}M)$ benötigt, aus Satz 2, der zweiten Fassung des grundlegenden Einschließungssatzes, hergeleitet. Hierzu werden Abschätzungen für die Norm der Lösung $\tilde{w}$ der Gleichung

$$M\tilde{w} = \tilde{h} \tag{12}$$

herangezogen, und zwar

a) Abschätzungen, die auf der monotonen Art von M beruhen (§7), und

b) a-priori-Abschätzungen (§8).

§7 Wenn M von monotoner Art ist, kann man im allgemeinen leicht obere Schranken für die Norm der Lösung $\tilde{w}$ der Gleichung (12) erhalten.

Satz 15

Voraussetzungen:

1. Auf D(M) sei eine Halbordnung erklärt, und zwar so, daß $(D(M), \leq)$ ein halbgeordneter linearer Raum ist und aus $f,g \in D(M)$, $f \geq 0$, $g \geq 0$ stets $\{f|g\} \geq 0$ folgt.
2. Auf H sei eine Halbordnung $\leq$ erklärt; M sei ein Operator monotoner Art, d. h. aus $f,g \in D(M)$ und $Mf \leq Mg$ folgt stets $f \leq g$.
3. Es seien $\tilde{w}, z \in D(M)$ und $\tilde{h} \in H$; es gelte $M\tilde{w} = \tilde{h}$ und $-Mz \leq \tilde{h} \leq Mz$.

Behauptung: $\|\tilde{w}\| \leq \|z\|$

Beweis: Aus $-Mz \leq \tilde{h} \leq Mz$ folgt zunächst $-z \leq \tilde{w} \leq z$ und weiter $0 \leq \{z-\tilde{w}|z+\tilde{w}\} = \{z|z\} - \{\tilde{w}|\tilde{w}\}$.

Kombiniert man diese Abschätzung mit Satz 2, so erhält man den folgenden Einschließungssatz:

Satz 16

Voraussetzungen und Bezeichnungen:

1. Die Voraussetzungen 1 und 2 von Satz 15 seien erfüllt.
2. Es seien $\varepsilon_*, \lambda_* \in \mathbb{R}$, $v \in D(M^{-1}N) \cap D(M)$ und $z \in D(M)$;

es gelte $0 \leq \varepsilon_* < 1$, $\lambda_* > 0$, $v \neq 0$, $-Mz \leq Mv - \lambda_* Nv \leq Mz$ und $\|z\| \leq \varepsilon_* \|v\|$.

Behauptung: Im Intervall $[\frac{\lambda_*}{1+\varepsilon_*}, \frac{\lambda_*}{1-\varepsilon_*}]$ liegt mindestens ein Eigenwert der Eigenwertaufgabe $M\phi = \lambda N\phi$.

Beweis: Setzt man in Satz 15 $\tilde{w} := v-\lambda_* M^{-1}Nv$ und $\tilde{h} := Mv-\lambda_* Nv$, so erhält man $\|v-\lambda_* M^{-1}Nv\| \leq \|z\| \leq \varepsilon_* \|v\|$; hieraus folgt nach Satz 2 die Behauptung.

Eine einfache Möglichkeit, Satz 16 auf Beispiel 3 aus §2 anzuwenden, sei kurz skizziert: D(M) und H werden mit der kanonischen Halbordnung versehen; $\lambda_* \in \mathbb{R}$ und $v \in C^\infty(\overline{\Omega})$ werden so gewählt, daß die Bedingungen $\lambda_* > 0$, $-\Delta v = \lambda_* v$ und $\int_\Omega dxdy \leq \int_\Omega v^2 dxdy$ erfüllt sind; ε_* und z werden durch $\varepsilon_* := \max_{(x,y)\in\partial\Omega} |v(x,y)|$ und $z(x,y) := \varepsilon_*$ für alle $(x,y) \in \overline{\Omega}$ definiert. Gilt nun $\varepsilon_* < 1$, so enthält das Intervall $[\frac{\lambda_*}{1+\varepsilon_*}, \frac{\lambda_*}{1-\varepsilon_*}]$ mindestens einen Eigenwert der betrachteten Aufgabe.

Die Idee, die Satz 16 zugrunde liegt, stammt von Fox, Henrici und Moler [17]; sie wurde von Nickel [43], Moler und Payne [42] sowie Donnelly [14] weiterentwickelt. Numerische Resultate liegen zu Membranproblemen für verschiedene Grundgebiete vor ([14], [17]).

§8 Kennt man eine positive Zahl γ derart, daß

$$\{f|f\} \leq \gamma\langle Mf|Mf\rangle \quad \text{für alle } f \in D(M) \tag{13}$$

gilt, so kann man leicht eine obere Schranke für die Norm der Lösung $\tilde{w}$ der Gleichung (12) angeben:

$$\|\tilde{w}\|^2 \leq \gamma\langle\tilde{h}|\tilde{h}\rangle \; . \tag{14}$$

Wenn man diese Abschätzung mit Satz 2 kombiniert, erhält man

Satz 17

Voraussetzungen und Bezeichnungen:

1. Es sei $\gamma \in \mathbb{R}$ mit $\gamma > 0$; für alle $f \in D(M)$ gelte $\{f|f\} \leq \gamma\langle Mf|Mf\rangle$.
2. $v_1,\ldots,v_n$ seien linear unabhängige Elemente von $D(M^{-1}N) \cap D(M)$.

3. Es seien $\lambda_* \in \mathbb{R}$ mit $\lambda_* > 0$,
$A := \gamma(\langle (M-\lambda_* N)v_i \mid (M-\lambda_* N)v_k \rangle)_{i,k=1,\ldots,n}$ und
$B := (\{v_i|v_k\})_{i,k=1,\ldots,n}$.
4. Für $i=1,\ldots,n$ sei μ_i der i-tkleinste Eigenwert der Matrix-Eigenwertaufgabe $Ax = \mu Bx$.

Behauptung: Für alle $r \in \mathbb{N}$ mit $r \leq n$ und $\mu_r < 1$ enthält das Intervall $\left[\frac{\lambda_*}{1+\sqrt{\mu_r}}, \frac{\lambda_*}{1-\sqrt{\mu_r}}\right]$ mindestens r Eigenwerte der Aufgabe $M\phi = \lambda N\phi$.

Beweis: Es sei $r \in \mathbb{N}$ mit $r \leq n$ und $\mu_r < 1$; $\tilde{V}$ sei der Unterraum des $\mathbb{R}^n$, der von denjenigen Eigenvektoren der Aufgabe $Ax = \mu Bx$ aufgespannt wird, die zu einem Eigenwert $\mu_i \leq \mu_r$ gehören. Dann gilt $\dim \tilde{V} \geq r$ und $c'Ac \leq \mu_r c'Bc$ für alle $c \in \tilde{V}$.
$V := \{ \sum_{i=1}^{n} c_i v_i : (c_1,\ldots,c_n)' \in \tilde{V}\}$ ist daher ein Unterraum von $D(M^{-1}N) \cap D(M)$ mit $\dim V \geq r$, und für alle $v \in V$ ergibt sich $\gamma\langle (M-\lambda_* N)v \mid (M-\lambda_* N)v\rangle \leq \mu_r\{v|v\}$.
Hieraus folgt mit Hilfe von (13) für alle $v \in V$ die Abschätzung $\|v - \lambda_* M^{-1}Nv\|^2 \leq \mu_r\|v\|^2$, aus der sich nach Satz 2 die Behauptung ergibt.

Viele der für Randwertaufgaben aufgestellten "expliziten a-priori-Abschätzungen" (s. [52]) können in die Form (13) überführt und dann gemäß Satz 17 zur Einschließung herangezogen werden. So läßt sich die Ungleichung
$$\int_\Omega f^2 dxdy \leq \gamma\int_\Omega (\Delta f)^2 dxdy + \gamma\omega \int_{\partial\Omega} f^2 ds \quad \text{für alle } f \in C^\infty(\overline{\Omega}) ,$$
in der γ und ω positive, nur von Ω abhängige Zahlen sind, zur Behandlung von Beispiel 3 aus §2 verwenden; hierzu muß man lediglich γ und ω ausrechnen (vgl. [52], S. 22).

Als nächstes wird eine Verallgemeinerung von Satz 2 bewiesen, die es ermöglicht, einige Voraussetzungen von Satz 17 abzuschwächen; hierdurch wird die Einschließung von Eigenwerten Stekloffscher Eigenwertaufgaben wesentlich erleichtert.

Satz 18

Voraussetzungen und Bezeichnungen:

1. $D(N_*)$ sei ein Unterraum von $D(S)$; $N_*: D(N_*) \to H$ sei eine li-

neare Abbildung; es gelte $D(N) \subset D(N_*)$ und $N_*f = Nf$ für alle $f \in D(N)$.

2. Für alle $f,g \in D(N_*)$ gelte $\langle Sf|N_*g\rangle = \langle Sg|N_*f\rangle$ und $\langle Sf|N_*f\rangle \geq 0$. Für alle $f \in D(N_*)$ sei $\|f\|_* := \sqrt{\langle Sf|N_*f\rangle}$.
3. M sei injektiv.
4. Es seien $\varrho, \tau \in \mathbb{R}$ mit $0 \leq \tau < |\varrho|$.
5. V_* sei ein r-dimensionaler Unterraum von $D(M^{-1}N_*) \cap D(N^{-1}N_*)$; für alle $v_* \in V_*$ mit $v_* \neq 0$ gelte $N_*v_* \neq 0$ und

   $$\|v_* - \frac{\varrho^2-\tau^2}{\varrho} M^{-1}N_*v_*\|_* \leq \frac{\tau}{|\varrho|}\|v_*\|_* .$$

Behauptung: Im Intervall $[\varrho-\tau, \varrho+\tau]$ liegen mindestens r Eigenwerte der Eigenwertaufgabe $M\phi = \lambda N\phi$.

Beweis (durch Zurückführung auf Satz 2): $V := \{N^{-1}N_*v_*: v_* \in V_*\}$ ist ein r-dimensionaler Unterraum von $D(M^{-1}N)$. Es sei $v \in V$; dann gibt es genau ein $v_* \in V_*$ mit $Nv = N_*v_*$, und es gilt $\|v_*\|_* = \|v\|$ sowie

$$\|v_* - (\varrho^2-\tau^2)\varrho^{-1}M^{-1}N_*v_*\|_* = \|v - (\varrho^2-\tau^2)\varrho^{-1}M^{-1}Nv\| .$$

V erfüllt somit die in Satz 2 genannten Bedingungen.

Satz 17 kann nun folgendermaßen verallgemeinert werden:

Satz 19

Voraussetzungen und Bezeichnungen:

1. Die Voraussetzungen 1 und 2 von Satz 18 seien erfüllt.
2. $D(M_*)$ sei ein Unterraum von $D(N_*)$; $M_*: D(M_*) \to H$ sei eine lineare Abbildung; es gelte $D(M) \subset D(M_*)$ und $M_*f = Mf$ für alle $f \in D(M)$.
3. Es sei $\gamma \in \mathbb{R}$ mit $\gamma > 0$; für alle $f \in D(M_*)$ gelte $\|f\|_*^2 \leq \gamma\langle M_*f|M_*f\rangle$.
4. $v_1,\dots,v_n$ seien Elemente aus $D(M^{-1}N_*) \cap D(N^{-1}N_*) \cap D(M_*)$.
5. Es seien $\lambda_* \in \mathbb{R}$ mit $\lambda_* > 0$,
   $A := \gamma(\langle(M_*-\lambda_*N_*)v_i \mid (M_*-\lambda_*N_*)v_k\rangle)_{i,k=1,\dots,n}$ und
   $B := (\langle Sv_i|N_*v_k\rangle)_{i,k=1,\dots,n}$; B sei positiv definit.
6. Für $i=1,\dots,n$ sei μ_i der i-tkleinste Eigenwert der Matrix-Eigenwertaufgabe $Ax = \mu Bx$.

Behauptung: Für alle $r \in \mathbb{N}$ mit $r \leq n$ und $\mu_r < 1$ enthält das Intervall $[\frac{\lambda_*}{1+\sqrt{\mu_r}}, \frac{\lambda_*}{1-\sqrt{\mu_r}}]$ mindestens r Eigenwerte der Aufgabe $M\phi = \lambda N\phi$.

Beweis: Es sei $r \in \mathbb{N}$ mit $r \leq n$ und $\mu_r < 1$; $\tilde{V}$ sei der Unterraum des $\mathbb{R}^n$, der von denjenigen Eigenvektoren der Aufgabe $Ax = \mu Bx$ aufgespannt wird, die zu einem Eigenwert $\mu_i \leq \mu_r$ gehören. Dann gilt $\dim \tilde{V} \geq r$ und $c'Ac \leq \mu_r c'Bc$ für alle $c \in \tilde{V}$.

$V_* := \{ \sum_{i=1}^{n} c_i v_i : (c_1,...,c_n)' \in \tilde{V}\}$ ist daher ein Unterraum von $D(M^{-1}N_*) \cap D(N^{-1}N_*) \cap D(M_*)$ mit $\dim V_* \geq r$, und für alle $v_* \in V_*$ mit $v_* \neq 0$ ergibt sich $N_* v_* \neq 0$ sowie

$\gamma\langle (M_*-\lambda_* N_*)v_* \mid (M_*-\lambda_* N_*)v_*\rangle \leq \mu_r \langle S v_* | N_* v_*\rangle$.

Hieraus folgt mit Hilfe von

$\| v_*-\lambda_* M^{-1}N_* v_* \|_*^2 \leq \gamma\langle M_* v_*-\lambda_* N_* v_* | M_* v_*-\lambda_* N_* v_*\rangle$

für alle $v_* \in V_*$ die Abschätzung

$\| v_*-\lambda_* M^{-1}N_* v_* \|_*^2 \leq \mu_r \| v_* \|_*^2$.

Setzt man $\varrho := \frac{\lambda_*}{1-\mu_r}$ und $\tau := \varrho\sqrt{\mu_r}$, so ergibt sich aus Satz 18 die Behauptung.

Der Beweis von Satz 19 verläuft weitgehend analog zu dem von Satz 17; in mancher Hinsicht weicht er aber auch von dem Schema ab, nach dem die Beweise der anderen Einschließungssätze verlaufen; so tritt beispielsweise die Halbnorm $\|\cdot\|_*$ an die Stelle der Norm $\|\cdot\|$. Natürlich kann man den grundlegenden Satz in §1 so allgemein fassen, daß sich alle Einschließungssätze in genau der gleichen Weise daraus herleiten lassen; man muß dann aber umständlichere Formulierungen in Kauf nehmen.

An Hand eines Beispiels soll nun erläutert werden, wie man Satz 19 auf Stekloffsche Eigenwertaufgaben anwenden kann.

Beispiel 4

Eigenwertaufgabe: $\Delta^2\phi = 0$ in Ω; $\phi = 0$, $\Delta\phi = \lambda\frac{\partial\phi}{\partial n}$ auf $\partial\Omega$

Voraussetzung: γ und ω seien positive Zahlen mit der Eigenschaft, daß

$$\int_{\partial\Omega} (\frac{\partial f}{\partial n})^2 ds \leq \gamma\omega\int_{\Omega} (\Delta^2 f)^2 dxdy + \gamma\int_{\partial\Omega} (\Delta f)^2 ds$$

für alle $f \in C^{\infty}(\overline{\Omega})$ mit $f = 0$ auf $\partial\Omega$ gilt (bezüglich γ und ω s. [30]).

<u>Einordnung:</u> $H := L_2(\Omega) \times L_2(\partial\Omega)$,

$$\left\langle \begin{pmatrix} f_1 \\ f_2 \end{pmatrix} \middle| \begin{pmatrix} g_1 \\ g_2 \end{pmatrix} \right\rangle := \omega\int_{\Omega} f_1 g_1 dxdy + \int_{\partial\Omega} f_2 g_2 ds \text{ für } \begin{pmatrix} f_1 \\ f_2 \end{pmatrix}, \begin{pmatrix} g_1 \\ g_2 \end{pmatrix} \in H,$$

$$D(M) := D(N) := \{f \in C^{\infty}(\overline{\Omega}) : \Delta^2 f = 0 \text{ in } \Omega,\ f = 0 \text{ auf } \partial\Omega\},$$

$$D(S) := D(M_*) := D(N_*) := \{f \in C^{\infty}(\overline{\Omega}) : f = 0 \text{ auf } \partial\Omega\},$$

$$M_* f := \begin{pmatrix} \Delta^2 f \\ \Delta f_{|\partial\Omega} \end{pmatrix},\ M := M_{*|D(M)},\ N_* f := Sf := \begin{pmatrix} 0 \\ \frac{\partial f}{\partial n} \end{pmatrix},\ N := N_{*|D(N)}.$$

Die zur Anwendung der Sätze 15 und 17 benötigten a-priori-Abschätzungen können im allgemeinen dann leicht aufgestellt werden, wenn (grobe) untere Schranken für den kleinsten Eigenwert gewisser Eigenwertaufgaben bekannt sind (vgl. z. B. [30], [32]).

Die ersten Sätze, in denen zur Eigenwerteinschließung explizite a-priori-Abschätzungen verwendet werden, sind von Kuttler und Sigillito [30], [31], [32] bewiesen worden; in [32] werden Eigenwertaufgaben $M\phi = \lambda\phi$ betrachtet, bei denen M ein symmetrischer Operator ist, während in [30] und [31] Einschließungssätze für zwei Stekloffsche Eigenwertaufgaben zu finden sind. Satz 19 enthält alle diese Resultate als Spezialfälle; er ermöglicht ferner die Behandlung von Eigenwertaufgaben, bei denen der Eigenwert sowohl in der Differentialgleichung als auch in den Randbedingungen auftritt. Während die Sätze von Kuttler und Sigillito nur aussagen, daß in bestimmten Intervallen mindestens e i n Eigenwert der betrachteten Aufgabe liegt, kann man nach Satz 19 auch Intervalle berechnen, die eine größere Mindestanzahl von Eigenwerten enthalten; dies dürfte für die Behandlung von Eigenwertaufgaben mit mehrfachen oder sehr nahe beieinander liegenden Eigenwerten von Interesse sein. - Erwähnt seien noch die Ergebnisse von McLaurin [40], [41], die als Vorläufer der Sätze von Kuttler und Sigillito aufgefaßt und ebenfalls aus dem grundlegenden Satz in §1 hergeleitet werden können. - Kuttler und Sigillito haben ihre Sätze mit gutem Erfolg an zwei Stekloffschen Eigenwertaufgaben [30], [31] und zahlreichen Schwingungsproblemen (Berechnung von Eigenfrequenzen rhombischer Membranen [32], orthotroper Rechteckplatten [34] sowie dreieckiger, trapezförmiger und rhombischer Platten [33], [35]) numerisch erprobt.

Die Tatsache, daß sich so viele bekannte Ergebnisse in einheitlicher Weise herleiten lassen, legt die Vermutung nahe, daß die geschilderte Methode auch zur Aufstellung neuer Einschließungssätze beitragen kann.

Die Verfasser danken der Deutschen Forschungsgemeinschaft für die großzügige Förderung und Herrn Dr. H.-E. Lahmann für die freundliche Unterstützung ihres Forschungsvorhabens.

Literatur

[1] Albrecht, J.: Verallgemeinerung eines Einschließungssatzes von L. Collatz. Z. Angew. Math. Mech. 48 (1968), T43 - T46

[2] Albrecht, J.: Quotienteneinschließungssätze für allgemeine Eigenwertaufgaben. Wiss. Z. Tech. Univ. Dresden 29 (1980), 424 - 426

[3] Albrecht, J.: Einschließung von Eigenwerten bei Schwingungen von Kreisbögen. Z. Angew. Math. Mech. 63 (1983), 387 - 389

[4] Andrushkiw, R.I.: On the approximate solution of K-positive eigenvalue problems Tu - λSu = O. J. Math. Anal. Appl. 50 (1975), 511 - 529

[5] Aubin, J.-P.: Approximation of elliptic boundary-value problems. New York - London - Sydney - Toronto: Wiley-Interscience 1972

[6] Bartsch, H.: Ein Einschließungssatz für die charakteristischen Zahlen allgemeiner Matrizen-Eigenwertaufgaben. Arch. Math. (Basel) 4 (1953), 133 - 136

[7] Bauer, F.L. und A.S. Householder: Moments and characteristic roots. Numer. Math. 2 (1960), 42 - 53

[8] Bazley, N.W. und D.W. Fox: Truncations in the method of intermediate problems for lower bounds to eigenvalues. J. Res. Nat. Bur. Standards Sect. B 65 (1961), 105 - 111

[9] Bredehöft, M. und W. Hauger: An iteration procedure and bounds for the eigenvalues for a class of non-selfadjoint eigenvalue problems. Mech. Res. Comm. 6 (1979), 105 - 111

[10] Collatz, L.: Einschließungssatz für die charakteristischen Zahlen von Matrizen. Math. Z. 48 (1942), 221 - 226

[11] Collatz, L.: Eigenwertaufgaben mit technischen Anwendungen. 2. Auflage. Leipzig: Akademische Verlagsgesellschaft Geest & Portig KG 1963

[12] Collatz, L.: Einschließungssatz für Eigenwerte bei partiellen Differentialgleichungen 2. und 4. Ordnung. Z. Angew. Math. Mech. 43 (1963), 277 - 280

[13] Dettmar, H.-K.: Symmetrisierbare Eigenwertaufgaben bei gewöhnlichen linearen Differentialgleichungen. Z. Angew. Math. Mech. 34 (1954), 284 - 287

[14] Donnelly, J.D.P.: Eigenvalues of membranes with reentrant corners. SIAM J. Numer. Anal. 6 (1969), 47 - 61

[15] Falk, S.: Einschließungssätze für die Eigenwerte normaler Matrizenpaare. Z. Angew. Math. Mech. 44 (1964), 41 - 55

[16] Fichera, G.: Numerical and quantitative analysis. London - San Francisco - Melbourne: Pitman 1978

[17] Fox, L., P. Henrici und C. Moler: Approximations and bounds for eigenvalues of elliptic operators. SIAM J. Numer. Anal. 4 (1967), 89 - 102

[18] Goerisch, F.: Über Quotienten-Einschließungssätze bei allgemeinen Eigenwertaufgaben. In: J. Albrecht, L. Collatz und G. Hämmerlin (Hrsg.): Numerische Behandlung von Differentialgleichungen mit besonderer Berücksichtigung freier Randwertaufgaben. International Series of Numerical Mathematics (ISNM), Vol. 39. Basel - Stuttgart: Birkhäuser 1978, 86 - 100

[19] Goerisch, F.: Weiterentwicklung von Verfahren zur Berechnung von Eigenwertschranken. Dissertation TU Clausthal 1978

[20] Goerisch, F.: Eine Verallgemeinerung eines Verfahrens von N.J. Lehmann zur Einschließung von Eigenwerten. Wiss. Z. Tech. Univ. Dresden 29 (1980), 429 - 431

[21] Goerisch, F.: Über die Anwendung einer Verallgemeinerung des Lehmann-Maehly-Verfahrens zur Berechnung von Eigenwertschranken. In: J. Albrecht und L. Collatz (Hrsg.): Numerische Behandlung von Differentialgleichungen, Band 3. International Series of Numerical Mathematics (ISNM) Vol. 56. Basel - Boston - Stuttgart: Birkhäuser 1981, 58 - 72

[22] Goerisch, F.: Ein Stufenverfahren zur Berechnung von Eigenwertschranken. Erscheint demnächst in den Nova Acta Leopoldina (N.F.)

[23] Goerisch, F. und J. Albrecht: Untere Schranken für die Eigenwerte Stekloffscher Eigenwertaufgaben. In: M. Gregus (Hrsg.): Equadiff 5. Proceedings. Teubner-Texte zur Mathematik, Bd. 47. Leipzig: Teubner 1982, 9 - 13

[24] Goerisch, F. und J. Albrecht: Die Monotonie der Templeschen Quotienten. Eingereicht bei der Z. Angew. Math. Mech.

[25] Goerisch, F. und H. Haunhorst: Eigenwertschranken für Eigenwertaufgaben mit partiellen Differentialgleichungen. Eingereicht bei der Z. Angew. Math. Mech.

[26] Hadeler, K.P.: Einschließungssätze bei normalen und bei positiven Operatoren. Arch. Rational Mech. Anal. 21 (1966), 58 - 88

Harazov, D.F. siehe Kharazov, D.F.

[27] Held, W.: Einschließungen von Eigenwerten. Dissertation TU Clausthal 1972

[28] Held, W.: Die Collatzschen Einschließungssätze für Eigenwerte bei Differentialgleichungen. In: L. Collatz und K.P. Hadeler (Hrsg.): Numerische Behandlung von Eigenwertaufgaben. International Series of Numerical Mathematics (ISNM), Vol. 24. Basel - Stuttgart: Birkhäuser 1974, 47 - 55

[29] Kharazov, D.F.: Estimates for the eigenvalues of certain operators with discrete spectrum. Differencial'nye Uravnenia 1 (1965), 1054 - 1069 (Englische Übersetzung: Differential Equations 1 (1965), 822 - 834)

[30] Kuttler, J.R.: Dirichlet eigenvalues. SIAM J. Numer. Anal. 16 (1979), 332 - 338

[31] Kuttler, J.R.: Bounds for Stekloff eigenvalues. SIAM J. Numer. Anal. 19 (1982), 121 - 125

[32] Kuttler, J.R. und V.G. Sigillito: Bounding eigenvalues of elliptic operators. SIAM J. Math. Anal. 9 (1978), 768 - 773

[33] Kuttler, J.R. und V.G. Sigillito: Upper and lower bounds for frequencies of clamped rhombical plates. Journal of Sound and Vibration 68 (1980), 597 - 607

[34] Kuttler, J.R. and V.G. Sigillito: Upper and lower bounds for the frequencies of clamped orthotropic plates. Journal of Sound and Vibration 73 (1980), 247 - 259

[35] Kuttler, J.R. and V.G. Sigillito: Upper and lower bounds for frequencies of trapezoidal and triangular plates. Journal of Sound and Vibration 78 (1981), 585 - 590

[36] Lange, E., Diplomarbeit TU Clausthal 1979

[37] Lehmann, N.J.: Beiträge zur numerischen Lösung linearer Eigenwertprobleme. I. Z. Angew. Math. Mech. 29 (1949), 341 - 356

[38] Lehmann, N.J.: Beiträge zur numerischen Lösung linearer Eigenwertprobleme. II. Z. Angew. Math. Mech. 30 (1950), 1 - 16

[39] Lehmann, N.J.: Optimale Eigenwerteinschließungen. Numer. Math. 5 (1963), 246 - 272

[40] McLaurin, J.: Bounding eigenvalues of clamped plates. Z. Angew. Math. Phys. 19 (1968), 676 - 681

[41] McLaurin, J.: Bounds for vibration frequencies and buckling loads of clamped plates. Dissertation ETH Zürich 1969

[42] Moler, C.B. und L.E. Payne: Bounds for eigenvalues and eigenvectors of symmetric operators. SIAM J. Numer. Anal. 5 (1968), 64 - 70

[43] Nickel, K.L.E.: Extension of a recent paper by Fox, Henrici and Moler on eigenvalues of elliptic operators. SIAM J. Numer. Anal. 4 (1967), 483 - 488

[44] Rektorys, K.: Variational methods in mathematics, science and engineering. Dordrecht - Boston: D. Reidel Publishing Company 1977

[45] Rektorys, K. und Vospěl, Z.: On a method of twosided eigenvalue estimates for elliptic equations of the form $Au - \lambda Bu = 0$. Apl. Mat. 26 (1981), 211 - 240

[46] Rieder, G.: Elementargeometrische Herleitung von Einschließungssätzen für Eigenwerte. Z. Angew. Math. Mech. 48 (1968), 207 - 210

[47] Ruppert, K.: Über die Anwendung von Quotienten-Einschließungssätzen für Eigenwerte. Dissertation TU Clausthal 1983

[48] Schäfke, F.W. und A. Schneider: S-hermitesche Rand-Eigenwertprobleme. I. Math. Ann. 162 (1965), 9 - 26

[49] Schäfke, F.W. und A. Schneider: S-hermitesche Rand-Eigenwertprobleme. II. Math. Ann. 165 (1966), 236 - 260

[50] Schellhaas, H.: Verfahren zur expliziten Berechnung unterer Eigenwertschranken beim allgemeinen Eigenwertproblem im Hilbertraum. Dissertation TH Darmstadt 1966

[51] Schellhaas, H.: Ein Verfahren zur Berechnung von Eigenwertschranken mit Anwendung auf das Beulen von Rechteckplatten. Ingenieur-Archiv 37 (1968), 243 - 250

[52] Sigillito, V.G.: Explicit a priori inequalities with applications to boundary value problems. London - San Francisco - Melbourne: Pitman 1977

[53] Stummel, F.: Rand- und Eigenwertaufgaben in Sobolewschen Räumen. Berlin - Heidelberg - New York: Springer 1969

[54] Temple, G.: The theory of Rayleigh's principle as applied to continuous systems. Proc. Roy. Soc. Ser. A 119 (1928), 276 - 293

[55] Velte, W.: Direkte Methoden der Variationsrechnung. Stuttgart: Teubner 1976

[56] Weinstein, A. und W. Stenger: Methods of intermediate problems for eigenvalues. New York - London: Academic Press 1972

[57] Wielandt, H.: Ein Einschließungssatz für charakteristische Wurzeln normaler Matrizen. Arch. Math. (Basel) 1 (1949), 348 - 352

Übersicht über die Herleitung der Einschließungssätze

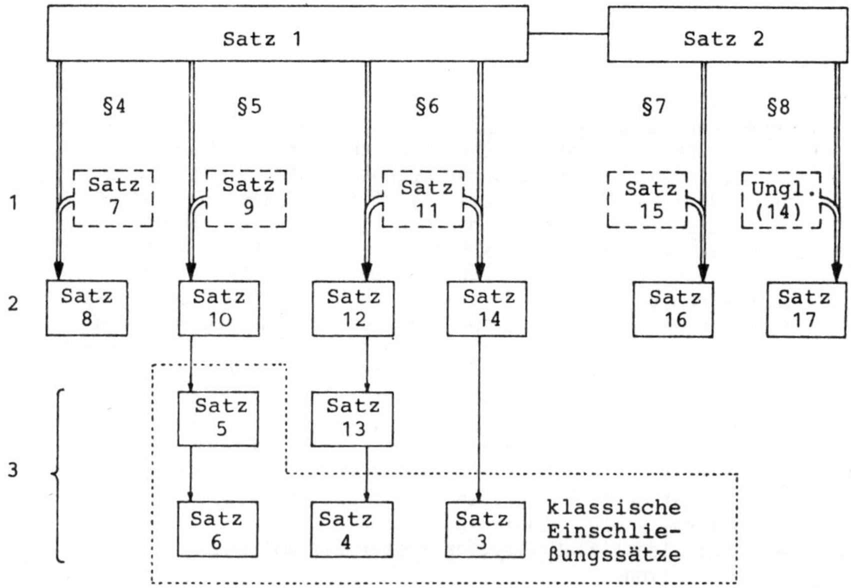

1 Normabschätzungen 2 Einschließungssätze 3 Spezialfälle

Dr. F. Goerisch, Prof. Dr. J. Albrecht
Institut für Mathematik der TU Clausthal
Erzstraße 1, D 3392 Clausthal-Zellerfeld

ON PARAMETER IDENTIFIABILITY

Guy CHAVENT
Université de Paris IX-Dauphine
75775 Paris Cédex 16
FRANCE and INRIA B.P. 105 78150 Le Chesnay Cédex
FRANCE

Abstract

Given a mathematical model of a physical process, the parameter identification problem is defined as that of determining the parameters of the model that, for a known given input, yield a given measured output. Identifiability is often defined as the injectivity of the parameter → output mapping defined by the mathematical model used. Though already this definition is difficult to check in practical situations, it is not guaranteed, when it is satisfied, that parameters can be determined in a unique and stable way from the measurement. So we will review some recent definitions as OLS-stability and OLS-identifiability: Two different ways of obtaining OLSI will be given: either by studying the injectivity of the linearized parameter → output mapping (the "sensitivity matrix"), which will yield OLSI for finite dimensional parameters, but will usually be of little help for infinite dimensional parameters, or by using a regularization technique, which will be shown to yield OLSI for a large enough regularization parameter.
Keywords Identifiability, parameter estimation, regularization.

1. Introduction

This paper is a theoretically oriented paper where different sufficient conditions for identifiability are studied. The emphasis is on developing tools which give insight into the properties of the system that are of interest in order to guarantee the identifiability of the parameter.

Let us denote by E the parameter space and by F the output and observation space. In practice the space F is a Hilbert space, and for simplicity we will suppose that the parameter space E is also a Hilbert space:

$$\begin{cases} E, F \text{ are Hilbert spaces with} \\ \text{Scalar products } \langle\, ,\rangle_E \text{ and } \langle\, ,\rangle_F \end{cases} \tag{1}$$

The parameter $x \in E$ is linked to the output $y \in F$ by a usually complicated sequence of equations, including a state equation and an observation operator, which selects the components of the state vector that can be observed (see Fig. 1). *The parameter estimation problem* consists of finding an estimated value $\hat{x}$ of the parameter x from the knowledge of the data z, of the parameter → output mapping ϕ, and of some *a priori* knowledge on the parameter contained in a set C of admissible parameters. Because of the measurement and model errors, the equation

$$\text{find } \hat{x} \in 0 \text{ such that } \phi(\hat{x}) = z \tag{2}$$

usually has no exact solution so that it has to be solved approximately using a least square formulation

$$\begin{cases} \text{find } \hat{x} \in C \text{ such that} \\ J(\hat{x}) \le J(x) \ \forall\, x \in C \end{cases} \tag{3}$$

where $J(x)$ is the output or measurement error criterion:

$$\forall\, x \in C \quad , \quad J(x) = \| \phi(x) - z \|_F^2 \tag{4}$$

The least square setting (3), (4) of the parameter estimation problem is now standard and widely used.

The identifiability problem consists then in studying the well-posedness of the parameter estimation problem. The early definitions of identifiability were all more or less intended to ensure the unique-

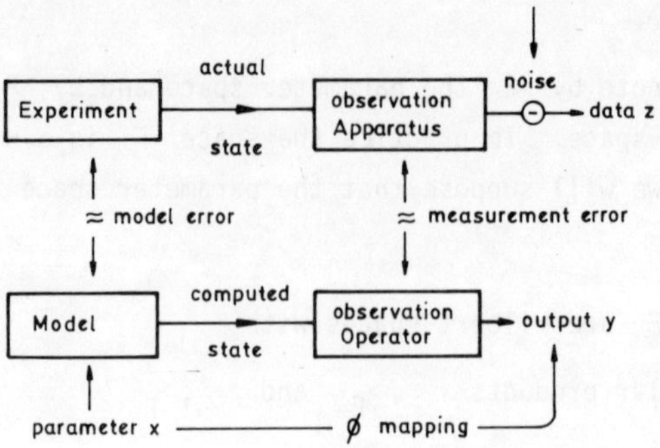

Fig. 1 An example of interpretation of the ϕ-mapping and of data z.

ness of the solution $\hat{x}$ of the formulation (2), for example:

Definition 1: The parameter x is identifiable at $\bar{x} \in C$ iff: $x \in C$ and $\phi(x) = \phi(\bar{x}) \Rightarrow x = \bar{x}$.

Definition 2: The parameter x is identifiable on C iff: $x, x' \in C$ and $\phi(x) = \phi(x') \Rightarrow x = x'$, i.e. iff ϕ is injective on C.

These definitions are not easy to use (proving the injectivity of the nonlinear mapping is a difficult task) and they suffer two inherent weakness: firstly, they do not say anything on the continuity of the solution $\hat{x}$ of (2) when the right-hand size z is changed (which is usually included in the definition of well-posedness); secondly, they do not say anything at all about the well-posedness of the least square formulation (3) of parameter estimation problem, which is the only effective setting for practical situations where measurements and model errors cannot be avoided.

So in the last years some new definitions of identifiability have been given, which apply essentially to the least square formulation (3), and consider stability and/or uniqueness of its solution.

First we give a definition recently introduced by Kunison (1984) for the stability of the local minimizers of the OLS problem:

<u>Definition 3</u>: Let $\hat{x} \in C$ be a (local or global) solution of the OLS problem (3), (4). Then x is called Output Least Square (OLS-) stable at $\hat{x}$ in C if there exists a neighborhood $\mathcal{Z}$ of z in F, a constant $r > 0$ and a nondecreasing continuous real valued function ρ with $\rho(\sigma) = 0$, such that for all $z' \in \mathcal{Z}$, there exists a local solution $\hat{x}'$ of the OLS problem (3), (4) (with z' instead of z) such that

$$\| \hat{x}' - \hat{x} \|_E \leq r \tag{5}$$

and if, for every such local solution satisfying (5):

$$\| \hat{x}' - \hat{x} \|_E \leq \rho(\| z' - z \|) \quad . \tag{6}$$

For ease of presentation, we have dropped from the original definition of Kunisch and Colonius the stability with respect to the right-hand side of the constraints defining the admissible set C. The (OLS)-stability property, which does not require any injectivity properties of the mapping ϕ, can be achieved by different techniques, for example by taking a finite dimensional parameter space, or by adding a regularization term. Let $\varepsilon > 0$ be the regularization parameter and $x_0 \in E$ be some *a priori* estimate of the unknown parameter (one can take $x_0 = 0 \ldots$). Then the OLS problem (3), (4) is replaced by the *Regularized OLS (ROLS) problem:*

$$\begin{cases} \text{find } \hat{x}_\varepsilon \in C \text{ such that} \\ J_\varepsilon(\hat{x}_\varepsilon) \leq J_\varepsilon(x) \qquad \forall\, x \in C \end{cases} \tag{7}$$

with

$$\forall\, x \in C \quad , \quad J_\varepsilon(x) = \| \phi(x) - z \|_F^2 + \varepsilon^2 \| x - x_0 \|_E^2 \quad . \tag{8}$$

Colonius and Kunisch (1984) introduced a stability concept adapted to this regularized problem:

<u>Definition 3 bis</u>: The parameter x is called Output Least Squares Stables by Regularization (ROLS-Stable) in C at $z \in F$ for

$\varepsilon \in J \subset (0, T\infty)$ if for every $\varepsilon \in J$ and for every (global) solution $\hat{x}_\varepsilon$ of the ROLS problem (7), (8), there exists a neighborhood $\mathcal{X}$ of z in F, a constant $r > 0$ and a continuous nondecreasing real valued function ρ with $\rho(0) = 0$ such that, for all $z' \in \mathcal{X}$, there exists a local solution $\hat{x}'$ of the ROLS problem (7), (8) (with z' instead of z) satisfying

$$\| \hat{x}'_\varepsilon - \hat{x}_\varepsilon \|_E \leq r \tag{9}$$

and if, for every such local solution:

$$\| \hat{x}'_\varepsilon - \hat{x}_\varepsilon \|_E \leq \rho(\| z' - z \|) \quad . \tag{10}$$

The same authors have shown that both stability properties of definition 3 and 3 bis holds for an infinite dimensional parameter estimation problem in a two point boundary value problem. Although this is a useful tool for infinite dimensional problems, it does not address the uniqueness problem. As one can guess, this uniqueness (together with stability of course) is important for the identifiability of the parameter; so we turn to one definition which takes into account these two properties:

Definition 4: Output Least Square Identifiability (OLSI, see Chavent, 1979). The parameter x is OLSI on C from the measurement of $\phi(x)$ iff there exists a neighborhood $\mathcal{U}$ of $\phi(C)$ in F such that, for every $z \in \mathcal{U}$, the least square problem (3), (4) has a unique solution $\hat{x}$ depending continuously on z.

In fact, OLSI is an essential condition in interpreting the solution $\hat{x}$ of (3), (4). Suppose that $\bar{x} \in C$ is the exact - but unknown - parameter, and let $z \in F$ be the measured output. When the measurement errors go to zero, the point z tends towards $\phi(\bar{x})$ and, in case of OLSI, the corresponding estimate $\hat{x}$ tends towards $\bar{x}$, so that the solution $\hat{x}$ of (3), (4) can be interpreted as an approximation of the true parameter $\bar{x}$. Moreover, the sufficient condition for OLSI, which will be given in paragraph 2, does not rely on a compactness argument, but on a completeness argument. This is similar to the one used for proving the existence of the minimum of a convex function on a closed convex set. Hence the condition guarantees that any minimizing sequence is converging, which is important for the actual computation of $\hat{x}$ using a minimization algorithm.

So the OLS-Identifiability is the property that one would wish any parameter estimation problem to satisfy. Unfortunately, OLSI is seldom achieved, as it supposes one is able to prove the injectivity of ϕ - and even more! So we will recall in paragraph 2 a sufficient condition for OLSI (Chavent 1983), which is stated in terms of the derivatives $\phi'(x)$ and $\phi''(x)$ of the parameter $\rightarrow$ output mapping ϕ.

The "sensitivity operator" $\phi'(x)$ is the usual tool for the study of the shape of the level lines of the function J near one global minimum $\hat{x}$. Here, the second derivative of J is given by

$$J''(\hat{x})(\delta x, \delta y) = 2 < \phi'^{t}(\hat{x})\phi'(\hat{x})\delta x, \delta y >_E$$

$$+ 2 < \phi''(\hat{x})(\delta x, \delta y), \phi(\hat{x}) - z >_F$$

or, when the second term is neglected using the fact that $\phi(\hat{x}) - z = 0$ for small measurement and model errors:

$$J''(\hat{x}) = 2\phi'^{t}(\hat{x})\phi'(\hat{x}) \quad .$$

Let:

$$\left\{ \begin{array}{l} 0 \leq \alpha^2(\hat{x}) \leq \delta^2(\hat{x}) \text{ be the smallest} \\ \text{and largest eigenvalues of } \phi'^{t}(\hat{x})\phi'(\hat{x}) \end{array} \right.$$

then the number $\delta(\hat{x})/\alpha(\hat{x})$ yields an estimation of the conditioning of the optimization problem (3), (4) near its local minimum $\hat{x}$. Of course this number cannot give any information about what occurs "far from $\hat{x}$": even if $J''(\hat{x})$ is positive definite, i.e.,

$$0 < \alpha(\hat{x}) \leq \delta(\hat{x}) \quad .$$

(which ensures that $\hat{x}$ is a strict local minimum of J), one cannot guarantee that J does not possess other local minimas, and hence $\hat{x}$ may not be the global minimum of J.

The idea underlying the sufficient condition for OLSI of paragraph 2 is that, by using information on $\phi'(x)$ and $\phi''(x)$ all over C, namely the numbers $\underline{\alpha}$, $\bar{\delta}$ and $\bar{\beta}$ defined by

$$\underline{\alpha} = \operatorname*{Inf}_{x \in C} \alpha(x) \qquad \bar{\delta} = \operatorname*{Sup}_{x \in C} \delta(x)$$

$$\bar{\beta} = \operatorname*{Sup}_{x \in C} \| \phi''(x) \|_{\mathscr{L}_2(E, F)}$$

one can obtain enough global information on ϕ to guarantee OLSI, provided that $\underline{\alpha}$ is strictly positive and that C is not too large. In some sense, this sufficient condition enables us to replace the study of the injectivity of the mapping ϕ by the expectedly easier study of the injectivity of the linearized problems for each parameter value $x \in 0$.

This sufficient condition will be shown in Section 3 to resume for finite dimensional parameters, to plain injectivity of $\phi'(x)$ over C.

Finally, we will devote the last two paragraphs to the cases where the condition $\bar{\alpha} > 0$ (stronger than injectivity) cannot be fulfilled. Such cases are important as they cover a large majority of the practical parameter estimation problems. In paragraph 4, we will study the influence of a weaker injectivity condition on ϕ' and on the uniqueness of the solution of the OLS problem (3), (4). In paragraph 5 we will see how to use the regularized problem (7), (8) to restore OLSI when injectivity is lacking, i.e. when $\underline{\alpha} = 0$. The main result there is that a *minimum amount* ε *of regularization* is required, in contrast to ROLS-stability where an arbitrary small amount of regularization may be enough. When the regularizing term is interpreted as the norm of the deviation from some *a priori* estimate of the unknown parameter, the results enable us to quantify the minimum amount of confidence one should have in this *a priori* estimate in order to get a well-posed parameter estimation problem.

2. A Sufficient Condition for OLSI

We recall here results of Chavent, 1983. We suppose that the set of admissible parameters C satisfy:

$$C \text{ is a bounded, closed and convex subset of } E \tag{11}$$

and that the parameter $\rightarrow$ output mapping ϕ satisfies

$$\phi \text{ is twice Gateau-differentiable over } C \tag{12}$$

with a derivative $\phi'(x)$ satisfying

$$0 < \underline{\alpha} \leq \bar{\delta} \quad , \quad 0 \leq \bar{\beta} \leq +\infty \tag{13}$$

where $\underline{\alpha}$, $\bar{\delta}$, and $\bar{\beta}$ are defined by (9), (10), or equivalently by:

$$\forall\, x \in C \quad , \quad \forall\, \delta x \in E : \underline{\alpha} \| \delta x \|_E \leq \| \phi'(x) \delta x \|_F \tag{14a}$$

and

$$\begin{cases} \forall\, x \in C \quad , \quad \forall\, \delta x \in E \\ \| \phi'(x) \cdot \delta x \|_F \leq \bar{\delta} \| \delta x \|_E \\ \| \phi''(x)(\delta x, \delta x) \|_F \leq \bar{\beta} \| \delta x \|_E^2 \end{cases} \tag{14b}$$

Of course, the most restrictive hypothesis is the condition $\underline{\alpha} > 0$ in (13), which means that the derivative $\phi'(x)$ is invertible in the least-square sense all over C.

We introduce now a condition on the size of the set C of admissible parameters

$$\bar{\beta} \text{ diam } C < 2\sqrt{2}\, \underline{\alpha} \tag{15}$$

(where $\text{diam } C : \sup_{x,y \in C} \| x - y \|_E$). This condition is a sufficient condition for the image set $\phi(C)$ not to overlap because of the curvature introduced by the mapping ϕ (the larger $\bar{\beta}$, the smaller C). When the size condition (15) is satisfied, one can define a strictly positive number γ by:

$$\gamma = \frac{\alpha^2}{\bar{\beta}} - \frac{\bar{\beta}}{8} (\text{diam } C)^2 > 0 \tag{16}$$

Then we have the

Theorem 1

Under hypothesis (1), (11) to (15), the parameter x is OLSI on C from the measurement of $\phi(x)$, with the neighborhood $\mathcal{V}$ of $\phi(C)$ given by

$$\mathcal{V} = \{z \in F \mid d(z,\phi(c)) < \gamma\} \tag{17}$$

Moreover, for z $\mathcal{V}$, any minimizing sequence of the OLS problem (3), (4) converges toward the unique solution $\hat{x}$.

This theorem reduces the study of OLSI to that of the "sensitivity operator" $\phi'(x)$ for all admissible parameters. This is of course still a large and difficult task, but we will see that for some cases it can be done.

From the formulae (15), (16), (17) we see that there is some sort of balance between the amount of *a priori* available information (represented by the diameter of C) and the largest measurement and model errors which still maintains the well-posedness of OLS problem (represented by the number γ): the closest diam C approaches $2\sqrt{2}\ \underline{\alpha}/\bar{\beta}$ from below, the smallest the admissible error γ.
If now the diameter of C is larger than $2\sqrt{2}\ \underline{\alpha}/\bar{\beta}$, then Theorem 1 does not allow any conclusion, as it is only a sufficient condition. The optimal constant in (15) is somewhere between $2\sqrt{2}$ and 2π, as it can be seen easily by considering the counter example $E = \mathbf{R}$, $C = [0, X] \subset \mathbf{R}$ with $X \geq 2\pi$, $F = \mathbf{R}^2$ and $\phi(x) = (\cos x, \sin x)$.

We will see now that the inverse problem is Lipschitz-stable under the sufficient condition of Theorem 1. Let z and z' be two data belonging to the neighborhood $\mathcal{V}$ defined in (17), and $\hat{x}$ and $\hat{x}'$ be the corresponding solution,

$$d = \text{Max}\{d(z, \phi(C)), d(z', \phi(C))\} < \gamma \tag{18}$$

and define a mapping $f: [OT] \to \mathbb{R}$ by

$$\forall\ \theta \in [0, T] \quad , \quad f(\theta) = \| \phi(\theta\hat{x} + (1-\theta)\hat{x}) + \theta z' + (1-\theta)z \|_F \tag{19}$$

Then we have

Theorem 2

Under the hypothesis of Theorem 1 and (18), (19), the following alternative holds:

i) either

$$\exists\ \theta \in]0, 1[\text{ s.t. } \quad f(\theta) > d \tag{20}$$

(we will say that "z and z' are on a concave side of $\phi(C)$") with

$$\| \hat{x}' - \hat{x} \|_E \leq \frac{2\bar{\delta}}{\bar{\beta}(\gamma - d)} \| z' - z \|_F \tag{21}$$

ii) or

$$\forall\ \theta \in]0, 1[\quad f(\theta) < d \tag{22}$$

(we will say that "z and z' are on a convex side of $\phi(C)$") with

$$\| \hat{x}' - \hat{x} \|_E \leq \frac{2\bar{\delta}}{\underline{\alpha}^2 - \bar{\beta}\gamma} \| z' - z \|_F \quad . \tag{23}$$

As one can reasonably expect, the Lipschitz constant in (21) on the "concave" side of $\phi(C)$ blows-up to infinity when z and z' approach the boundary of the neighborhood $\mathcal{V}$ (i.e. when $d \to \gamma$). On the "convex" side of $\phi(C)$, the Lipschitz constant in (23) takes the value 2 when $E = F$ and $\phi = $ Identity, whereas the optimal value would be 1!

3. Finite Dimensional Parameters

We consider in this paragraph the case of a finite number of real parameters:

$$E = \mathbb{R}^n \tag{24}$$

and where one has been able to prove, by theoretical or numerical considerations, that the sensitivity operator $\phi'(x)$ (which is matrix if F is finite dimensional too) is injective over C:

$$\begin{cases} \forall\, x \in C \\ \phi'(x) \cdot \delta x = 0 \Rightarrow \delta x = 0 \end{cases} \tag{25}$$

Because of hypothesis (24), both the set C satisfying (11) and the unit sphere $S = \{\delta x \in E \mid \|\delta x\|_E = 1\}$ of E are compact. Hence the continuous functions from $C \times S$ in $\mathbb{R}$ defined by $(x, \delta x) \to \|\phi'(x) \cdot \delta x\|_F$ and $(x, \delta x) \to \|\phi''(x) \cdot (\delta x, \bar{\delta} x)\|_F$ both attain their minima and maxima over $C \times S$:

$$\begin{cases} \exists\, x_j \in C \quad \text{and} \quad \delta x_j \in S \quad , \quad j = 0,1,2 \quad \text{such that} \\ \forall\, x \in C \quad \text{and} \quad \forall\, \delta x \in S \\ \underline{\alpha} = \|\phi'(x_0)\delta x_0\|_F \le \|\phi'(x)\delta x\|_F \le \|\phi'(x_1)\delta x_1\|_F = \bar{\delta} \\ \|\phi''(x_2)(\delta x_2, \delta x_2)\|_F = \bar{\beta} \end{cases} \tag{26}$$

From (25) we get then

$$\underline{\alpha} > 0 \tag{27}$$

so that $\underline{\alpha}$, $\bar{\delta}$ and $\bar{\beta}$ defined by (26) satisfy the hypothesis (13) of Theorem 2. Hence we get

<u>Corollary</u>: Under hypothesis (1), (24), (25), (15), (16), the results of Theorems 1 and 2 hold.

Of course, if for example the finite dimensional parameter has been obtained by discretization of an infinite dimensional problem for which one has been unable to prove the strong injectivity condition (13), (14a) ($\underline{\alpha} > 0$), it is to be expected that, even if the plain injectivity condition (25) holds for this finite dimensional problem, the number $\underline{\alpha}/\bar{\beta}$ obtained by (25) will be very small, so that the size of the corresponding set of admissible parameter will be very small. Even in truly finite dimensional problems with injective derivative, the maximum size of C given by (15) may be too small for a practical use, as the majorations done for its derivation are very rough. More refined estimations are under development.

4. A Weaker Condition on the Derivative

When the parameter space E is infinite dimensional, the plain injectivity of $\phi'(x)$ over C does not imply the existence of $\underline{\alpha} > 0$ satisfying (14a). So we will see in this paragraph what happens if one replaces (14a) with $\underline{\alpha} > 0$ by the weaker condition

$$\begin{cases} \forall\, x \in C \quad , \quad \forall\, \delta x \in E \\ \alpha(\|\delta x\|_E)\,\|\delta x\|_E \leq \|\phi'(x)\cdot\delta x\|_F \end{cases} \tag{28}$$

where

$$\alpha : \mathbf{R}^+ \to \mathbf{R}^+ \quad \text{is continuous with} \quad \alpha(0) = 0 \tag{29}$$

This kind of hypothesis has been shown to be satisfied for the estimation of the diffusion coefficient of a parabolic equation from a boundary measurement in a paper by Pagani (1982), under adequate - and complicated - choice of the parameter space E and of the set C of admissible parameters. Depending on that choice, the functions α exhibited in this paper are $\alpha(t) = \exp(-1/t)$ and $\alpha(t) = t^{1/p}$ $(p \geq 0)$.
We will require that the function $t \to \alpha(t)$ does not vanish too quickly when $t \to 0$; more precisely, we will suppose that

$$\alpha(t)/t \to +\infty \qquad \text{when} \qquad t \to 0 \quad . \tag{30}$$

The condition (15) on the size of C will be replaced by

$$\begin{cases} \exists \eta > 0 \text{ such that } \forall\, t \in [0, \text{diam } C]: \\ \beta t < \sqrt{1-\eta} \times 2\sqrt{2}\,\alpha(t) \end{cases} \tag{31}$$

which can always be satisfied for diam C small enough using (30). Then we have the following theorem.

Theorem 3

When C is convex and satisfies (31), with ϕ' satisfying (28), (29), (30) and ϕ'' satisfying $\|\phi''(x)\| \leq \bar{\beta}\ \forall\, x \in C$, the diameter Δ of the set of solutions of the OLS problem (3), (4), supposed to be non empty, is bounded by

$$\eta\alpha^2(\Delta) \leq \bar{\beta} d(z, \phi(C)) \tag{32}$$

In particular, ϕ is injective over C.

Proof: Let $\hat{x}$ and $\hat{x}'$ be two solutions of the OLS problem (3), (4), and let:

$$w = \hat{x}' - \hat{x}$$

Define a function $f: [OT] \to \mathbf{R}$ by

$$\forall\, \theta \in [0,1] \quad , \quad f(\theta) = \| \phi(\theta\hat{x}' + (1-\theta)\hat{x}) - z\|^2 \quad .$$

Differentiating f yields $(x_\theta = \theta\hat{x}' + (1-\theta)\hat{x})$:

$$f'(\theta) = 2 < \phi'(x_\theta) \cdot w,\ \phi(x_\theta) - z >_F \quad ,$$

$$f''(\theta) = 2\| \phi'(x_\theta) \cdot w\|_F^2 + 2 < \phi''(x_\theta)(w, w),\ \phi(x_\theta) - z >_F \quad . \tag{33}$$

Hence:

$$|f'(\theta)| \leq 2\sqrt{f(\theta)}\, \| \phi'(x_\theta) \cdot w\|_F$$

and

$$\frac{f''(\theta)}{f(\theta)^{1/2}} - \frac{f'(\theta)^2}{2f(\theta)^{3/2}} + 2\bar{\beta}\|w\|_E \geq 0 \quad .$$

This proves that the mapping $\theta \to 2\sqrt{f(\theta)} + \bar{\beta}(\theta^2 - \theta)\| w\|_E^2$ has a positive second derivative and hence is convex. This yields

$$\sqrt{f(\theta)} \le d(z, \phi(c)) + \frac{\bar{\beta}}{8} \| w \|_E^2 \qquad \forall\ \theta \in [0,1] \tag{34}$$

on the other hand, by definition of $\hat{x}$ and $\hat{x}'$, the function $f(\theta)$ satisfies always

$$f(\theta) \ge f(0) = f(1) = d(z, \phi(C))^2 \qquad \forall\ \theta \in [0,1]$$

so that there necessarily exists some $\theta_0 \in [0, 1]$ such that:

$$f''(\theta_0) \le 0$$

which yields, using (33) and hypothesis (29):

$$2\alpha^2(\| w \|_E) \| w \|_E^2 \le 2\bar{\beta} \| w \|_E^2 \sqrt{f(\theta_0)}$$

i.e.

$$\frac{\alpha^2(\| w \|)_E}{\bar{\beta}} \le \sqrt{f(\theta_0)} \tag{35}$$

Comparing (34) and (35) gives:

$$\frac{\alpha^2(\| w \|)_E}{\bar{\beta}} \le d(z, \phi(C)) + \frac{\bar{\beta}}{8} \| w \|_E^2 \quad . \tag{36}$$

But from (31) we get, as $t = \| w \|_E \in [0, \text{diam } C]$:

$$\frac{\bar{\beta}}{8} \| w \|_E^2 < (1 - \eta) \frac{\alpha^2(\| w \|_E)}{\bar{\beta}}$$

which, together with (36), yields

$$\eta \frac{\alpha^2(\| w \|_E)}{\bar{\beta}} < d(z, \phi(C))$$

which proves (32) as $\hat{x}$ and $\hat{x}'$ are any two solutions of the OLS

problem (3), (4). This ends the proof of Theorem 3.
Theorem 3 shows that, though ϕ is injective over C, the OLS problem (3), (4) may have many solutions if measurement or model errors are present (i.e. $d(z, \phi(C)) > 0$). However, the size of the set of solutions tends towards zero: one checks easily that (32) and (30) imply that

$$\Delta / \sqrt{d(z, \phi(C))} \to 0 \quad \text{when} \quad d(z, \phi(c)) \to 0 \tag{37}$$

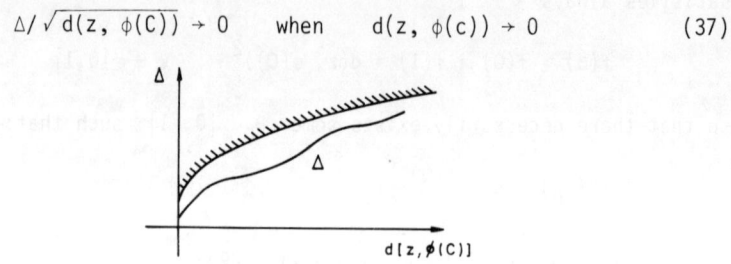

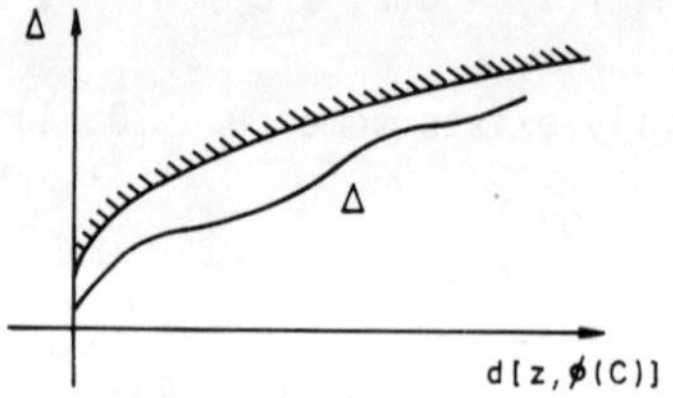

5. <u>OLS-Identifiability by Regularization</u>

We consider here the very common case where all the hypothesis of the sufficient condition for OLSI of Section 2 are satisfied, except for *the sensitivity operator* $\phi'(x)$ *which is not supposed to be injective* over C, in other terms, we may have in (13):

$$\underline{\alpha} = 0 \tag{38}$$

Then the OLS problem (3), (4) may have no or many solutions, and nothing can be said about its well-posedness. The standard approach in that case is to replace the problem (3), (4) by the *regularized problem (7), (8)*. Depending on one's inclination, one can think of the regularizing term either as a pure mathematical artifact (x_0 usually taken to be zero) to restore good mathematical properties of the optimization problem, or as corresponding to supplementary *a priori* information on the parameter (then x_0 is some *a priori* estimation of the parameter x, usually $x_0 \in C$, and ε is related to the confidence one has in this *a priori* estimate). In order to study the OLS-Identifiability of the regularized problem (7), (8), we introduce a mapping ϕ_ε from $C \to F \times E$, and $z_\varepsilon \in F \times E$, defined by:

$$\forall\ x \in C \qquad \phi_\varepsilon(x) = (\phi(x),\ \varepsilon x) \tag{39}$$

$$z_\varepsilon = (z,\ \varepsilon x_0) \tag{40}$$

and we equip $F \times E$ with the Max norm. One checks then easily that the bounds $\underline{\alpha}_\varepsilon$ and $\bar{\beta}_\varepsilon$ associated to ϕ_ε are given by:

$$\underline{\alpha}_\varepsilon = \varepsilon > 0 \quad , \quad \bar{\beta}_\varepsilon = \bar{\beta} \tag{41}$$

so that the sufficient condition of Section 2 can be used for this regularized problem. For this we first evaluate the distance of z_ε to $\phi_\varepsilon(C)$:

$$d(z_\varepsilon,\ \phi_\varepsilon(C)) = \underset{x \in C}{\text{Min}}\ \text{Max}\{\|\phi(x) - z\|_F,\ \varepsilon\|x - x_0\|_E\}$$

If we define the radius of C seen from x_0 by

$$\text{Rad}_{x_0}(C) = \sup_{x \in C} \|x - x_0\|_E$$

Then we get:

$$d(z_\varepsilon,\ \phi_\varepsilon(C)) \le d(z,\ \phi(C)) + \varepsilon \text{Rad}_{x_0}(C) \tag{42}$$

So we see that <u>*if* $\varepsilon > 0$ *is large enough*</u> so that:

$$\gamma = \varepsilon^2/\bar{\beta} - \varepsilon \text{Rad}_{x_0}(C) - \beta/8\ \text{diam}\ (C)^2 > 0 \quad . \tag{43}$$

It is obvious that C satisfies

$$\bar{\beta}_\varepsilon\ \text{diam}\ C \le 2\sqrt{2}\ \underline{\alpha}_\varepsilon \tag{44}$$

Moreover, if we choose z such that

$$d(z,\ \phi(C))_F < \gamma \tag{45}$$

we get from (42), (43) that

$$d(z_\varepsilon, \phi_\varepsilon(C)) \leq \varepsilon^2/\bar{\beta} - \bar{\beta}/8 \times \mathrm{diam}(C)^2 = \gamma_\varepsilon \quad . \tag{46}$$

Hence Theorem 1 applies and yields:

Theorem 4

Under hypothesis of Theorem 1, with $\underline{\alpha} = 0$, and if the regularization parameter ε is chosen large enough to satisfy (43), then the ROLS problem (7), (8) has a unique solution $\hat{x}_\varepsilon$ depending continuously on z as soon as $d(z, \phi(C)) < \gamma$ (defined in (43)); in other terms, the parameter x is OLSI over C from the measurement of $\phi(C)$ and the regularizing term $\varepsilon^2 \| x - x_0 \|_E^2$. Moreover, any minimizing sequence converges toward $\hat{x}_\varepsilon$.

A result similar to Theorem 2 can also be obtained, with explicit Lipschitz constants for the $z \to \hat{x}_\varepsilon$ mapping.

References

CHAVENT, G. (1979) "About the stability of the optimal control solution of inverse problem". Mathematical and Numerical Methods of Inverse and Improperly Posed Problems, ed. by G. ANGER, Akademie Verlag, Berlin.

CHAVENT, G. (1983). "Local Stability of the output least square parameter estimation technique". Matematicada Aplicada e Computacional, V. 2, no. 1, pp. 3-22.

COLONIUS, F. and KUNISCH, K. (1984). "Stability for Parameter estimation in Two Point Boundary Value Problems", Technische Universität GRAZ Report Nr 50 - 1984.

PAGANI, C.D. (1982). "Determining a coefficient of a parabolic equation". Applicable Analysis, Vol. 14.

Inclusion of solutions of certain types of boundary value problems

L. Collatz, Hamburg

Abstract
The paper describes methods for inclusion of unknown solutions of linear and nonlinear ordinary and partial differential equations, of integral equations and some other types of functional equations in not too complicated cases. These methods are based on approximation theory, iteration and optimization techniques, monotonicity principles and fixed point theorems of functional analysis, and are often the only possible methods which gives bounds for the absolute value of the error one can guarantee. Recently (mostly in the last year calculated) numerical examples illustrate the theorems.

Introduction

The most frequently used numerical methods for solving boundary value problems, integral- and functional equations, have been in the last decades the discretization methods (difference equations, finite element methods and their improvements) and variational methods (Ritz-Galerkin procedures and variants); the computer printes the results usually with many decimals, say f.i. 10 decimals, and then it is the problem for the mathematician to decide, how many of these decimals one can guarantee, whether f.i. the second decimal is sure or not. This question is often too difficult for the mathematician, but in the last decades one has had progress and to day one can answer this question in many cases, which appeared to be hopeless say 10 years ago. Therefore one hopes that in the futute many more types of problems become accessible for a rigorous mathematical decision about the reached accuracy - supposed that more and more mathematicians will work on these difficult and important problems.

Many problems in numerical mathematics can be written in one of the following forms:

(1.1) $$Tu = \theta,$$

(1.2) $$Tu = r,$$

(1.3) $$Tu = u.$$

Here u is an unknown (wanted) element in a Banach space R and T a given linear or nonlinear operator which maps a certain domain D of definition (which is a subset of R) into a Banach space $\hat{R}$ with the zero element θ; and r is a given element of $\hat{R}$; if $R=\hat{R}$, one can ask for fixed point elements u with (1.3).

A numerical procedure calculates an approximate solution z for u; one is interested on bounds for the error

(1.4) $$\varepsilon = z - u.$$

I. Different types of error considerations

1. Qualitative error-estimates; these are often of the form

(1.5) $$||\varepsilon|| = O(h^p)$$

where $||\cdot||$ is a norm, h a value which is characteristic for the computation, f.i. the stepsize, and p an order for the accuracy.

(1.5) may have f.i. the meaning

(1.6) $|\varepsilon| \leq K \cdot h^p$,

where K is often a constant the value of which is unknown or difficult to determine. Therefore usually one can seldom guarantee certain decimals with aid of (1.5).

2. Quantitative error bounds

Let us consider all quantities as real, the spaces R, $\hat{R}$ as partially ordered, and the elements of R and $\hat{R}$ as realvalued vectors or functions. The sign $\leq$ may mean the classical ordering of real numbers. Let as consider as an important case the set C(B) of continuous functions f(x) defined in an open connected set B, which may be a subset of the space R^n of n-dimensional vectors $x=(x_1,\ldots,x_n)$. One can introduce as ordering $f<g$ for a pair $f,g \in C(B)$:

(1.7) $f \leq g$ means $f(x) \leq g(x)$ for all $x \in B$ (We write shortly $f \leq g$)

Then we can define in the case $v < w$ an "interval" $J = [v,w]$ as the set of functions $h(x) \in C(B)$:

(1.8) $$J = [v,w] = \{h \in C(B),\ v < h < w\}$$

(J means the "strip" between v and w, fig. 1)

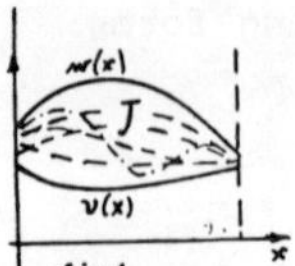

fig. 1

It is a central problem in numerical analysis to determine for an unknown function u(x) an interval [v,w] for which one can guarantee that it contains u, and which is small enough for the wanted accuracy (compare Kantorowitsch-Akilow [64]). We describe two methods for calculating such intervals.

II. Operators T of monotonic type

An Operator T (with notations as in I) is called "of monotonic type", if (compare Collatz [57],[66], Bohl [74], Schröder [80])

(2.1) $\quad Tv \leq Tw$ for all $v,w \in D$ implies $v \leq w$.

Let us consider a boundary value problems

$$(2.2)\quad \begin{cases} Lu = r(x) & \text{on } B \\ Su = \gamma(x) & \text{on } \partial B \end{cases}$$

with a linear or nonlinear given differential operator L, a given boundary operator S, which may be the identical operator (f.i. Dirichlet condition) or may contain derivatives (f.i. Neumann condition). $r(x)$ and $\gamma(x)$ are given functions. We define T as the vector $T = (L,S)$. For wide classes of boundary value problems with elliptic and parabolic equations one can prove the condition (2.1), (compare Collatz [66] § 23 and 24).

Numerical procedure (Approximation and Optimization

One chooses a class $W = w(x,a_\nu) = w(x,a_1,a_2,\ldots,a_p)$ of functions w which depend on certain parameters a_ν (compare the examples); one determines values $\bar{a}_\nu$ for the parameters a_ν such that for the corresponding function $\bar{w} = w(x,\bar{a}_\nu)$ holds $T\bar{w} \geq Tu$, and one gets the upper bound $\bar{w} \geq u$. Then one asks for other value a_ν such that $T\underline{w} \leq Tu$ holds for $\underline{w} = w(x,\underline{a}_\nu)$; this gives the inclusion

$$(2.3)\qquad \underline{w} \leq u \leq \bar{w} \quad \text{or} \quad u \in [\underline{w},\bar{w}]$$

(so called twosided Chebychev Approximation), compare Meinardus [67] a.o.).

We wish to get good error bounds and write the approximation as semi-infinite optimization problem with restrictions: (compare Collatz [81] a.o.)

$$(2.4)\quad \begin{matrix} 0 \leq \bar{w} - \underline{w} \leq \delta, & \delta = \text{Min} \\ T\underline{w} \leq (r,\gamma) \leq T\bar{w} & \end{matrix}$$

Here $\delta,\underline{a}_\nu,\bar{a}_\nu$ are the variables of the optimization problem, written now in a form for which subroutines for computers are often available.

Numerical Example 1

A function

$u(\varphi,t)$ should satisfy the nonlinear heat conduction equation

$$(2.5)\qquad Tu = \frac{\partial u}{\partial t} - 2\frac{\partial^2 u}{\partial \varphi^2} - \frac{1}{2}(u+u^3) = 0$$

with the initial boundary conditions

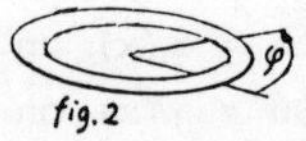

fig.2

$$(2.6)\qquad \begin{cases} u(\varphi,0) = \cos\varphi \text{ for } 0 \leq \varphi \leq 2\pi, \\ u(\varphi,t) = u(\varphi + 2\pi,t) \text{ for all } \varphi \text{ and } t \geq 0. \end{cases}$$

u may be interpreted as temperature in a ring, fig. 2; by symetry we can substitute the boundary condition (fig. 3) by

$$(2.7)\qquad \begin{cases} u(\varphi,0) = \cos\varphi \text{ for } |\varphi| \leq \frac{\pi}{2},\ u(\varphi,t) = 0 \text{ for } |\varphi| = \frac{1}{2}\pi \\ \text{and all } t > 0 . \end{cases}$$

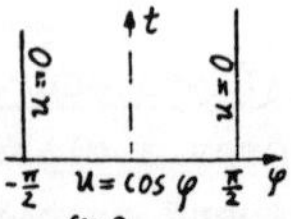

fig.3

We approximate u by

$$u \approx v = (\cos\varphi)\, e^{-kt}.$$

Then v satisfies the condition (2.7); the monotonicity principle (compare Collatz [66], Redheffer [67], Walter [70], Werner [75] a.o.) with v = u on the boundary ∂B holds:

$$(2.8)\qquad Tv \leq Tu \text{ in } B \text{ implies } v \leq u \text{ in } |\varphi| \leqq \frac{\pi}{2},\ t \geq 0.$$

One calculates immedeately $Tv = v \cdot \Phi$ with

$$\Phi = -k + 2 - \frac{1}{2} - \frac{1}{2}(\cos\varphi)^2 e^{-2kt} .$$

We have $\Phi \geq 0$ for $k = 1$ and $\Phi \leq 0$ for $k = \frac{3}{2}$; therefore we have got the inclusion for the solution $u(\varphi,t)$

$$(\cos\varphi)\, e^{-\frac{3}{2}t} \leq u(\varphi,t) \leq (\cos\varphi)\, e^{-t}$$

Of course one can improve the accuracy by introducing more parameters instead of k.

Numerical Example 2 Ideal flow of water between chanals; B may be the domain in the x-y-plane:

$$B = \{(x,y);\ 0 < x < 1,\ |y| < \infty \quad \text{and} \quad x \geq 1,\ |y| < 1\} \quad \text{fig. 4;}$$

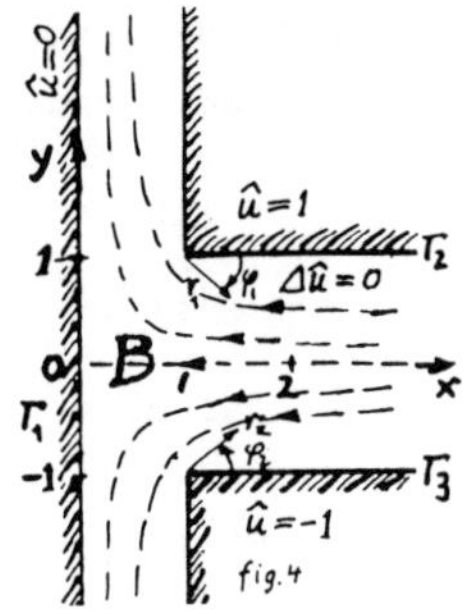

The boundary ∂B consists of three parts Γ_1 (with $x = 0$) Γ_2 (in the quadrant $x > 0$, $y > 0$) and Γ_3 (in the quadrant $x > 0$, $y < 0$), fig. 4; we look for a function $\hat{u}(x,y)$ with

(2.9) $\Delta\hat{u} = \dfrac{\partial^2\hat{u}}{\partial x^2} + \dfrac{\partial^2\hat{u}}{\partial y^2} = 0$ in B, $\hat{u} = 0$ on Γ_1, $\hat{u} = 1$ on Γ_2, $\hat{u} = -1$ on Γ_3.

The curves $\hat{u}$ = const can be interpreted as streamlines; we restrict ourselves on a quarter Q of B:

$$Q = \{(x,y);\ (x,y) \in B \text{ with } 0 < y < 1,\ y < x\} \text{ fig. 5;}$$

the boundary ∂Q consists of three parts Γ_4 (with $y = 1$), Γ_5 (with $x = y$) and Γ_6 (with $y = 0$). We have for $u = \hat{u} - y$ the mixed boundary value problem, fig. 5

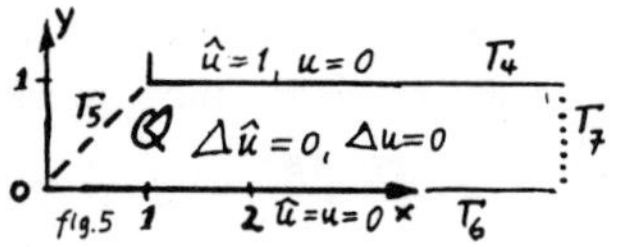

(2.10) $\Delta u = 0$ in Q, $u = 0$ on Γ_4 and Γ_6,
$\dfrac{\partial \hat{u}}{\partial n} = 0$ on Γ_5,

where n means the auter normal.

$$u \approx w(x,y) = b_0 w_0 + \sum_{\nu=1}^{p} a_\nu w_\nu + \sum_{\mu=1}^{q} c_\mu P_\mu$$
$$w_0 = r_1^{2/3} \sin\frac{2\varphi_1}{3} - r_2^{2/3} \sin\frac{2\varphi_2}{3}$$
$$w_\nu = \sin(\nu\pi y)\exp(-\nu\pi x)$$
Polynomials P_μ with $\Delta P_\mu = 0$.

The monotonicity principle insures for a function $\psi(x,y)$:

(2.11) $\{-\Delta\psi \geq 0$ in Q, $\psi \geq 0$ on Γ_4 and Γ_6, $\dfrac{\partial\psi}{\partial n} \geq 0$ on $\Gamma_5\}$ implies
$\psi \geq 0$ in Q.

Therefore we have again the possibility for an inclusion of the wanted function $u(x,y)$. Introducing polarcoordinates r_1, φ_1, r_2, φ_2 as in fig.4; the lines $\hat{u}$ = const in fig. 4 have been calculated on a computer by taking a finite domain (boundary part Γ_7, $x = x_0$ = const, $u = 0$, fig.5) by approximating u by a function w of the form $w(x,y)$ (see above), and determining the parameters b_0, a_ν, c_μ by optimization as in (2.4).

III. Monotonically decomposible operators and fixed point theorems

We define for an Operator T (with notations as in I)

$$(3.1)\begin{cases} T \text{ is called syntone,} & \text{if } f \le g \text{ implies } Tf \le Tg \text{ for all } f,g \in D \\ T \text{ is called antitone,} & \text{if } f \le g \text{ implies } Tf \ge Tg \text{ for all } f,g \in D. \end{cases}$$

Example 1: The linear integral-Operator

$$(3.2) \qquad T f(x) = \int_B K(x,s)\, f(s) ds$$

with a given realvalued (continuous) kernel $K(x,s)=K(x_1,\ldots,x_n,s_1,\ldots s_n)$ is syntone, if $K(x,s) \ge 0$ in $B \times B$, and antitone for $K(x,s) \le 0$. In the first case we write symbolically $T\nearrow$, in the second case $T\searrow$.

Definition: The operator T is called monotonically decomposible (compare Bohl [74], Schröder [80] a.o.), if one can write

$$(3.3) \qquad T = T_1\nearrow + T_2\searrow \text{ with } T_1\nearrow \text{ (syntone) and } T_2\searrow \text{ (antitone).}$$

Example 2: An example for (3.3) is the operator T in (3.2), if the kernel is real (not necessarely of fixed sign). Any real kernel K can be written as

$$(3.4) \qquad K = K_1 + K_2 \text{ with } K_2 \ge 0 \text{ in } B \times B, \;\; K_2 \le 0 \text{ in } B \times B$$

Then one can define in (3.3): $T_j(f(x)) = \int_B K_j(x,s)\, f(s)\, ds$ for $(j=1,2)$.

Example 3: Every Hammerstein-Operator

$$(3.5) \qquad T f(x) = \int_B K(x,s)\, \varphi(f(s)) ds$$

with a real kernel $K(x,s)$ and a realvalued function $\varphi(z)$ of bounded variation is monotonically decomposible (compare f.i. Collatz [81] Bohl [74], J. Schröder [80]), because $\varphi(z)$ can be written as

$$(3.6) \qquad \varphi(z) = \varphi_1(z) + \varphi_2(z) \text{ with } \varphi_1\nearrow \text{ and } \varphi_2\searrow.$$

One has

$$(3.7) \qquad T_1(u) = \int_B (K_1\varphi_1(u) + K_2\varphi_2(u)) ds, \qquad \text{and}$$

$$T_2(u) = \int_B (K_1\varphi_2(u) + K_2\varphi_1(u)) ds.$$

Iterations procedure: The operator T may satisfy (3.3). Let the iteration procedure

$$(3.8) \begin{cases} v_{n+1} = T_1 v_n + T_2 w_n \\ w_{n+1} = T_1 w_n + T_2 v_n \end{cases} \qquad (n = 0,1,2,\ldots)$$

start with elements $v_0, w_0 \in D$ with

(3.9) $\quad v_0 \leq v_1;\ w_1 \leq w_0;\quad v_0 \leq w_0\,.$

Then one can proove

$$v_n \leq v_{n+1},\ w_{n+1} \leq w_n,\ v_n \leq w_n \quad (m = 0,1,2,\ldots)$$

If the operator T is furthermore completely continuous, then the fixed point theorem of Schauder gives the existence of at least one solution u of (1.3) $u = Tu$ in every interval $M_n = [v_n, w_n]$ $(n = 0,1,2,\ldots)$.

<u>Numerical Example 3</u>: For illustration we take a simple example of a non-linear integral equation of Urysohn-type

(3.10) $$u(x) = \lambda \int_0^1 \frac{dt}{1+x+u(t)} = T\,u(x)\,.$$

Let be λ a given positive constant; we look on a positive solution $u(x) > 0$ in $[0,1]$; the operator T is antitone, $T = T_2$ and (3.8) reduces to

(3.11) $\quad v_{n+1} = Tw_n,\quad w_{n+1} = Tv_n \quad (n = 0,1,2,\ldots)$

The first trial $v_0(x) \equiv 0$ gives $w_1 = Tv_0 = \frac{\lambda}{1+x}$ with $w_1(x) \leq \lambda$; we try $w_0(x) = \lambda$ and get $v_1 = Tw_0 = \frac{\lambda}{1+x+\lambda}$. The conditions (3.9) are satisfied, fig. 6, and by the Schauders theorem we have the existence of at least one solution $u(x)$ with

$$\frac{\lambda}{1+x+\lambda} \leq u(x) \leq \frac{\lambda}{1+x}$$

The existence is assured for arbitrately great λ; this shows the effectiveness of Schauders theorem.

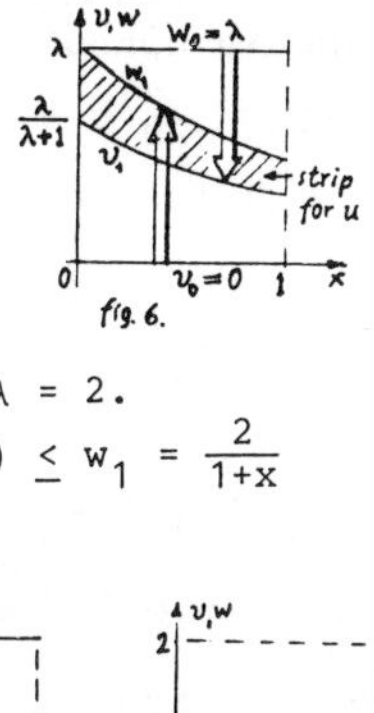

fig. 6.

Let us take a little more in detail the special value $\lambda = 2$.

1) $v_0 = 0$, $w_0 = 2$ gives the strip, fig. 7 $v_1 = \frac{2}{3+x} \leq u(x) \leq w_1 = \frac{2}{1+x}$

2) Better constants are $v_0 = \frac{\sqrt{5}-1}{2}$, $w_0 = \sqrt{5}-1$ with

$$v_1 = Tw_0 = \frac{2}{\sqrt{5}+x},\quad w_1 = Tv_0 = \frac{4}{\sqrt{5}+1+2x}$$

with the strip in fig. 7

3) Even better results one gets with linear functions for v_0, w_0:

$$\varphi(x) = \alpha - \beta x \text{ gives } T\varphi(x) = \frac{\lambda}{\beta}\,\ln\left(\frac{1+x+\alpha}{1+x+\alpha-\beta}\right).$$

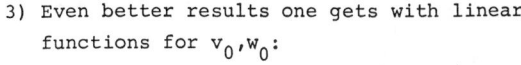

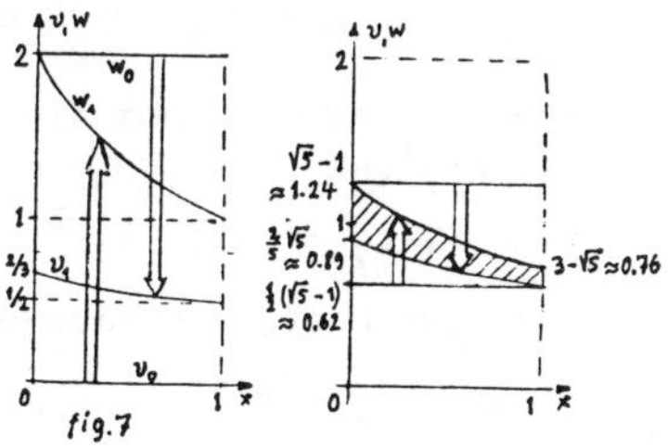

fig.7

One gets with $v_0(x) = 1 - 0.35\,x$ and $w_0(x) = 1.1 - 0.39x$ the strip in fig. 8. Of course it is easy to improve the inclusion-strip for $u(x)$.

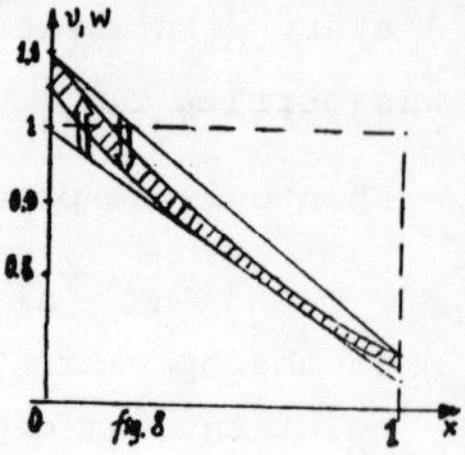

IV. Singular boundary value problems

The mentioned methods have been applied in the last years to many types of problems f.i. with free boundaries and moving boundaries, to delay equations, to eigenvalue problems, to problems with unbounded domains a.o.

We select here the case of singularities along the boundary, f.i. singular boundary points. Then one has to know the type of singularities otherwise one cannot expect good numerical results (Whiteman [85], Tolksdorf [85], Dobrowolski [85], Collatz [85] a.o.). Recently one has treated also threedimensional singularities. We restrict ourselves to mention some different types of singular problems for which it was possible to get for the solution inclusion-intervals one can guarantee.

Numerical Example 4: (A Crack-Problem) A function $u(x,y)$ in the x-y-plane is schould satisfy the Laplace-Equation

(4.1) $\quad -\Delta u = 1$ in B, $\quad u = 0$ on ∂B

where B is the domain

$$B = \{(x,y),\ y > 0,\ r < 1 \setminus x = 0,\ 0 < y \le \tfrac{1}{4}\} \text{ with } r^2 = x^2 + y^2.$$

B is a half-circle with a slit, fig. 9.

One approximates u by

$$u \approx v_{p,q} = -\frac{r^2}{4} + \sum_{k=1}^{q} a_k \psi_k + \sum_{j=1}^{p} b_j \zeta_j$$

where ζ_j are polynomials with $\Delta\zeta_j = 0$ which satisfy the symetry of the problem f.i.

$\zeta_1 = 1,\ \zeta_2 = y,\ \zeta_3 = x^2 - y^2,\ \zeta_4 = 3x^2y - y^3, \ldots$

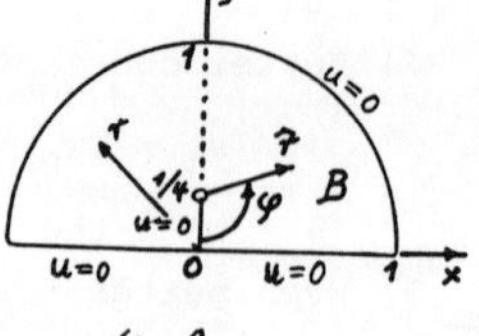

fig. 9

and ψ_k with $\Delta\psi_k = 0$ have the singularity:

(4.2) $\quad \psi_k = \hat{r}^{(k-\frac{1}{2})} \sin\left((k-\tfrac{1}{2})\varphi\right),\ \hat{r}^2 = x^2 + (y-\tfrac{1}{4})^2$, ($\hat{r}, \varphi$ polarcoordinates at the point $x = 0,\ y = \frac{1}{4}$. The optimization

(4.3) $\quad -\delta \leq v_{p,q}(x,y) \leq \delta, \quad (x,y) \in \partial B, \ \delta = \text{Min}$

was curried out on a computer, I thank Mr. Zheng, Qinghua for the calculation.

For $p = 11$, $q = 3$ one gets $\delta = 0.000424$; one has the inclusion for u one can guarantee

$$|v_{11,3} - u| \leq 0.000424.$$ More details are given in Collatz [85].

Numerical Example 5: (Unbounded domain from electrostatics)

A function $u(x,y,z)$ should satisfy

(4.5) $\quad \Delta u = u_{xx} + u_{yy} + u_{zz} = 0$ in B: $|z| < 2$, $r > 1$; $u = 0$ for $|z| = 2$, $u = 1$ for $r = 1$.

with $r^2 = x^2 + y^2 + z^2$, fig. 10.

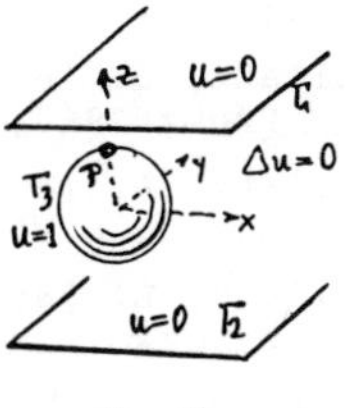

fig.10

One asks for the derivative $\eta = \frac{\partial u}{\partial z}(P)$ at the point $P = (0,0,1)$ (greatest gradient). By approximating with singular functions of the form $\frac{1}{r_j}$, where r_j is the distance between the variable point (x,y,z) from $(0,0,d_j)$ one gets with 17 values d_j [8 values > 3, 8 values < -3 and one value $= 0$] the inclusion

$-1.358016 \leq \eta \leq -1.358022$; for details see Collatz (Lect. Conf. Diff. Equ. Brno, C.S.S.R., 26.8.1985).

Numerical Example 6 (A nonlinear problem)

A function $u(x,y)$ should satisfy

$Tu = -\Delta u + 1 + \frac{1}{2} u^2 = 0$ in B; $u = 0$ on ∂B

$B = \{(x,y),\ |x| < 1,\ 0 < y < 2 \setminus x = 0; 0 < y \leq \frac{1}{2}\}$

B is a squere with a slit fig. 11; u (x,y) may be interpreted as temperature in a room with a wall.

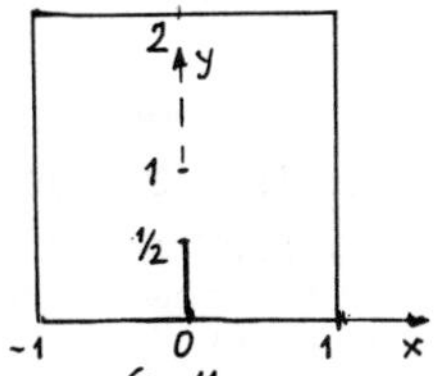
fig. 11

One approximates u by

$$v = \sum_{\nu=1}^{q} a_\nu \psi_\nu(x,y) + \sum_{j=1}^{p} b_j \zeta_j$$

with $\zeta_1 = 1$, $\zeta_2 = y$, $\zeta_3 = x^2$, $\zeta_4 = y^2$, $\zeta_5 = x^2 y, \ldots$

and ψ_k as in (4.2), but with $r^2 = x^2 + (y - \frac{1}{2})^2$.

Now one has to approximate on the boundary ∂B and in the interior B (so called sinultaneous approximation). One looks for a lower bound $\underline{v}$ with parameters $\underline{a}$,$\underline{b}_j$ and an upper bound $\bar{v}$ with parameters $\overline{a_\nu}$, $\overline{b}_j$ and has the optimization

$$-\delta \leq \bar{v} - \underline{v} \leq \delta, \; \delta = \text{Min}, \; T\underline{v} \leq 0 \leq T\bar{v} \text{ in } B, \; \underline{v} \leq 0 \leq \bar{v} \text{ on } \partial B.$$

I thank Mr. E. Greiner for the calculation on a computer; he got für $p = 6$, $q = 1$ the error bound $\delta = 0.087$ and for $p = 9$, $q = 2$ the error bound $\delta = 0.014$.

References

Bohl, E. [74] Monotonie, Lösbarkeit und Numerik bei Operator-gleichungen Springer, 1974, 255 S.

Collatz, L. [52] Aufgaben monotoner Art, Arch. Math. Anal. Mech. 3 (1952) 366-376.

Collatz, L. [66] Functional Analysis and Numerical Mathematics, Acad. Press 1966, 473 p.

Collatz, L. [81] Anwendung von Monotoniesätzen zur Einschließung der Lösungen von Gleichungen, Jahrb. Überbl. Math., 1981, 189-225.

Collatz, L. [85] Inclusion of regular and singular solutions of certains types of integral equations, Intern.Ser.Num.Math. 73 (1985) 93-102.

Dobrowolski, M. [85] Vortrag Tagung: Singularities and Constructive Methods for their treatment. Proc. Oberwolfach 21.-26. November 1983 and a paper 1985 to appear.

Kantorowitsch, L.W. - G. P. Akilow [64] Funktionalanalysis in normierten Räumen, Akad. Verl. Berlin, 1964, 622 p.

Meinardus, G. [67] Approximation of Functions, Theory and Numerical Methods, Springer Verlag, 1967, 198 p.

Redheffer, R.M. [67] Differentialungleichungen unter schwachen Voraussetzungen, Abhandl. Math. Sem. Univ. Hbg. 31 (1967) 33-50.

Schröder, J. [80] Operator Inequalities, Acd. Press, 1980, 367 p.

Tolksdorf P. [85] Vortràg Tagung: Singularities and Constructive Methods for their treatment. Proc. Oberwolfach 21.-26. November 1983 and a paper 1985 to appear

Walter, W. [70] Differential and Integral Inequalities, Springer, 1970, 352 p.

Werner, B. [75] Monotonie und finite Elemente bei elliptischen Differentialgleichungen, Intern.Ser.Num.Math. 27 (1975), 393-401.

Whiteman, J. R. [85] Vortrag Tagung: Singularities and Constructive Methods for their treatment. Proc. Oberwolfach 21.-26. Nov. 1983, and a paper 1985 to appear.

CANONICAL DIFFERENCE SCHEMES FOR HAMILTONIAN CANONICAL DIFFERENTIAL EQUATIONS

FENG KANG
Computing Center, Academia Sinica
Beijing, CHINA

The present program[1] that the author and his group have started is a systematic study, within the framework of symplectic geometry, of the numerical methods for the solution of differential equations of mathematical physics expressed in Hamiltonian formalism. As is well known, Hamiltonian canonical systems serve as the basic mathematical formalism, for diverse areas of physics, mechanics, engineering, as well as pure and applied mathematics, e.g., geometrical optics, analytical dynamics, non-linear PDE's of first order, group representations, WKB asymptotics, pseudodifferential and Fourier integral operators, electro-dynamics, plasma physics, elasticity, hydrodynamics, relativity, control theory, etc. It is generally accepted that all real physical processes with negligible dissipation could be expressed, in some way or other, in suitable Hamiltonian forms.

We consider the canonical system in finite dimensions

$$\frac{dp_i}{dt} = -H_{qi} \quad , \quad \frac{dq_i}{dt} = H_{pi} \quad , \quad i = 1,2,\dots,n \quad , \tag{1}$$

with Hamiltonian $H(p_1,\dots,p_n, q_1,\dots,q_n)$.

In the following, vectors are always represented by column matrices, matrix transpose is denoted by a prime '. Let $z = (z_1,\dots,z_n, z_{n+1},\dots, z_{2n})' = (p_1,\dots,p_n, q_1,\dots,q_n)'$, (1.1) can be written as

$$\frac{dz}{dt} = J^{-1}H_z \quad , \quad J = \begin{pmatrix} 0 & In \\ -In & 0 \end{pmatrix} \tag{2}$$

defined in phase space R^{2n} with a standard symplectic structure given by the non-singular anti-symmetric closed differential 2-form $\omega = \Sigma\, dz_i \wedge dz_{n+i} = \Sigma\, dp_i \wedge dq_i$. The Fundamental Theorem on Hamiltonian Formalism says that the solution of the canonical system (1.2) can be generated by a one-parameter group G_t of canonical transformations of R^{2n} (locally in t and z) such that

$$G_{t_1} G_{t_2} = G_{t_1+t_2}$$

$$z(t) = G_t z(0) \quad .$$

A transformation $z \to \hat{z}$ of R^{2n} is called canonical if it is a local diffeomorphism whose Jacobian $\frac{\partial \hat{z}}{\partial z} = M$ is symplectic everywhere, i.e.

$$M'JM = J \quad , \quad \text{i.e.} \quad M \in S_p(2n) \quad .$$

Linear canonical transformations are simply symplectic transformations.

The canonicity of G_t implies the preservation of 2-form ω, 4-form $\omega \wedge \omega, \ldots,$ 2n-form $\omega \wedge \omega \ldots \wedge \omega$. They constitute the class of conservation laws of phase area of even dimensions for the Hamiltonian system (1.2).

Moreover, the Hamiltonian systems possesses another class of conservation laws related to the energy $H(z)$. A function $\phi(z)$ is said to be an invariant integral of (1.2) if it is invariant under (1.2)

$$\phi(z(t)) \equiv \phi(z(0))$$

which is equivalent to

$$\{\phi, H\} = 0 \quad ,$$

where the Poisson Bracket for two functions $\phi(z)$, $\psi(z)$ are defined as

$$\{\phi, \psi\} = \phi_z' J^{-1} \psi_z \quad .$$

H itself is always an invariant integral. See, e.g., Refs. 5 and 6.

For the numerical study, we are less interested in (1.2) as a general system of ODE per se, but rather as a specific system with Hamiltonian structure. It is natural to look for those discretization systems which preserve as much as possible the characteristic properties and inner symmetries of the original continuous systems. To this end the transition $\hat{z} \to z$ from the k-th time step $z^k = \hat{z}$ to the next (k+1)-th time step $z^{k+1} = z$ should be canonical for all k and, moreover, the invariant integrals of the original system should remain invariant under these transitions. We try to conceive, design, analyse and evaluate difference schemes and algorithms specifically within the framework of symplectic geometry. The approach proves to be quite successful as one might expect, we actually derive in this way numerous "unconventional" difference schemes.

We give a brief survey of some of our preliminary results[1-4].

Consider at first the case for which the Hamiltonian is a quadratic form

$$H(z) = \frac{1}{2} z'Sz \quad , \quad S' = S \quad , \quad H_z = S(z) \quad . \tag{3}$$

Then the canonical system is linear

$$\frac{dz}{dt} = Bz \quad , \tag{4}$$

where $B = J^{-1}S$ is infinitesimally symplectic, i.e.

$$B'J + JB = 0 \quad .$$

The solution is

$$z(t) = G_t z(0) \quad ,$$

where $G(t) = \exp tB$, as the exponential transform of infinitesimally symplectic tB, is symplectic.

<u>Theorem 0</u>. The weighted Euler scheme

$$\frac{z^{k+1} - z^k}{\tau} = B(\alpha z^{k+1} + (1-\alpha)z^k) \qquad (5)$$

for the linear system (4) is symplectic if and only if $\alpha = 1/2$, i.e. it is the case of time-centered Euler Scheme with the transition.

For the time-centered case, the transition matrix

$$z^{k+1} = F_\tau z^k \quad , \quad F_\tau = \phi(\tau B) \quad , \quad \phi(\lambda) = \frac{1 + \frac{\lambda}{2}}{1 - \frac{\lambda}{2}} \qquad (6)$$

F_τ, as the Cayley transform of infinitesimally symplectic τB, is symplectic.

In order to generalize the time-centered Euler scheme, we need, apart from the exponential or Cayley transforms, other matrix transforms carrying infinitesimally symplectic matrices into symplectic ones.

<u>Theorem 1</u>. Let $\psi(\lambda)$ be a function of complex variable λ satisfying

(I) $\psi(\lambda)$ is analytic with real coefficients in a neighborhood D of $\lambda = 0$

(II) $\psi(\lambda)\psi(-\lambda) \equiv 1$ in D,

(III) $\psi(0) \neq 0$.

A is a matrix of order 2n, then $(\psi(\tau B))' A \psi(\tau B) = A$ for all τ with sufficiently small $|\tau|$, if and only if

$$B'A + AB = 0 \quad .$$

If, moreover, $\exp\lambda - \psi(\lambda) = O(|\lambda|^{m+1})$, then

$$z^{k+1} = \psi(\tau B) z^k \qquad (7)$$

considered as an approximative scheme for the canonical system (4) is symplectic, of m-th order of accuracy and has the property that $z'Aw$ is invariant under $\psi(\tau B)$ if and only if it is invariant under G_τ of (4).

Remark 1. The last property is remarkable in the sense that, all the bilinear invariants of the system (4), no more and no less, are kept invariant under the scheme (7), in spite of the fact that the latter is only approximate.

Remark 2. The approximative scheme in Theorem 1 becomes difference schemes only when $\psi(\lambda)$ is a rational function. As a concrete application for the construction of symplectic difference schemes, take the diagonal Padé approximants to the exponential function

$$\exp \lambda - \frac{P_m(\lambda)}{P_m(-\lambda)} = O(|\lambda|^{2m+1})$$

Theorem 2. The difference schemes

$$z^{k+1} = \frac{P_m(\tau B)}{P_m(-\tau B)} z^k \quad , \qquad m = 1,2,\ldots \tag{8}$$

for the system (4) are symplectic, A-stable, of 2m-th order of accuracy, and having the same set of bilinear invariants as that of system (4), the case $m = 1$ is the centered Euler scheme.

For the general non-linear canonical system (1.2), the time-centered Euler scheme is

$$\frac{1}{\tau}(z^{k+1} - z^k) = J^{-1}H_z(\tfrac{1}{2}(z^{k+1} + z^k)) \quad . \tag{9}$$

The transition $z^{k+1} \to z^k$ is canonical with Jacobian

$$F_\tau = \left[I - \frac{\tau}{2} J^{-1}H_{zz}(\tfrac{1}{2}(z^{k+1} + z^k))\right]^{-1}\left[I + \frac{\tau}{2} J^{-1}H_{zz}(\tfrac{1}{2}(z^{k+1} + z^k))\right]$$

symplectic everywhere. However, unlike the linear case the invariant integrals $\varphi(z)$ of system (2), including $H(z)$, are conserved only approximately

$$\varphi(z^{k+1}) - \varphi(z^k) = O(\tau^3) \quad .$$

For non-linear separable system for which $H(p, q) = U(p) + V(q)$ where U, V are kinetic and potential energies respectively, we define the staggered scheme

$$\frac{1}{\tau}(p^{k+1} - p^k) = -V_q(q^{k+\frac{1}{2}}) \quad ,$$

$$\frac{1}{\tau}(q^{k+1+\frac{1}{2}} - q^{k+\frac{1}{2}}) = U_p(p^{k+1}) \quad . \tag{10}$$

The p's are set at integer times $t = k\tau$, q's at half-integer times $t = (k + \frac{1}{2})\tau$. The transition

$$w^k = \begin{pmatrix} p^k \\ \\ q^{k+\frac{1}{2}} \end{pmatrix} \longrightarrow \begin{pmatrix} p^{k+1} \\ \\ q^{k+1+\frac{1}{2}} \end{pmatrix} = w^{k+1} = F_\tau w^k \quad ,$$

$$F_\tau = \begin{pmatrix} I & 0 \\ \\ -\tau M & I \end{pmatrix}^{-1} \begin{pmatrix} I & -\tau L \\ \\ 0 & I \end{pmatrix} \quad ,$$

$$M = U_{pp}(p^{k+1}) \quad , \qquad L = V_{qq}(q^{k+\frac{1}{2}}) \quad ,$$

is symplectic, of 2nd order accuracy and practically explicit. Since p, q are computed at different times, we need synchronization, e.g., using

$$q^k = \frac{1}{2}(q^{k-\frac{1}{2}} + q^{k+\frac{1}{2}})$$

to compute the invariant integrals $\varphi(p, q)$,

$$\varphi(p^{k+1}, q^{k+1}) - \varphi(p^k, q^k) = O(\tau^3) \quad .$$

For the comparison of stability for the linear system (4) with separable Hamiltonian with the canonical schemes in Theorems 1, 2 the staggered scheme and the application of the latter to the wave equation see Ref. 1. In Ref. 1, a class of energy conservative schemes was constructed using the differencing of the Hamiltonian function, the symplectic property is not satisfactory. The problem of compatibility of energy conservation with phase area conservation in difference schemes is solved successfully for linear canonical systems (Theorem 1), it seems to be difficult, however, for the general non-linear systems.

In order to develop a general method of construction of canonical difference schemes we first give a constructive generalization of the classical theory of generating functions and Hamilton-Jacobi equations. Our approach is in part inspired by the early works of Siegel[7] and Hua[8]. Every $4n \times 2n$ matrix

$$A = \begin{pmatrix} A_1 \\ A_2 \end{pmatrix} \in M(4n, 2n) \quad , \quad A_1, A_2 \in M(2n) \quad , \quad \text{rank } A = 2n$$

defines in R^{4n} a 2n-dim. Subspace $\{A\}$ spanned by its column vectors. $\{A\} = \{B\}$ if and only if $A \sim B$, i.e.

$$AP = B \quad , \quad \text{i.e.,} \begin{pmatrix} A_1P \\ A_2P \end{pmatrix} = \begin{pmatrix} B_1 \\ B_2 \end{pmatrix} \quad , \quad \text{for some } P \in GL(2n) \quad .$$

The spaces of symmetric and symplectic matrices of order $2n$ will be denoted by $S_m(2n)$, $S_p(2n)$ resp. Let

$$J_{4n} = \begin{pmatrix} 0 & I_{2n} \\ -I_{2n} & 0 \end{pmatrix} \quad , \quad \tilde{J}_{4n} = \begin{pmatrix} -J_{2n} & 0 \\ 0 & +J_{2n} \end{pmatrix}$$

$$X = \begin{pmatrix} X_1 \\ X_2 \end{pmatrix} \quad , \quad Y = \begin{pmatrix} Y_1 \\ Y_2 \end{pmatrix} \in M(4n, 2n) \quad , \quad \text{of rank } 2n.$$

Subspace $\{X\} \subset R^{4n}$ is called J_{4n}-Lagrangian (and $\begin{pmatrix} X_1 \\ X_2 \end{pmatrix}$ is called a symmetric pair) if

$$X'J_{4n}X = O_{2n} \quad , \quad \text{i.e.} \quad X_1'X_2 - X_2'X_1 = O_{2n} \quad .$$

If, moreover $|X_2| \neq 0$, then $X_1X_2^{-1} = N \in S_m(2n)$ and $\begin{pmatrix} X_1 \\ X_2 \end{pmatrix} \sim \begin{pmatrix} N \\ I \end{pmatrix}$, N is determined uniquely by the subspace $\{X\}$. Similarly, subspace $\{Y\} \subset R^{4n}$ is called $\tilde{J}_{4n}$-Lagrangian (and $\begin{pmatrix} Y_1 \\ Y_2 \end{pmatrix}$ is called a symplectic pair) if

$$Y'\tilde{J}_{4n}Y = O_{2n} \quad , \quad \text{i.e.} \quad Y_1'J_{2n}Y_1 - Y_2'J_{2n}Y_2 = O_{2n} \quad .$$

If, moreover, $|Y_2| \neq 0$, then $Y_1Y_2^{-1} = M \in S_p(2n)$ and $\begin{pmatrix} Y_1 \\ Y_2 \end{pmatrix} \sim \begin{pmatrix} M \\ I \end{pmatrix}$, M is determined uniquely by the subspace $\{Y\}$.

A 2n-dim. submanifold $U \subset R^{4n}$ is called J_{4n}-Lagrangian (resp. $\tilde{J}_{4n}$-Lagrangian) if the tangent plane of U is a J_{4n}-Lagrangian (resp. $\tilde{J}_{4n}$-Lagrangian) subspace of the tangent space at each point of U.

Let $z \to \hat{z} = g(z)$ be a canonical transformation in R^{2n}, with Jacobian $g_z = M(z) \in S_p(2n)$. The graph

$$V = \left\{ \begin{pmatrix} \hat{z} \\ z \end{pmatrix} \in R^{4n} \,\middle|\, \hat{z} = g(z) \right\}$$

of g is a $\tilde{J}_{4n}$-Lagrangian submanifold, whose tangent plane is spanned by the symplectic pair $\begin{pmatrix} M(z) \\ I \end{pmatrix}$.

Similarly, let $w \to \bar{w} = f(w)$ be a gradient transformation in R^{2n}, the Jacobian $f_w = N(w)$ $S_m(2n)$. This is equivalent to the (local) existence of a scalar function $\phi(w)$ such that $f(w) = \phi_w(w)$. The graph

$$U = \left\{ \begin{pmatrix} \hat{w} \\ w \end{pmatrix} \in R^{4n} \middle| \hat{w} = f(w) \right\}$$

of f is a J_{4n}-Lagrangian submanifold with tangent planes spanned by the symmetric pair $\begin{pmatrix} N(w) \\ I \end{pmatrix}$. The following algebraic theorem is crucial for our construction.

<u>Theorem 3</u>. $T \in GL(4n)$ carries every $\tilde{J}_{4n}$-Lagrangian submanifolds into J_{4n}-Lagrangian submanifolds if and only if

$$T'J_{4n}T = \mu\tilde{J}_{4n} \quad , \quad \text{for some} \quad \mu \neq 0 \quad ,$$

i.e.

$$A_1 = -\mu^{-1}J_{2n}C' \quad , \quad B_1 = \mu^{-1}J_{2n}A' \quad , \quad T = \begin{pmatrix} A & B \\ C & D \end{pmatrix} \quad ,$$

$$T^{-1} = \begin{pmatrix} A_1 & B_1 \\ C_1 & D_1 \end{pmatrix} \quad , \quad C_1 = \mu^{-1}J_{2n}D' \quad , \quad D_1 = -\mu^{-1}J_{2n}B' \quad .$$

The totality of T's in Theorem 3 will be denoted by $CS_p(\tilde{J}_{4n}, J_{4n})$. Canonical transformations can in some way be expressed in implicit form, as gradient transformations with generating functions via suitable linear transformations. The graphs of canonical and gradient transformations in R^{4n} are $\tilde{J}_{4n}$-Lagrangian and J_{4n}-Lagrangian submanifolds respectively. Theorem 3 leads to the existence and construction of the generating functions, under certain transversality condition, for the canonical transformations.

<u>Theorem 4</u>. Let $T = \begin{pmatrix} A & B \\ C & D \end{pmatrix}$, $T^{-1} = \begin{pmatrix} A_1 & B_1 \\ C_1 & D_1 \end{pmatrix}$, $T \in CS_p(\tilde{J}_{4n}, J_{4n})$, they define linear transformations

$$\hat{w} = A\hat{z} + Bz \quad , \quad \hat{z} = A_1\hat{w} + B_1 w \quad ,$$
$$w = C\hat{z} + Dz \quad , \quad z = C_1\hat{w} + D_1 w \quad .$$

Let $z \to \hat{z} = g(z)$ be a canonical transformation in (some neighborhood of R^{2n}, with Jacobian $g_z = M(z) \in S_p(2n)$ and graph

$$V^{2n} = \left\{ \begin{pmatrix} \hat{z} \\ z \end{pmatrix} \in R^{4n} \middle| \hat{z} - g(z) = 0 \right\} .$$

If (in some neighborhood of R^{4n}) the transversality condition

$$|CM + D| \neq 0 \quad , \tag{11}$$

holds, then there exists in (some neighborhood of) R^{2n} a gradient transformation $w \to \hat{w} = f(w)$ with Jacobian $f_w = N(w) \in S_m(2n)$ and graph

$$U^{2n} = \left\{ \begin{pmatrix} \hat{w} \\ w \end{pmatrix} \in R^{4n} \middle| \hat{w} - f(w) = 0 \right\}$$

and a scalar function — generating function — $\phi(w)$ such that

(1) $f(w) = \phi_w(w)$

(2) $N = (AM + B)(CM + D)^{-1} \quad , \quad M = (NC - A)^{-1}(B - ND)$

(3) $T(V^{2n}) = U^{2n} \quad , \quad V^{2n} = T^{-1}(U^{2n}) \quad .$

This corresponds to the fact that, under non-exceptional condition (3.1)

$$[\hat{w} - \phi_w(w)]_{\hat{w}=A\hat{z}+Bz,\ w=C\hat{z}+Dz} = 0$$

gives the implicit representation of the canonical transformation $\hat{z} = g(z)$ via linear transformation T and generating function φ.

For the time-dependent canonical transformation, related to the time-evolution of the solutions of a canonical system (2) with Hamiltonian function $H(z)$, we have the following general theorem on the existence and construction of time-dependent generating function and Hamilton-Jacobi equation depending on T and H under the transversality condition.

Theorem 5. Let T be such as in Theorems 3 and 4. Let $z \to \hat{z} = g(z, t)$ be a time-dependent canonical transformation (in some neighborhood) of R^{2n} with Jacobian $g_z(z, t) = M(z, t) \in S_p(2n)$ such that

(a) $g(*, 0)$ is a linear canonical transformation, $M(z, 0) = M_0$ be independent of z.

(b) $g^{-1}(*, 0)g(*, t)$ is the time-dependent canonical transformation carrying the solution $z(t)$ at moment t to $z(0)$ at moment $t = 0$ for the canonical system (2). If

$$|CM_0 + D| \neq 0 \quad .$$

then for sufficiently small $|t|$, the transversality condition (11) holds and there exists and in (some neighborhood of) R^{2n} a time-dependent gradient transformation $w \to \hat{w} = f(w, t)$ with Jacobian $f_w(w, t) = N(w, t) \in S_m(2n)$ and a time-dependent generating function $\phi(w, t)$ such that

(1) $$[\hat{w} - f(w, t)]_{\hat{w}=A\hat{z}+Bz,\ w=C\hat{z}+Dz} = 0$$

is the implicit representation of the canonical transformation $\hat{z} = g(z, t)$

(2) $$N = (AM + B)(CM + D)^{-1} \quad , \quad M = (NC - A)^{-1}(B - ND)$$

(3) $$\phi_w(w, t) = f(w, t)$$

(4) $$\phi_t(w, t) = -\mu H(C_1\phi_w(w, t) + D_1 w) \quad , \quad w = C\hat{z} + Dz \quad .$$

The last equation is the most general Hamilton-Jacobi equation for the canonical system with Hamiltonian $H(z)$ and linear transformation

$T \in CS_p(\tilde{J}_{4n}, J_{4n})$.

By recursions we can determine explicitly the power series expansions for all possible time-dependent generating functions for analytic Hamiltonians.

<u>Theorem 6</u>. Let $H(z)$ depends analytically on z, then $\phi(w, t)$ in Theorem 5 is expressible as convergent power series in t for sufficiently small $|t|$:

$$\phi(w, t) = \sum_{k=0}^{\infty} \phi^{(k)}(w) t^k \quad ,$$

$$\phi^{(0)}(w) = \frac{1}{2} w' N_0 w \quad , \quad N_0 = (AM_0 + B)(CM_0 + D)^{-1}$$

$$\phi^{(1)}(w) = -\mu H(E_0 w) \quad , \quad E_0 = (CM_0 + D)^{-1}$$

$$\phi^{(k+1)}(w) = -\frac{\mu}{k+1} \sum_{m=1}^{k} \frac{1}{m!} \sum_{i_1,\ldots,i_m=1}^{2n} H_{z_{i_1},\ldots,z_{i_m}}(E_0 w) \sum_{\substack{k_1+\ldots+k_m=k \\ k_j \geq 1}}$$

$$\times (C_1 \phi_w^{(k_1)}(w))_{i_1} \ldots (C_1 \phi_w^{(k_m)}(w))_{i_m} \quad , \quad k \geq 1 .$$

Examples of special types of generating functions:

(I)

$$T = \begin{pmatrix} -I_n & 0 & 0 & 0 \\ 0 & 0 & I_n & 0 \\ 0 & I_n & 0 & 0 \\ 0 & 0 & 0 & I_n \end{pmatrix} \quad , \quad \mu = +1 \quad , \quad M_0 = J_{2n} \quad ,$$

$$|CM_0 + D| \neq 0$$

$$w = \begin{pmatrix} \hat{q} \\ q \end{pmatrix} \quad , \quad \phi = \phi(\hat{q}, q, t) \quad ,$$

$$\hat{w} = \begin{pmatrix} -\hat{p} \\ p \end{pmatrix} = \begin{pmatrix} \phi_{\hat{q}} \\ \phi_q \end{pmatrix} \quad , \quad \phi_t = -H(\phi_q, q) \quad .$$

This is the generating function and H.J. equation of the first kind for the "free" canonical transformations[5].

(II)

$$T = \begin{pmatrix} I_n & 0 & 0 & 0 \\ 0 & 0 & 0 & -I_n \\ 0 & I_n & 0 & 0 \\ 0 & 0 & I_n & 0 \end{pmatrix} \quad , \quad \mu = +1 \quad , \quad M_0 = I_{2n} \quad ,$$

$$|CM_0 + D| \neq 0 \quad .$$

$$w = \begin{pmatrix} \hat{q} \\ p \end{pmatrix} \quad , \quad \phi = \phi(\hat{q}, p, t) \quad ,$$

$$w = \begin{pmatrix} -\hat{p} \\ -q \end{pmatrix} = \begin{pmatrix} \phi_{\hat{q}} \\ \phi_p \end{pmatrix} \quad , \quad \phi_t = -H(p, -\phi_p) \quad .$$

This is the generating function and H.J. equation of the second kind[5].

(III)

$$T = \begin{pmatrix} J_{2n} & -J_{2n} \\ \frac{1}{2} I_{2n} & \frac{1}{2} I_{2n} \end{pmatrix} \quad , \quad \mu = +1 \quad , \quad M_0 = I_{2n} \quad ,$$

$$|CM_0 + D| \neq 0 \quad .$$

$$w = \begin{pmatrix} \frac{1}{2}(\hat{p} + p) \\ \frac{1}{2}(\hat{q} + q) \end{pmatrix} = \begin{pmatrix} \bar{p} \\ \bar{q} \end{pmatrix} = \text{mean values} \quad , \quad \phi = \phi(\bar{p}, \bar{q}, t) \quad ,$$

$$\hat{w} = \begin{pmatrix} q - \hat{q} \\ \\ \hat{p} - p \end{pmatrix} = \begin{pmatrix} \phi_{\bar{p}} \\ \\ \phi_{\bar{q}} \end{pmatrix} \quad , \quad \phi_t = -H(\bar{p} + \frac{1}{2}\phi_{\bar{q}}, \bar{q} - \frac{1}{2}\phi_{\bar{p}}) \quad .$$

This is a new type of generating functions and H.J. equations, not encountered in the classical literature. $\phi(\bar{p}, \bar{q}, t)$ is an odd function in t.

Generating functions play the central role for the construction of canonical difference schemes for Hamiltonian systems. The general methodology for the latter is as follows: Choose some suitable type of generating function (Theorem 5) with its explicit expression (Theorem 6). Truncate it or approximate it in some way, take the gradient of this approximate generating function, then we get automatically the implicit representation of some canonical transformation for the transition of the difference scheme. In this way one can get an abundance of canonical difference schemes. This methodology is unconventional in the ordinary sense, but natural from the point of view of symplectic geometry. As an illustration we construct a generalization of centered Euler scheme with arbitrarily high order of accuracy.

Theorem 7. Choose the generating function of Example (III) with its m-th truncation

$$\phi(\bar{p}, \bar{q}, \tau) = \sum_{k=0}^{\infty} \phi^{(2k+1)}(\bar{p}, \bar{q})\tau^{2k+1} \quad ,$$

$$\phi_m(\bar{p}, \bar{q}, \tau) = \sum_{k=0}^{m-1} \phi^{(2k+1)}(\bar{p}, \bar{q})\tau^{2k+1} \quad , \quad m = 1, 2, \ldots,$$

with $\tau > 0$ small enough as the time-step, $\phi(\tau) - \phi_m(\tau) = O(\tau^{2m})$. Let $\hat{z} = z^k$, $z = z^{k+1}$, $\bar{z} = \frac{1}{2}(z^k + z^{k+1})$, then

$$\hat{q} - q = \sum_{k=0}^{m-1} \phi_{\bar{p}}^{(2k+1)}(\bar{p}, \bar{q})\tau^{2k+1} \quad ,$$

$$p - \hat{p} = \sum_{k=0}^{m-1} \phi_{\bar{q}}^{(2k+1)}(\bar{p}, \bar{q})\tau^{2k+1}$$

is a canonical difference scheme of 2m-th order of accuracy, $m = 1$ is the centered Euler scheme (9).

References

1. Feng Kang, "On difference schemes and symplectic geometry", Prof. of 1984 Beijing International Symposium on Differential Geometry and Differential Equations, "Computation of Partial Differential Equations", August 1984, Beijing, ed. Feng Kang, pp. 42-58, Science Press, Beijing, 1985.

2. Feng Kang, "Difference schemes for Hamiltonian formalism and symplectic geometry", Computational Math. 3 (1986).

3. Feng Kang, Wu Hua-mo, Qing Meng-zhou, "Symplectic difference schemes for linear Hamiltonian canonical systems", to appear.

4. Feng Kang, Wu Hua-mo, Qing Meng-zhou, "Construction of canonical difference schemes for Hamiltonian formalism via generating functions", to appear.

5. V.I. Arnold, Mathematical Methods of Classical Mechanics, New York, 1978.

6. R. Abraham, J. Marsden, Foundations of Mechanics, 2nd ed., Mass., 1978.

7. C.L. Siegel, "Symplectic geometry", Amer. J. Math. 65 (1943) 1-86.

8. L.K. Hua, "On the theory of automorphic functions of a matrix variable I, II", Amer. J. Math. 66 (1944) 470-488, 531-563.

Blow up of solutions of nonlinear evolution equations

Avner Friedman

In part I of the talk I shall study nonlinear heat equations

$$u_t - \Delta u = f(u) \qquad (x \in \Omega,\ t > 0)$$

$$u = 0 \quad \text{if} \quad x \in \partial\Omega,\ t > 0,$$

$$u(x,u) = \phi(x) \quad \text{if} \quad x \in \Omega$$

with $\phi \geq 0$, $f \geq 0$, f superlinear. Let T be the maximal time for which the solution exists. We shally analyze the behavior of $u(x,t)$ as $t \uparrow T$; in particular, we shall study the set of blow-up points, and the growth of u at these points as $t \uparrow T$.

In part II we shall study the nonlinear wave equations

$$u_{tt} - \Delta u = f(u)$$

with Cauchy data, and analyze the surface (in the (x,t)-space) where the solution blows up.

OPTIMAL CONTROL OF SPECIAL DISTRIBUTED PARAMETER SYSTEMS

Abstract

Optimal control problems were presented, where the state is governed partial differential equations involving free boundaries. The costfunctions contain the state and control functions as well as the free boundary itself. The relations to improperty posed problems are discussed. Finally we prove existence and uniqueness results and present a numerical scheme, which is based on finite element techniques. The numerical results will be represented using modern computer graphics.

(Prof. Dr. K.-H. Hoffmann)

The Solution of Soliton Equation and Darboux Transformation*

Li Yi-shen
(Department of Mathematics,
University of Science and Technology of China, Hefei, China)

1. Introduction

Since the discovery of soliton and the inverse scattering method for solving the KdV equation, a lot of soliton equations has been found[1]-[3]. Associated with a given equation there exist at least two interelated aspects of fundamental interest : (1) The development of methods for finding special solutions. (2) The investigation of various "algebraic" properties of the equation. In this paper, we shall propose a method for finding special solutions of some soliton equations. The source of this method lies upon the old theorem proved by Darboux for Schrödinger (Strum-Liouville) equation

$$-\phi_{xx}(x,\lambda)+u(x)\,\phi(x,\lambda)=\lambda\phi(x,\lambda) \tag{1.1}$$

Darboux's theorem says that for an eigenvalue $\lambda=\lambda_0$ and a solution $f=f(x,\lambda_0)$ of (1.1) we can construct all the solution of the new equation

$$-\Phi_{xx}(x,\lambda)+u_1(x)\Phi(x,\lambda)=\lambda\Phi(x,\lambda) \tag{1.2}$$

by the transformation

$$\Phi(x,\lambda)=\phi_x(x,\lambda)-\frac{f_x}{f}\phi(x,\lambda) \tag{1.3}$$

$$u_1(x)=u(x)-2(\ln f)_{xx} \tag{1.4}$$

It is interesting that equation (1.1) is "invariant" under the action $\phi\to\Phi$, $u\to u_1$. We call (1.3) and (1.4) the Darboux Transformation (abv. as DT hereafter) of (1.1).

Recently, we have extended the Darboux's idea for finding special solutions of some soliton equations, we shall summarize our main results in the following sections.

---

* Projects Supported by the Science Fund of the Chinese Academy of Science.

2. KdV type equations[4],[5]

Consider the eigenvalue problem (E.P) (1.1) with coefficient $u(x) = u(x,t)$, and associated (1.1) with a time evolution

$$\phi_t = A\phi + B\phi_x \tag{2.1}$$

where

$$A = -\frac{\alpha(t)}{4} u_x , \quad B = \frac{\alpha(t)}{4} (2u+4\lambda)$$

suppose $\lambda_t=0$, $\phi_{xxt} = \phi_{txx}$ implies

$$u_t = \frac{\alpha(t)}{4} (-u_{xxx} + 6uu_x) \tag{2.2}$$

when $\alpha(t)=4$, (2.2) reduced to the KdV equation

$$u_t - 6uu_x + u_{xxx} = 0 \tag{2.3}$$

In [4],[5] we have proved the following theorem:

Theorem 1: If $u(x,t)$ is a solution of (2.2), and ϕ is a solution of (1.1) and (2.1), then Φ defined by (1.3) satisfies equation (1.2) and

$$\Phi_t = \bar{A}\Phi + \bar{B}\Phi_x \tag{2.4}$$

with

$$\bar{A} = -\frac{\alpha(t)}{4} u_{1x} , \quad \bar{B} = \frac{\alpha(t)}{4}(2u_1 + 4\lambda)$$

(This means that (2.1) is also "invariant" under the Darboux Transformation) and u_1 defined by (1.4) is another solution of equation (2.2).

Now, we can generate a series of solutions of (2.2) from the Darboux Transformation (1.3) and (1.4).

First,we find a solution $u(x,t)$ of the equation (2.2) as "seed", then find the general solution of (1.1) and (2.1) for a fixed eigenvalue $\lambda = \lambda o$, $u_1(x,t) = u(x,t) - 2(\ln f)_{xx}$ is a new solution of (2.2)

Then, we begin with $u_1(x,t)$, we can get Φ from (1.3) (we need'nt to solve (1.2) and (2.4)) for another eigenvalue $\lambda = \lambda_1$,using (1.4) we get another new solution $u_2=u_1 - 2(\ln \Phi)_{xx}$, this procedure can be continued.

Using a special solution and the eigenfunction of the coresponding eigenvalue problem, the DT enable us to generate an infinte many special solutions in purely an algebraic way.

As example, we consider the KdV equation (2.3) and given a "seed"

$$u(x,t) = b = \text{const.} \tag{2.5}$$

(1.1) and (2.1) are reduced to

$$-\phi_{xx} = (\lambda - b)\phi \tag{2.6}$$

$$\phi_t = (2b + 4\lambda)\phi_x \tag{2.7}$$

i) when $\lambda = b$

$$\phi = x + 6bt \tag{2.8}$$

$$u_1 = b + 2/(x + 6bt)^2 \tag{2.9}$$

ii) when $\lambda \neq b$

$$\phi = c_1 e^{\alpha z} + c_2 e^{-\alpha z} \tag{2.10}$$

where c_1, c_2 are constants and

$$\alpha = (b - \lambda)^{\frac{1}{2}} \qquad z = x + \beta t \qquad \beta = 2b + 4\lambda \tag{2.11}$$

$$u_1 = b - 2\lambda^2 \frac{4c_1c_2}{[c_1 e^{\alpha z} + c_2 e^{-\alpha z}]^2} \tag{2.12}$$

when $b - \lambda_0 > 0$ $c_1 = \frac{1}{c_2} = e^{\delta}$ u_1 is reduced to

$$u_1 = b - 2\alpha_0^2 \operatorname{sech}^2(\alpha_0 x + \alpha_0\beta_0 t + \delta) \tag{2.13}$$

where $\alpha_0 = (b - \lambda_0)^{\frac{1}{2}}$ $\beta_0 = 2b + 4\lambda_0$ (2.14)

Let us begin with $u_1 = b + \dfrac{2}{(x + 6bt)^2}$

i) $\lambda = b$

$$\phi = c_1[(x + 6bt)^2 + \frac{12t + c_2}{x + 6bt}] \tag{2.15}$$

$$u_2 = u_1 + 2(\ln\phi)_{xx}$$

$$= b - \frac{2}{(x+6bt)^2} + \left(4(x+6bt)^2 - \frac{12b(12t+c)}{x+6bt}\right) / \left[(x+6bt)^3-(12t+c)\right] \quad (2.16)$$

ii) $\lambda \neq b$

$$\phi = c_1\left(\alpha + \frac{1}{x+6bt}\right)e^{\alpha z} + c_2\left(-\alpha + \frac{1}{x+6bt}\right)e^{-\alpha z}$$

$$u_2 = u_1 - 2(\ln\phi)_{xx}$$

$$= b - \frac{2}{x+6bt} + 2\,\phi_x^2(x,\lambda_1)/\phi^2(x,\lambda_1) \quad (2.17)$$

when $b \to 0$ the coresponding solution have been obtained in[6].

We can extend this method to nonisopectral case, for simplicity, consider the case $\lambda_t = k(t)\lambda$, and associate (1.1) with another time evolution

$$\phi_t = \left[g(t) + \frac{\alpha(t)}{4}u_x\right]\phi + \left[\frac{k(t)\,x}{2} + \frac{\alpha(t)}{4}(2u+4\lambda)\right]\phi_x \quad (2.18)$$

where $g(t)$ is an arbitrary function of t.
From the compatibility condition $\phi_{xxt} = \phi_{txx}$, we find:

$$u_t = \frac{\alpha(t)}{4}\left[-u_{xxx} + 6uu_x\right] + k(t)\left[\tfrac{1}{2}xu_x + u\right] \quad (2.19)$$

This is a nonlinear equation with variational coefficients, when $u=-q+x/12t$ $\alpha(t)=4$, $k(t)=-\frac{1}{t}$ we get the c-kdv equation

$$q_t + q_{xxx} + 6q_xq + \frac{q}{2t} = 0 \quad (2.20)$$

We have also proved that[1] if u is a solution of (2.19), ϕ is a solution of (1.1) and (2.18), then u_1 defined by (1.4) is also a solution of (2.19).

For example, consider the c-kdv equation (2.20), define "seed" $q = x/12t$ and put $\lambda=0$ then (1.1) and (2.18) are reduced to

$$-\phi_{xx} = 0 \quad (2.21)$$

$$\phi_t = g(t)\phi - \frac{x}{12t}\phi_x \quad (2.22)$$

$$\phi = e^{\int g(t)dt}[c_1 x t^{-\frac{1}{2}} + c_2] \tag{2.23}$$

$$q_1 = \frac{x}{12t} - \frac{2c_1^2 t^{-1}}{(c_1 x t^{-\frac{1}{2}} + c_2)^2} \tag{2.24}$$

3. AKNS System[7]

In this section we deduce the Darboux Transformation for AKNS E.P (or: generalized Zaharov-Shabat E.P)

$$\begin{pmatrix}\phi_1\\ \phi_2\end{pmatrix}_x = M \begin{pmatrix}\phi_1\\ \phi_2\end{pmatrix} \qquad M = \begin{pmatrix}-i\xi & q(x)\\ r(x) & i\xi\end{pmatrix} \tag{3.1}$$

$$\begin{pmatrix}\Phi_1\\ \Phi_2\end{pmatrix}_x = M_1 \begin{pmatrix}\Phi_1\\ \Phi_2\end{pmatrix} \qquad M_1 = \begin{pmatrix}-i\xi & q_1(x)\\ r_1(x) & i\xi\end{pmatrix} \tag{3.2}$$

If the transformation

$$\begin{pmatrix}\Phi_1\\ \Phi_2\end{pmatrix} = T \begin{pmatrix}\phi_1\\ \phi_2\end{pmatrix} \tag{3.3}$$

where $T = \begin{pmatrix}-2i & 0\\ 0 & 2i\end{pmatrix}\xi + \begin{pmatrix}a(x) & b(x)\\ c(x) & d(x)\end{pmatrix}$

map (3.1) to (3.2). Then T must satisfy

$$T_x = M_1 T - TM \tag{3.4}$$

We have

$$b = (q_1 + q) \qquad c = (r_1 + r) \tag{3.5}$$

$$a_x = q_1 c - rb \qquad d_x = r_1 b - qc \tag{3.6}$$

$$b_x = q_1 d - qa \qquad c_x = r_1 a - rd \tag{3.7}$$

Let $\phi = (\phi_{ij})$, $\Phi = (\Phi_{ij})$ $i\,j = 1,2$ be the solution matrix of (3.1) and (3.2) respectively. Since $\mathrm{Tr}M = \mathrm{Tr}M_1 = 0$, $\det\phi$ and $\det\Phi$ do not depend on x, from $\Phi = T\phi$, we get

$$\det T = \det\Phi / \det\phi = \text{const.} \tag{3.8}$$

The quadratic polynomial det T can be factorized as follows

$$\det T = 4(\xi - \xi_1)(\xi - \xi_2) \qquad \xi_i = \text{const} \quad i=1,2 \tag{3.9}$$

So:

$$\det \Phi(x, \xi_j) = \det T(x, \xi_i) \det\phi(x, \xi_j) = 0 \qquad j=1\ 2$$

i.e.

$$\begin{vmatrix} (-2i\xi_j + a)\phi_{11} + b\,\phi_{21} & (-2i\xi_j + a)\phi_{12} + b\phi_{22} \\ \phi_{11} + (2i\xi_j + d)\phi_{21} & c\phi_{12} + (2i\xi_j + d)\phi_{22} \end{vmatrix} = 0 \tag{3.10}$$

There exist two constants k_1, k_2 such that:

$$\begin{aligned} &(-2i\xi_1 + a)\phi_1 + b\phi_2 = 0 \qquad \phi_1 + (2i\xi_1 + d)\phi_2 = 0 \\ &(-2i\xi_2 + a)\psi_1 + b\,\psi_2 = 0 \qquad \psi_1 + (2i\xi_2 + d)\psi_2 = 0 \end{aligned} \tag{3.11}$$

where

$$\begin{aligned} &\phi_1 = \phi_{11}(x_1, \xi_1) + k_1\phi_{12}(x, \xi_1) \qquad \psi_1 = \phi_{11}(x, \xi_2) + k_2\,\phi_{12}(x, \xi_2) \\ &\phi_2 = \phi_{21}(x, \xi_1) + k_1\,\phi_{22}(x, \xi_1) \qquad \psi_2 = \phi_{21}(x, \xi_2) + k_2\,\phi_{22}(x, \xi_2) \end{aligned} \tag{3.12}$$

Solving (3.11) we get

$$\begin{aligned} &a = (2i\xi_1\alpha\beta - 2i\xi_2)/\Delta \qquad b = 2i(\xi_2 - \xi_1)\alpha/\Delta \\ &c = 2i(\xi_2 - \xi_1)\,\beta/\Delta \qquad d = (2i\xi_1 - 2i\xi_2\alpha\beta)/\Delta \end{aligned} \tag{3.13}$$

where

$$\Delta = \alpha\beta - 1 \qquad \alpha = \phi_1/\phi_2 \qquad \beta = \psi_2/\psi_1 \tag{3.14}$$

Using (3.5) and (3.13) we have

$$\begin{aligned} q_1 &= -q + 2i(\xi_2 - \xi_1)\,\alpha/\Delta \\ r_1 &= -r + 2i(\xi_2 - \xi_1)\,\beta/\Delta \end{aligned} \tag{3.15}$$

we call the formula (3.3) and (3.15) the DT for (E.P) (3.1)**

** Similar formulas have been deduced indenpendently in[8] in different way

Associate with (3.1) a time evolution

$$\begin{pmatrix}\phi_1\\ \phi_2\end{pmatrix}_t = N\begin{pmatrix}\phi_1\\ \phi_2\end{pmatrix} \qquad N = \begin{pmatrix}A, & B\\ C, & -A\end{pmatrix} = \begin{bmatrix}\sum_{j=0}^{n} a_j\,\xi^{n-j} & \sum_{j=0}^{n} b_j\,\xi^{n-j}\\ \sum_{j=0}^{n} c_j\,\xi^{n-j} & -\sum_{j=0}^{n} a_j\,\xi^{n-j}\end{bmatrix} \tag{3.16}$$

when $\xi_t = 0$ and set

$$a_o = -i\,\alpha_0(t) \qquad a_j = D^{-1}(qc_j - rb_j) - i\,\alpha_j(t) \tag{3.17}$$

$$\begin{pmatrix}c_j\\ b_j\end{pmatrix} = (\alpha_0(t)L^{j-1} + \alpha_1(t)L^{j-2} + \dots \alpha_{j-1}(t))\begin{pmatrix}r\\ q\end{pmatrix} \tag{3.18}$$

where $\alpha_j(t)$ $j = 0,1,2,\dots n$ are arbitrary function of t

$$L = \frac{1}{2i}\begin{pmatrix}D - 2rD^{-1}q & 2rD^{-1}r\\ -2qD^{-1}q & -D + 2qD^{-1}r\end{pmatrix} \tag{3.19}$$

here D denote $\frac{\partial}{\partial x}$ and $D^{-1}D = DD^{-1} = I$.

The compatibility of (3.1) and (3.16) gives[9]

$$M_t - N_x + MN - NM = 0$$

i.e.

$$\begin{pmatrix}r\\ -q\end{pmatrix}_t = 2i\sum_{j=0}^{n} \alpha_j(t)L^{n-j}\begin{pmatrix}r\\ q\end{pmatrix} \tag{3.20}$$

We have proved the following theorem:

Theorem 2 [10],[11] : If $q(x,t)$, $r(x,t)$ is a solution of (3.20), Φ is a fundamental solution matrix of (3.1) and (3.16), define a,b,c,d and Φ by (3.13) and (3.3) respectively, then Φ satisfying $\Phi_x = M_1\Phi$,

$\Phi_t = N_1\Phi$, where M_1, N_1, are same type as M,N defined by (3.1) and (3.16) with $q_1(x,t)$, $r_1(x,t)$ instead of $q(x,t)$, $r(x,t)$, and the relation:

$$(M_t - N_x + MN - NM) = T^{-1}(M_{1t} - N_{1x} + M_1N_1 - N_1M_1)T \tag{3.21}$$

yields that q_1, r_1 defined by (3.15) is another solution of (3.20).

For example[7],[12], we put

$$M = \begin{pmatrix} -i\xi & q \\ -q^* & i\xi \end{pmatrix}, \quad N = \begin{pmatrix} -2i & 0 \\ 0 & 2i \end{pmatrix}\xi^2 + \begin{pmatrix} 0 & 2q \\ -2q^* & 0 \end{pmatrix}\xi + \begin{pmatrix} i|q|^2 & iq_x \\ iq_x^* & -i|q|^2 \end{pmatrix}. \tag{3.22}$$

(3.20) is reduced to nonlinear Schrödinger equation

$$iq_t + q_{xx} + 2|q|^2 q = 0 \tag{3.23}$$

It is well known that $q = e^{i(t-x)}$ is a periodic solution of (3.23). Let it be the "seed" solution, for fixed eigenvalue $\xi_1 = \xi_2^* = \frac{1}{2} + \frac{\sqrt{3}}{2} i$, solving (3.1) and (3.16) we get

$$\begin{aligned} \phi_1 &= e^{i(1+\xi_1)t} - i\xi_1 e^{-(x+\xi_1 t)} & \phi_2 &= i\xi_1 e^{i(x+\xi_1 t)} - e^{-i(1+\xi_1)t} \\ \psi_1 &= e^{i(1+\xi_1^*)t} + i\xi_1^* e^{-i(x+\xi_1^* t)} & \psi_2 &= i\xi_1^* e^{i(x+\xi_1^* t)} + e^{-i(1+\xi_1^*)t} \end{aligned} \tag{3.24}$$

substituting (3.24) into (3.15) we get a new periodic solution

$$q_1 = e^{i(t-x)}\left[1 - \frac{i\sqrt{3}\,\mathrm{sh}\sqrt{3}t + \mathrm{ch}\sqrt{3}t}{\sqrt{3}\cos(x+2t) + 2\mathrm{ch}\sqrt{3}t}\right] \tag{3.25}$$

In fig.1 we plot the evolution of the modulas, real and imaginary part of the solution. We have discussed some properties of this new periodic solutions in [13]. It is interesting that, under DT, the main spectrum remains unchanged, but two sets of the nondegenerated auxiliary spectrum ν_i, η_i are changed.

4. Some other Eigenvalue Problems

In[14],[15], we have extended this method to the following E.P:

i) Kaup-Newell E.P.[16]

$$\phi_x = M\phi \qquad M = \begin{pmatrix} -i\xi^2 & \xi q \\ \xi r & i\xi^2 \end{pmatrix} \tag{4.1}$$

we get DT as follows:

$$\Phi = T\phi \qquad T = \begin{pmatrix} \xi^2 a + 1 & \xi b \\ \xi c & \xi^2 d + 1 \end{pmatrix} \tag{4.2}$$

$$a = \left(\frac{1}{\xi_2} - \frac{\alpha\beta}{\xi_1}\right)/(\xi_1\alpha\beta - \xi_2) \qquad c = \beta\left(\frac{\xi_1}{\xi_2} - \frac{\xi_2}{\xi_1}\right)/(\xi_2\alpha\beta - \xi_1)$$

$$b = \alpha\left(\frac{\xi_2}{\xi_1} - \frac{\xi_1}{\xi_2}\right)/(\xi_1\alpha\beta - \xi_2) \qquad d = \left(\frac{1}{\xi_1} - \frac{\alpha\beta}{\xi_2}\right)/(\xi_2\alpha\beta - \xi_1) \tag{4.3}$$

$$q_1 = q + b_x \qquad r_1 = r + c_x \tag{4.4}$$

ii) Eigenvalue Problem[17]:

$$\phi_x = -i\xi S\phi \qquad S = \begin{pmatrix} w & u \\ v & -w \end{pmatrix} \qquad w^2 + uv = 1 \tag{4.5}$$

we get DT as follows

$$\Phi = T\phi \ , \qquad T = \begin{pmatrix} \xi a+1 & \xi b \\ \xi c & \xi d+1 \end{pmatrix} \tag{4.6}$$

$$a = \left(-\frac{1}{\xi_1} + \frac{1}{\xi_2}\alpha\beta\right)/(1-\alpha\beta) \qquad c = \beta\left(\frac{1}{\xi_1} - \frac{1}{\xi_2}\right)/(\alpha\beta - 1)$$

$$b = \alpha\left(\frac{1}{\xi_1} - \frac{1}{\xi_2}\right)/(1-\alpha\beta) \qquad d = \left(-\frac{1}{\xi_2}\alpha\beta + \frac{1}{\xi_1}\right)/(\alpha\beta - 1) \tag{4.7}$$

$$u_1 = u + ib_x \qquad v_1 = v + ic_x \qquad u_1v_1 + w_1^2 = 1 \tag{4.8}$$

iii) Jaulent, Miodek E.P [18]

$$\phi_x = M\phi \qquad M = \begin{pmatrix} -i\xi & u+iv \\ 1 & i\xi \end{pmatrix} \tag{4.9}$$

we get DT as follows:

$$\Phi = T\phi \qquad T = \begin{pmatrix} a_1\xi + a_o & b_1\xi + b_o \\ c_o & d_1\xi + d_o \end{pmatrix} \tag{4.10}$$

when $a_1d_1 = 1$ we define:

$$f_1 = \beta(\xi_2 - \xi_1)/(\alpha\beta - 1) \qquad f_2 = (\xi_1 - \xi_2\alpha\beta)/(\alpha\beta - 1)$$

then

$$a = (1-2if_1)^{-\frac{1}{2}} \qquad d_1 = (1-2if_1)^{\frac{1}{2}}$$

$$c_o = d_1 f_1 \qquad d_o = d_1 f_2 \tag{4.11}$$

$$b_1 = d_{1x} + vc_o \qquad a_o = c_{ox} + d_o \qquad b_o = d_{ox} + ac_o$$

$$v_1 = a_{1x} + b_1/c_o \qquad u_1 = b_{ox} + ua_o/d_o \tag{4.12}$$

iv) Boit-Tu E.P[19]:

$$\phi_x = M\phi \qquad M = \begin{pmatrix} -i\xi + \frac{is}{\xi} & u + \frac{iv}{\xi} \\ u - \frac{iv}{\xi} & i\xi - \frac{is}{\xi} \end{pmatrix} \tag{4.13}$$

we deduce DT as follows:

$$\Phi = T\phi \qquad T = \begin{pmatrix} \xi + a & -ib \\ ib & \xi - a \end{pmatrix} \tag{4.14}$$

where

$$a = \xi_1(\phi_1^2 + \phi_2^2)/(\phi_2^2 - \phi_1^2) \qquad b = -2i\xi_1\phi_1\phi_2/(\phi_2^2 - \phi_1^2) \tag{4.15}$$

$$u_1 = 2b + u$$

$$v_1 = -\frac{1}{\xi_1^2}(2isab - vb^2 + va^2) \tag{4.16}$$

$$s_1 = -\frac{1}{\xi_1^2}(-2ivab + sb^2 - sa^2) \qquad s_1^2 + v_1^2 = s^2 + v^2$$

When $\xi_t = 0$, we associate with each E.P a linear lime evolution. We can deduce four classes of nonlinear evolution equations which include many physical interesting equations such as ferromegnetic Heisenberg Chain equation, derivative nonlinear Schrödinger equation, the massive thirring model equation and Sine-Gordon equation with an external potential etc.

We can establish theorems as theorem 1 and theorem 2, this kinds of theorems made it is true that the above mentioned DT can be used to generate a series of solutions of the coresponding nonlinear evolution equation.

5. A System of Nonlinear Klein-Gordon Equations

Our method can be extended to $n\times n$ 1st order E.P,. Now we only consider a special case[20]

$$\phi_x = M\phi \qquad M = (M_{j\ell}) = \theta_{jx}\delta_{j\ell} + \xi\delta_{j,\ell-1} \qquad j,\ell = 1,2,\ldots n \tag{5.1}$$

$$\phi_t = N\phi \qquad N = (N_{j\ell}) = \frac{1}{\xi}\exp(\theta_j - \theta_{j-1})\,\delta_{j,\ell+1} \qquad j,\ell = 1,2,\ldots n \tag{5.2}$$

where θ_j are functions of x and t, the compatibility condition $\phi_{xt} = \phi_{tx}$ yields the klein-Gordon equation:

$$\theta_{j,xt} = e^{\theta_j - \theta_{j-1}} - e^{\theta_{j+1} - \theta_j} \qquad j = 1,2,\ldots n \ (\mathrm{mod}\ n) \tag{5.3}$$

We get D.T as follows

$$\Phi = T\phi \qquad T = \xi\delta_{j,\ell} + a_j\delta_{j,\ell+1} \qquad j,\ell = 1,2,\ldots n \ (\mathrm{mod}\ n) \tag{5.4}$$

where

$$a_j = -\alpha\psi_j / \psi_{j-1} \tag{5.5}$$

$$\psi_j = \sum_{\ell=1}^{n} \phi_{j,\ell}(x,\alpha)k_\ell \tag{5.6}$$

α , k_ℓ ($\ell=1,2,\ldots n$) are constants (Take one of k_j is 1)

$$(\bar{\theta}_j) = \ln\frac{\psi_j}{\psi_{j-1}} + \theta_{j-1} \tag{5.7}$$

We have proved that, if $\{\theta_j\}$ is a set of solution of (5.3) then $\{\bar{\theta}_j\}$, defined by (5.7), is also a set of solution of (5.3).

As an example, let us begin from $\theta_j = kx$ $j=1,2,\ldots,$ here k is a constants, $w_j = \exp(i(j-1)\pi/n)$ $j=1,2,\ldots n$, the solution of (5.1) and (5.2) is

$$\phi_{j,\ell} = (w_\ell)^{j-1} e^{(k+\alpha w_\ell)x + \frac{t}{\alpha w_\ell}} \qquad \ell, j = 1,2,\ldots n \tag{5.8}$$

$$\psi_j = \sum_{\ell=1}^{n} [(w_\ell)^{j-1} e^{(k+\alpha w_\ell)x + \frac{t}{\alpha w_\ell}}]k_\ell \qquad j = 1,2,\ldots n \tag{5.9}$$

$$(\bar{\theta})_j = kx + \ln\psi_j - \ln\psi_{j-1} \; . \qquad (5.10)$$

when n=2, put $\theta = \theta_1 - \theta_2$, the equation (5.3) is reduced to Sinh-Gordon equation:

$$\theta_{xt} = 4\mathrm{sh}\theta \qquad (5.11)$$

(5.10) is reduced to

$$\bar{\theta} = 2\ln\frac{\psi_j}{\psi_{j-1}} = \frac{k_1 e^{2\alpha x + \frac{2}{\alpha}t} + k_2}{k_1 e^{2\alpha x + \frac{2}{\alpha}t} - k_2} \qquad (5.12)$$

6. Kadomtsev - Petviashvili (KP) Equation[21]

In this section, we extend our method to a 1+2 dimentional case[22].

Consider the E.P

$$\phi_{xx} = (\lambda - u)\phi + \phi_y \qquad (6.1)$$

$$\phi_t = A\phi + B\phi_x - 4\phi_{xy} \qquad (6.2)$$

where

$$A = u_x - 3w_y \qquad B + -2u - 4\lambda \qquad w_x = u \; .$$

The compatibility condition $\phi_{xxt} = \phi_{txx}$ yields the KP equation

$$u_t + u_{xxx} + 6uu_x + 3w_{yy} = 0 \qquad (6.3)$$

We get DT as follows

$$\Phi = \phi_x + a\phi \qquad (6.4)$$

$$u_1 = u - 2a_x \qquad (6.5)$$

where $\quad a = \phi_x(x,y,t,\lambda_0) / \phi(x,y,t,\lambda_0) \qquad (6.6)$

We have proved the following theorem

Theorem 3: If u is a solution of (5.3) ϕ satisfies (5.1) and (6.2) then Φ defined by (6.4) satisfies the equation

$$\begin{aligned} \Phi_{xx} &= (\lambda - u_1)\Phi + \Phi_y \\ \Phi_t &= A_1\Phi + B_1\Phi_x - 4\Phi_{xy} \end{aligned} \qquad (6.7)$$

where

$$A_1 = (u_{1x} - 3w_{1y}) \qquad B_1 = -2u_{1x} - 4\lambda \qquad w_{1x} = u_1 \tag{6.8}$$

and u_1 defined by (6.5) is another solution of the KP equation (6.3).

For example, let us begin from $u=b$ and $\lambda = \lambda_0 = 0$, the equation (6.1) (6.2) are reduced to

$$\phi_{xx} = \phi_y \tag{6.9}$$

$$\phi_t = -6b\,\phi_x - 4\,\phi_{xy} \tag{6.10}$$

i) Solution $\phi = x^2 + 2y - 12bxt + 96b^2t$ (6.11)

$$u_1 = b + 4 / (x^2 + 2y - 12bx + 96b^2t) - 2(2x - 12bt)^2 / (x^2 + 2y - 12bxt + 96b^2t)^2 \tag{6.12}$$

ii) Solution $\phi = 1 + e^{az} \qquad z = x + ay - (6b + 4a^2)t$ (6.13)

$$u_1 = b + \frac{1}{2}a^2 \operatorname{sech}^2 \frac{a}{2}z\ . \tag{6.14}$$

Reference

1. В.Е.ЗахароВ, С.В. Манаков, С.П. Новиков. Л.П. Питаевский. "Теория солитонов, метод обратной задачи." ИЗД «НАУКА» 1980.

2. M.J.Ablowitz and H. Segur, "Solitons and Inverse Scattering Transform" SIAM Philadelphia (1981) .

3. F.Calogero and A Degasperis "Spectral Transform and Solitons I" Studies in Mathematics and Its Application vol 13 (1982).

4. Li Yi-shen "KdV type equations and Darboux transformation" preprint (1984).

5. Huang Jia-wu "Darboux transformation and KdV equation" preprint (1985) Chinese .

6. CAu, P.C.W.Fung "A KdV soliton propergating with varying velocity" J.Math. Phys. 19(1984) p1364-69 .

7. Li Yi-shen "Darboux transformation for AKNS eigenvalue problem" a lecture in seminar 1983 .

8. D.Levi, O.Ragnisco, A.Sym,"Dressing method vs classical Darboux transformation" Preprint 1983.

9. Li Yi-shen "A class of evolution equations and the spectral deformation" Scientia Sinica (Series A) Vol XXV (1982) 9 p911-917.

10. Cheng Yi "Darboux transformation and AKNS System" Preprint (1984) (Chinese).

11. Zou Mao Rong "Darboux transformation for AKNS and Kaup-Newell eigenvalue problem" Preprint (1985) (Chinese).

12. Li Yi-shen, Wang Cun-qi: "A new periodic solution of nonlinear schrödinger equation I" to be published in Kexue Tongbao.

13. Wang Cun-qi, Li Yi-shen " A new periodic solution of nonlinear schrödinger equation II " preprint (1985) (Chinese).

14. Li Yi-shen "Bäcklund transformation and Darboux transformation" to be submitted to Chin. Ann. Math. (1984) (Chinese).

15. Li Yi-shen, Gu Xin-shen "Darboux transformation associated with Boita-Tu eigenvalue problem" to be submitted to Chin. Ann. of Math. (1984) Chinese.

16. D.T.Kaup and A.C.Newell "An exact solution for a derivative nonlinear schrödinger equation" J. Math. Phys. 19(1978) p789-801.

17. D.Y.Chen, Li Yi-shen "The transformation operator V" to be published in Acta. Math. Sinica.

18. M.Jaulent, I Miodek "Nonlinear evolution equations associated with energy-dependent schrödinger potential" Lett. in Math. Phys. 1(1976) p243-250.

19. M.Boiti, G.Z.Tu "A simple approach to the Hamiltonian structure of soliton equation III" IL. Nuovo cimento 75 B(1983) p145-160.

20. Li Yi-shen "The Darboux transformation method for finding the solutions of a system of nonlinear Klein-Gordon equation" to be submitted to J. .. Math. Phys. (1984) (Chinese).

21. Tian Chou "Lax pair and Bäcklund transformation of K.P equation preprint (1985).

22. Li Yi-shen "Darboux transformation and K-P equation" preprint (1985).

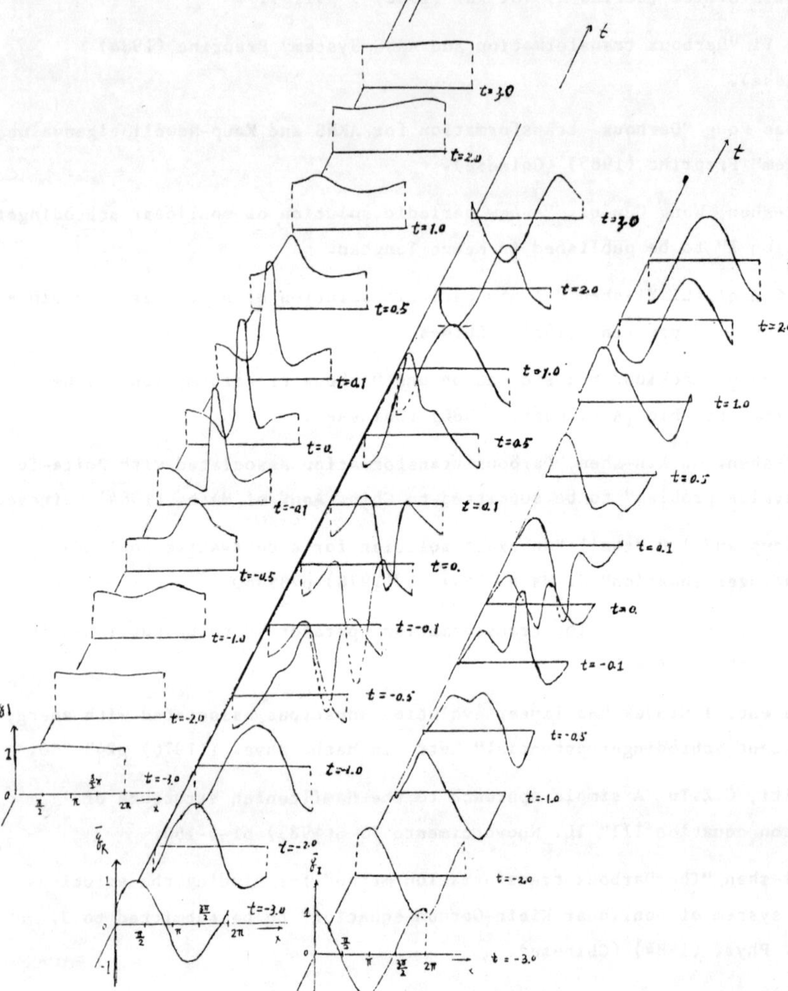
t
t=3.0
t=2.0
t=1.0
t=0.5
t=0.1
t=-0.1
t=-0.5
t=-1.0
t=-2.0
t=-3.0
δI
2
0
π/2
π
3π/2
2π
x
t
t=3.0
t=2.0
t=1.0
t=0.5
t=0.1
t=0.
t=-0.1
t=-0.5
t=-1.0
t=-2.0
t=-3.0
q̄R
1
0
-1
t
t=3.0
t=2.0
t=1.0
t=0.5
t=0.1
t=0.
t=-0.1
t=-0.5
t=-1.0
t=-2.0
t=-3.0
q̄I
1
0
-1

Fig. (1)

NECESSARY AND SUFFICIENT CONDITIONS FOR THE SOLVABILITY OF NONLINEAR EQUATIONS THROUGH THE DUAL LEAST ACTION PRINCIPLE

Jean Mawhin

Université de Louvain
Institut Mathématique
B-1348 Louvain-la-Neuve
BELGIUM

1. INTRODUCTION TO THE DUAL LEAST ACTION PRINCIPLE

A number of applications involving differential or other equations lead to abstract equations of the form

$$Lu = F'(u) \tag{1}$$

in a real Hilbert space H, where $L: D(L) \subset H \to H$ is self-adjoint with closed range and $F: H \to \mathbb{R}$ is convex, continuous and Gâteaux-differentiable. When the spectrum $\sigma(L)$ of L is unbounded from below and from above, the direct method of the calculus of variations can be difficult to apply to the functional naturally associated to (1), namely

$$f: H \to \mathbb{R} \quad , \quad u \mapsto (1/2)(Lu, u) - F(u) \quad ,$$

because, in general, f will also be indefinite, so that $\sup_H f = +\infty$ and $\inf_H f = -\infty$. The critical points u of f (i.e. the ones at which $f'(u) = 0$) will be obtained only through involved minimax arguments.

However, if $F': H \to H$ is bijective, one can associate as follows the solutions of (1) to the critical points of another function which can be more tractable. If $K = (L_{|D(L) \cap R(L)})^{-1} : R(L) \to R(L)$ denotes the right inverse of L, then, letting

$$v = Lu \in R(L) \quad ,$$

which is equivalent to

$$u \in Kv + N(L) \quad , \tag{2}$$

we see that if u satisfies (1), we have

$$v = F'(u) \tag{3}$$

or equivalently

$$u = (F')^{-1}(v) = (F^{\star})'(v) \quad , \tag{4}$$

where $F^{\star}$ is the Legendre transform of F defined by

$$F^{\star}(v) = (u, v) - F(u)$$

with u expressed in terms of v by (3). Eliminating u between (2) and (4), we get the relation to be satisfied by the new unknown v, namely

$$- Kv + (F^{\star})'(v) \in N(L) \quad ,$$

which is nothing but the Euler equation associated to the functional

$$g : R(L) \to \mathbb{R} \quad , \quad v \mapsto (1/2)(-Kv, v) + F^{\star}(v) \quad .$$

Reversing the argument, we easily see that u will satisfy (1) if and only if $v = Lu$ is a critical point on $R(L)$ of the functional g. This is the idea of the dual least action principle initiated by Clarke and Ekeland in the study of periodic solutions of Hamiltonian systems and then developed by Brézis, Coron, Nirenberg, Willem and others (see [4] for a survey and references).

The benefit in dealing with g instead of f lies in the fact that K is a bounded operator (so that $r\|v\|^2 \leq (Kv, v) \leq s\|v\|^2$) and, in many applications, K is even compact, making its contribution to g weakly sequentially continuous. When $r < 0$, its lack of positivity can be compensated through $F^{\star}$ under reasonable and natural assumptions upon F. To be more explicit, let us first notice that, for each fixed v, the mapping

$$F_v : H \to \mathbf{R} \quad , \quad u \mapsto (u, v) - F(u)$$

is concave and that (3) is equivalent to

$$(F_v)'(u) = 0 \quad ,$$

so that F_v achieves its maximum at the u given by (3). We have therefore the (Fenchel) formula

$$F^*(v) = \sup_{u \in H} [(u, v) - F(u)] \quad , \tag{5}$$

which can therefore be taken as the definition of an extension of the Legendre transform to situations where F' is no more bijective, and is then called the <u>Fenchel transform</u> of F. When F is convex and lower semi-continuous (l.s.c.), the same will be true for F^*. When F is Gâteaux-differentiable, F^* need not be so, but the Legendre identity (4) can be replaced by the relations

$$v = F'(u) \leftrightarrow u \in \partial F^*(v) \leftrightarrow F^*(v) + F(u) = (u, v)$$

where $\partial F^*(v)$ is the subdifferential of F^* at v, i.e. the set

$$\{w \in H : F^*(z) \geq F^*(v) + (w, z - v) \quad \text{for all} \quad z \in H\} \quad .$$

(see e.g. [8] for more details). Consequently, if we assume that

$$(a/2)\|u\|^2 - c \leq F(u) \leq (b/2)\|u\|^2 + c$$

for some $0 < a \leq b$, $c \geq 0$ and all $u \in H$, then, by (5) and easy calculations we shall have

$$(1/2a)\|v\|^2 + c \geq F^*(v) \geq (1/2b)\|v\|^2 - c$$

for all $v \in H$. Therefore, if we assume that there is some $\lambda > 0$ such that $\sigma(L) \cap]0, \lambda[= \emptyset$, we shall have

$$(-Kv, v) \geq -\lambda^{-1}\|v\|^2 \quad , \quad v \in R(L) \quad ,$$

and hence

$$g(v) \geq (1/2)(b^{-1} - \lambda^{-1})\|v\|^2 - c$$

for all $v \in R(L)$. Consequently, g will be coercive (i.e. $g(v) \to +\infty$ as $\|v\| \to \infty$) when

$$b < \lambda \quad .$$

Then, g will have a minimum if it is weakly lower semi-continuous (w.l.s.c.) and this is already the case for F^*. For the other term, let $\{P_s : s \in \mathbb{R}\}$ be the spectral resolution of $-L$,

$$P^- = \int_{-\infty}^{-\lambda/2} dP_s \quad , \quad P^+ = \int_{-\lambda/2}^{+\infty} dP_s \quad ,$$

so that P^- and P^+ are orthogonal projectors commuting with L and K, and such that $H = R(P^+) \oplus R(P^-)$, $-KP^+$ is semi-positive-definite and $-KP^-$ is semi-negative-definite. Thus,

$$(-Kv, v) = (-KP^+v, v) + (-KP^-v, v)$$

and the first term, convex and continuous, is w.l.s.c. If we finally assume that $\sigma(L) \cap]0, \infty[$ is made of isolated eigenvalues having finite multiplicity $\lambda_1 < \lambda_2 < \ldots$ (so that we can take $\lambda = \lambda_1$), then KP^- will be compact and $v \mapsto (-KP^-v, v)$ will be weakly sequentially continuous, making g w.l.s.c.. Thus, g has a minimum on $R(L)$ and there will exist $v \in R(L)$ and $w \in N(L)$ such that

$$w + Kv \in \partial F^*(v) \quad . \tag{6}$$

Setting

$$u = w + Kv \quad ,$$

so that $Lu = v$, we deduce from the duality relations that

$$v = F'(u) \quad ,$$

i.e.

$$Lu = F'(u) \quad .$$

We have therefore proved the following result, due to Willem [18].

Proposition 1. *Let* H *be a real Hilbert space,* $L : D(L) \subset H \to H$ *be a self-adjoint operator with closed range such that* $\sigma(L) \cap \,]0, \infty[$ *is made of eigenvalues* $\lambda_1 < \lambda_2 < \dots$ *having finite multiplicity, and let* $F : H \to \mathbb{R}$ *be a Gâteaux-differentiable, continuous and convex function such that*

$$(a/2)\|u\|^2 - c \leq F(u) \leq (b/2)\|u\|^2 + c \tag{7}$$

for all $u \in H$, *some* $c \geq 0$ *and some*

$$0 < a \leq b < \lambda_1 \quad .$$

Then the problem

$$Lu = F'(u)$$

has at least one solution u *such that* $v = Lu$ *minimizes*

$$g : v \mapsto (1/2)(-Kv, v) + F^*(v)$$

on $R(L)$, *with* F^* *the Fenchel transform of* F.

Conditions of the type (7) were first introduced by Dolph [6] for nonlinear integral equations.

2. AN EXISTENCE THEOREM FOR SEMILINEAR EQUATIONS AT RESONANCE

When $F(u) = (\lambda/2)\|u\|^2 + (h, u)$, with $h \in H$, condition (6) is equivalent to

$$0 < \lambda < \lambda_1$$

and hence Proposition 1 constitutes a nonlinear generalization of the *non-resonance* situation for the forced linear equation

$$Lu - \lambda u = h \quad , \tag{8}$$

i.e. to the case where (8) is solvable for every $h \in H$ in the Fredholm alternative. We shall now use the dual least action principle to prove an existence theorem for (1) which constitutes a nonlinear version of the situation of resonance for the linear problem (8). The statement that we shall give here is a slightly more abstract version of the basic theorem in [13].

Theorem 1. _Let_ H _be a real Hilbert space_, $L : D(L) \subset H \to H$ _a self-adjoint operator with closed range and non-trivial null-space such that_ $\sigma(L) \cap]0, \infty[$ _is made of eigenvalues_ $\lambda_1 < \lambda_2 \ldots$ _having finite multiplicity, and let_ $F : H \to \mathbb{R}$ _be a Gâteaux-differentiable, continuous and convex function satisfying the following properties_:

1. _There exist_ $c \geq 0$ _and_ $0 < b < \lambda_1$ _such that_

$$F(u) \leq (b/2)\|u\|^2 + c$$

for all $u \in H$.

2. $F(w) \to +\infty$ _if_ $\|w\| \to \infty$ _in_ $N(L)$.

Then the problem

$$Lu = F'(u)$$

has at least one solution u _such that_ $v = Lu$ _minimizes_

$$g : v \mapsto (1/2)(-Kv, v) + F^*(v)$$

_on_ $R(L)$.

The proof of Theorem 1 is based upon a sequence of lemmas.

Lemma 1. _There exists_ $\varepsilon_0 > 0$ _such that, for each_ $\varepsilon \in]0, \varepsilon_0]$, _the equation_

$$Lu = \varepsilon u + F'(u) = F_\varepsilon'(u)$$

(_with_ $F_\varepsilon(u) = (\varepsilon/2)\|u\|^2 + F(u)$) _has a solution_ u_ε _such that_ $v_\varepsilon = Lu_\varepsilon$ _minimizes_ $g_\varepsilon : v \mapsto (1/2)(-Kv, v) + F_\varepsilon^*(v)$ _on_ $R(L)$.

Proof. By assumptions, F_ε is strictly convex and such that $F_\varepsilon(u)/\|u\| \to +\infty$ as $\|u\| \to \infty$. Thus, F^* is its Legendre transform.

Let $\varepsilon_0 > 0$ be such that $b + \varepsilon_0 < \lambda_1$. Then, for each $0 < \varepsilon \leq \varepsilon_0$, F_ε satisfies the conditions of Proposition 1, with b replaced by $b + \varepsilon$, because F is bounded below by an affine mapping. Also, the coercivity of g_ε is independent of ε, as

$$g_\varepsilon(v) \geq (1/2)((b+\varepsilon)^{-1} - \lambda_1^{-1})\| v \|^2 - c \geq (1/2)((b+\varepsilon_0)^{-1} - \lambda_1^{-1})\| v \|^2 - c$$

$$= (\eta/2)\| v \|^2 - c \quad . \tag{9}$$

Thus, by Proposition 1, there will exist for each $\varepsilon \in]0, \varepsilon_0]$ a solution u_ε of

$$Lu_\varepsilon = \varepsilon u_\varepsilon + F'(u_\varepsilon)$$

such that $v_\varepsilon = Lu_\varepsilon$ minimizes g_ε on $R(L)$.

We now obtain _a posteriori_ estimates on our approximate solutions u_ε.

Lemma 2. _There exist_ $C \geq 0$ _and_ $D \geq 0$ _such that, for each_ $\varepsilon \in]0, \varepsilon_0]$, _one has_

$$\| v_\varepsilon \| = \| Lu_\varepsilon \| \leq C \quad , \quad \| u_\varepsilon \| \leq D \quad .$$

Proof. Assumption 2 and the w.l.s. continuity of F implies that F achieves its minimum on $N(L)$ and hence there exists some $w_0 \in N(L)$ such that

$$v_0 = F'(u_0) \in R(L) = (N(L))^\perp \quad .$$

By duality, this implies that

$$F^*(v_0) = (w_0, v_0) - F(v_0) < +\infty \quad ,$$

and obviously,

$$F_\varepsilon^*(v_0) \leq F(v_0) \qquad (\varepsilon > 0) \quad .$$

Now, as v_ε minimizes g_ε on $R(L)$, we have, using (9),

$$(\eta/2)\| v_\varepsilon\|^2 - c \le g_\varepsilon(v_\varepsilon) \le g_\varepsilon(v_0) \le g(v_0)$$

$$= (-Kv_0, v_0) + F^*(v_0) = (-Kv_0, v_0) + (w_0, v_0) - F(v_0) = C_1 \quad ,$$

and hence

$$\| v_\varepsilon\| = \| Lu_\varepsilon\| \le C \qquad (\varepsilon \in]0, \varepsilon_0]) \quad .$$

Moreover, if we write $u_\varepsilon = \bar{u}_\varepsilon + \tilde{u}_\varepsilon$ with $\bar{u}_\varepsilon \in N(L)$, $\tilde{u}_\varepsilon \in R(L)$, we deduce

$$\| \tilde{u}_\varepsilon\| = \| KLu_\varepsilon\| \le \| K\| C \quad .$$

Now, by the convexity of F, we get

$$F(\bar{u}_\varepsilon/2) = F((u_\varepsilon/2) - (\tilde{u}_\varepsilon/2)) \le (1/2)F(u_\varepsilon) + (1/2)F(-\tilde{u}_\varepsilon)$$

$$\le (1/2)[F(0) + (F'(u_\varepsilon), u_\varepsilon) + (b/2)\| \tilde{u}_\varepsilon\|^2 + c]$$

$$\le (1/2)[F(0) + (Lu_\varepsilon, \tilde{u}_\varepsilon) + (b/2)\| \tilde{u}_\varepsilon\|^2 + c] \le C_2 \quad .$$

Hence, assumption 2 implies the existence of C_3 such that, for all $\varepsilon \in]0, \varepsilon_0]$, one has

$$\| \bar{u}_\varepsilon\| \le C_3 \quad ,$$

so that $\| u_\varepsilon\| \le C_3 + \| K\| C = D$, and the proof is complete.

Proof of Theorem 1. By the results of Lemma 2, there exists $u \in H$, $v \in R(L)$ and a sequence (ε_k) in $]0, \varepsilon_0]$ converging to 0 such that

$$u_k = u_{\varepsilon_k} \to u \quad , \quad Lu_k \to v$$

as $k \to \infty$. The weak closedness of the graph of L implies then that $u \in D(L)$ and $v = Lu$. By the convexity of F, F' is a monotone operator and hence, for all $y \in H$ and all $k \in \mathbb{N}^*$, we shall have

$$(F'(u_k) - F(y), u_k - y) \ge 0 \quad .$$

Consequently, by the definition of u_k,

$$(Lu_k - \varepsilon_k u_k - F'(y),\ u_k - y) \geq 0 \quad , \quad k \in \mathbb{N}^* \quad , \tag{10}$$

which implies that

$$(LP^- u_k,\ P^- u_k) + (LP^+ u_k,\ P^+ u_k) - (Lu_k - \varepsilon_k u_k,\ y) - (F'(y),\ u_k - y) \geq 0 \ ,$$
$$k \in \mathbb{N}^* \quad . \tag{11}$$

From $P^- u_k = P^- K L u_k = K P^- L u_k$ and the compactness of KP^-, we deduce that

$$P^- u_k \to P^- u \quad \text{as} \quad k \to \infty \quad ,$$

and hence that

$$(LP^- u_k,\ P^- u_k) \to (LP^- u,\ P^- u) \quad \text{as} \quad k \to \infty \quad .$$

We therefore deduce from (11) that

$$(LP^- u,\ P^- u) - (Lu,\ y) - (F'(y),\ u - y) \geq \liminf_{k \to \infty} (-LP^+ u_k,\ P^+ u_k)$$
$$\geq (-LP^+ u,\ P^+ u) \quad ,$$

i.e.

$$(Lu - F'(y),\ u - y) \geq 0 \quad \text{for all} \quad y \in H \quad .$$

Taking $y = u - tz$ with $t > 0$ and $z \in H$ (Minty's trick), we obtain

$$(Lu - F'(u - tz),\ z) \geq 0 \quad \text{for all} \quad t > 0 \quad \text{and} \quad z \in H \quad ,$$

and hence, if $t \to 0$, using the fact that if F is convex and Gâteaux-differentiable, then F' is demicontinuous, we get

$$(Lu - F'(u),\ z) \geq 0 \quad \text{for all} \quad z \in H \quad ,$$

i.e.

$$Lu = F'(u) \quad .$$

It only remains to prove that $v = Lu$ minimizes g on $R(L)$. For all

$k \in \mathbb{N}^*$ and all $h \in R(L)$, we have, if $v_k = Lu_k$,

$$g_{\varepsilon_k}(v_k) \le g_{\varepsilon_k}(h) \le g(h) \quad ,$$

and hence,

$$g(h) \ge \liminf_{k\to\infty} g_{\varepsilon_k}(v_k) = \liminf_{k\to\infty} (1/2)(KP^- v_k, v_k) + (1/2)(KP^+ v_k, v_k)$$
$$+ F^*_{\varepsilon_k}(v_k) \ge (1/2)(KP^- v, v)$$
$$+ (1/2)(KP^+ v, v) + \liminf_{k\to\infty} F^*_{\varepsilon_k}(v_k) \quad .$$

Now, for all $k \in \mathbb{N}^*$, by definition of the Fenchel transform, we have

$$F^*_{\varepsilon_k}(v_k) \ge (v_k, u) - F_{\varepsilon_k}(u) = (v_k, u) - (\varepsilon_k/2)\|u\|^2 - F(u) \quad ,$$

so that

$$\liminf_{k\to\infty} F^*_{\varepsilon_k}(v_k) \ge (v, u) - F(u) \quad .$$

Now, by duality and the fact that $v = F'(u)$, we have

$$F^*(v) = (v, u) - F(u)$$

and hence

$$\liminf_{k\to\infty} F^*_{\varepsilon_k}(v_k) \ge F^*(v) \quad ,$$

which, introduced in (12), implies that

$$g(h) \ge (1/2)(Kv, v) + F^*(v) = g(v)$$

for all $h \in R(L)$, and completes the proof of Theorem 1.

Assumption 2 in Theorem 1 is fairly sharp indeed, as shown by the special case where F is strictly convex and $N(L)$ finite-dimensional. Theorem 2 is a slightly more abstract version of a result of [13].

Theorem 2. Let L be like in Theorem 1 with moreover $\dim N(L) < \infty$ and $F : H \to \mathbb{R}$ be Gâteaux-differentiable, continuous and strictly convex.

Then the following conditions are equivalent and necessary for the solvability of (1).

(a) There exists $u \in D(L)$ such that $F'(u) \in R(L)$;

(b) There exists $w \in N(L)$ such that $F'(w) \in R(L)$;

(c) $F(w) \to +\infty$ if $\|w\| \to \infty$ in $N(L)$.

Proof. That (a) is necessary for the solvability of (1) is trivial, as well as (b)$\Rightarrow$(a). Let us show now that (a)$\Rightarrow$(c). Let $u \in D(L)$ satisfying (a) and let us write $u = \bar{u} + \tilde{u}$ with $\bar{u} \in N(L)$ and $\tilde{u} \in R(L)$; let us define on $N(L)$ the strictly convex function $F_{\tilde{u}}$ by

$$F_{\tilde{u}}(w) = F(w + \tilde{u}) \quad .$$

By condition (a), $(F_{\tilde{u}})'(u) = 0$ and, as $\dim N(L) < \infty$ and $F_{\tilde{u}}$ is strictly convex, this implies that

$$F_{\tilde{u}}(w) \to +\infty \qquad \text{if} \qquad \|w\| \to \infty \quad .$$

But, by convexity,

$$F_{\tilde{u}}(w) \leq (1/2)F(2w) + (1/2)F(2\tilde{u}) = (1/2)F(2w) + C \quad ,$$

so that

$$F(w) \to +\infty \qquad \text{as} \qquad \|w\| \to \infty \quad \text{in} \quad N(L) \quad .$$

Finally, the strict convexity of F and the fact that $\dim N(L) < \infty$ imply that (b)$\Leftrightarrow$(c), which completes the proof.

If we combine Theorem 1 and Theorem 2, we obtain a generalization of results of Berger and Schechter [3] who proved them by the method of natural constraints.

Remark 1. When $H \subset L^2(\Omega; \mathbb{R}^N)$ for some open set $\Omega \subset \mathbb{R}^m$ and

$$F(u) = \int_\Omega f(x, u(x))dx$$

for some Caratheodory function $f : \Omega \times \mathbb{R}^N \to \mathbb{R}$ which is convex in u

and satisfies suitable regularity conditions, the assumption 2 in Theorem 1 can be sharpened into nonuniform non-resonance conditions (with respect to x) which allow $\limsup_{|u| \to \infty} 2f(x, u)/|u|^2$ to be equal to λ_1 on large subsets of Ω. The reader can consult [13] for more details.

Remark 2. Theorem 1 obviously generalizes Proposition 1.

3. APPLICATIONS TO NONLINEAR BOUNDARY VALUE PROBLEMS

We list here some applications of Theorem 1 to boundary value problems for ordinary or partial differential equations. The reader will consult the indicated references for details of proofs and more general statements. For the sake of simplicity, all nonlinear terms in the differential equation will be assumed to be continuous in all variables, although less regularity is of course sufficient.

a) Periodic solutions of Hamiltonian systems

Let $F: [0, T] \times \mathbb{R}^{2N} \to \mathbb{R}$, with $F(t, 0)$ convex for each $t \in [0, T]$ be a given function, and

$$J = \begin{pmatrix} 0_N & I_N \\ -I_N & 0_N \end{pmatrix}$$

be the symplectic matrix. Let us assume that there exist $c \geq 0$ and

$$0 < b < \omega = 2\pi/T$$

such that

$$F(t, u) \leq (b/2)|u|^2 + c$$

for all $(t, u) \in [0, T] \times \mathbb{R}^{2N}$ and, for some fixed $m \in Z$, let us consider the periodic problems

$$\begin{aligned} &Ju'(t) - m\omega u(t) = D_u F(t, u(t)) \quad , \quad t \in \left]0, T\right[\quad , \\ &u(0) = u(T) \end{aligned} \tag{13}$$

or

$$-Ju'(t) - m\omega u(t) = D_u F(t, u(t)) \quad , \quad t \in]0, T[$$
$$u(0) = u(T) \quad . \tag{14}$$

The application of Theorem 1 to this problem implies that if

$$\int_0^T F(t, (\cos m\omega t)c + (\sin m\omega t)Jc)dt \to +\infty$$
$$(\text{resp. } \int_0^T F(t, (\cos m\omega t)c - (\sin m\omega t)Jc)dt \to +\infty) \tag{15}$$

as $|c| \to \infty$ in $\mathbb{R}^{2N}$, then the problem (13) (resp. (14)) has at least one solution. In particular, if $m = 0$ (so that F is the Hamiltonian), the conditions (15) reduce to

$$\int_0^T F(t, c)dt \to +\infty \quad \text{as} \quad |c| \to \infty \quad \text{in} \quad \mathbb{R}^{2N} \quad .$$

This result, which generalizes earlier ones of Clarke and Ekeland [5, 7], is basic for proving a number of existence and multiplicity results for periodic solutions of non-autonomous and autonomous Hamiltonian systems (see e.g. [14] for more details).

b) Periodic-Dirichlet problem for a system of semilinear dispersive wave equations

Let $\sigma = \{m^2 - j^2 : m \in \mathbb{N},\ j \in \mathbb{N}^*\} = \{\ldots\mu_{-2} < \mu_{-1} < \mu_0 = 0 < \mu_1 < \mu_2 < \ldots\}$ be the spectrum of the abstract realization in $L^2((0, 2\pi) \times (0, \pi))$ of the wave operator

$$\Box : u \mapsto D_t^2 u - D_x^2 u$$

with the periodic-Dirichlet boundary conditions

$$u(0, x) - u(2\pi, x) = D_t u(0, x) - D_t(2\pi, x) = 0 \quad , \quad x \in]0,\pi[\ ,$$
$$u(t, 0) = u(t, \pi) = 0 \quad , \quad t \in]0, 2\pi[\quad .$$

Let $F : [0, 2\pi] \times [0, \pi] \times \mathbb{R}^N \to \mathbb{R}$ with $F(t, x, \cdot)$ convex for each

$(t, x) \in [0, 2\pi] \times [0, \pi]$ be such that, for some $k \in Z \setminus \{0\}$, $c \geq 0$ and

$$0 < b < \mu_{k+1} - \mu_k \quad ,$$

one has

$$F(t, x, u) \leq (b/2)|u|^2 + c$$

for all $(t, x, u) \in [0, 2\pi] \times [0, \pi] \times \mathbb{R}^N$. If F is moreover such that

$$\int_0^{2\pi}\int_0^{\pi} F\Bigg(t,x, \sum_{r=1}^{N} \sum_{\substack{j\in\mathbb{N}^* \\ m\in\mathbb{N} \\ m^2-j^2=\mu_k}} (a_{jmr}\cos jt \sin mx + b_{jmr}\sin jt \sin mx)e_r\Bigg)$$

$$\times\, dtdx \to +\infty \qquad \text{as} \qquad \Sigma\Sigma(|a_{jmr}|^2 + |b_{jmr}|^2) \to \infty \quad , \qquad (16)$$

then the periodic-Dirichlet problem for

$$D_t^2 u - D_x^2 u - \mu_k u = F(t, x, u)$$

has at least one weak solution. This is a partial generalization of a result of Rabinowitz [17] which only needs $F(t, x, u) + (\mu_k/2)|u|^2$ convex in u but requires $D_u F$ bounded. Conditions of type (16) were introduced first, for a semilinear elliptic problem with bounded nonlinearity, by Ahmad, Lazer and Paul [1] who have shown that they generalize the classical Landesman-Lazer conditions for semi-linear problems at resonance.

c) Periodic-Dirichlet problem for a system of semi-linear beam equations

Let $F : [0, 2\pi] \times [0, \pi] \times \mathbb{R}^N \to \mathbb{R}$ be such that $F(t, x, \cdot)$ is convex for every $(t, x) \in [0, 2\pi] \times [0, \pi]$ and such that for some $c \geq 0$ and

$$0 < b < 1 \qquad (\text{resp. } 0 < b < 3)$$

and all $(t, x, u) \in [0, 2\pi] \times [0, \pi] \times \mathbb{R}^N$, one has

$$F(t, x, u) \leq (b/2)|u|^2 + c \quad . \tag{17}$$

Let us assume moreover that

$$F(t, x, u) \to +\infty \quad \text{as} \quad |u| \to \infty \quad \text{in} \quad \mathbb{R}^N \tag{18}$$

uniformly in $(t, x) \in [0, 2\pi] \times [0, \pi]$. Then the problem

$$D_t^2 u + D_x^4 u = D_u F(t, x, u) \quad (\text{resp.} \quad -D_t^2 u - D_x^4 u = D_u F(t, x, u))$$

$$u(0, x) - u(2\pi, x) = D_t u(0, x) - D_t u(2\pi, x) = 0 \quad , \quad x \in]0, \pi[\quad ,$$

$$u(t, 0) = u(t, \pi) = D_x^2 u(t, 0) = D_x^2 u(t, \pi) = 0 \quad , \quad t \in]0, 2\pi[\quad ,$$

has at least one weak solution. This result generalizes substantially in several directions a theorem of Bahri and Sanchez [2]. It is unknown if conditions (17) and (18) give the same existence result for the periodic-Dirichlet problem for the semi-linear equation of section b with $\mu_k = 0$. Partial results in this direction were obtained by Bahri and Brézis (see [4] for details and references).

d) Dirichlet and Neumann problems for semilinear elliptic equations

Let $\Omega \subset \mathbb{R}^m$ be a bounded open domain with sufficiently smooth boundary, $f : \mathbb{R} \to \mathbb{R}$ a continuous non-decreasing function and $h \in L^q(\Omega)$, and let us consider the problem

$$\Delta u + \mu_1 u + f(u) = h(x) \quad \text{in} \quad \Omega \quad ,$$

$$Bu = 0 \quad \text{on} \quad \partial\Omega \quad , \tag{19}$$

where $Bu = u$ or $\partial u/\partial \nu$ and $\mu_1 < \mu_2 < \ldots$ denote the eigenvalues of the corresponding linear problem ($\mu_1 = 0$ for the Neumann boundary conditions and $\mu_1 > 0$ in the Dirichlet case). Let $F(u) = \int_0^u f(s)ds$ be such that

$$F(u) \leq (b/2)u^2 + c$$

for some $c \geq 0$, $0 < b < \mu_2 - \mu_1$ and all $u \in \mathbb{R}$. Then, Theorem 1 can be used to prove that problem (19) has at least one weak solution if and only if the real function

$$\bar{F} : \mathbb{R} \to \mathbb{R} \quad , \quad c \mapsto \int_\Omega [F(c\varphi(x)) - ch(x)\varphi(x)]dx$$

has at least one critical point, where φ spans the eigenspace of μ_1, (so that $\varphi \equiv 1$ in the Neumann case), when h satisfies the regularity condition $q > m/2$ for $m \geq 4$ and $q = 2$ for $m = 1,2,3$ in the Neumann case, and $q > m$ in the Dirichlet case. See [11, 12, 13, 15, 16] for details. Notice that $\bar{F}$ has a critical point if and only if the equation

$$\int_\Omega [f(c\varphi(x)) - h(x)]\varphi(x)dx = 0$$

has a solution, i.e. if and only if

$$\int_\Omega h(x)\varphi(x)dx \in \text{Range}\left[\int_\Omega f(\cdot\varphi(x))\varphi(x)dx\right] \quad .$$

in particular, for $f \equiv 0$, we find the classical Fredholm condition

$$\int_\Omega h(x)\varphi(x)dx = 0$$

for the linear problem

$$\Delta u + \mu_1 u = h(x) \quad \text{in} \quad \Omega \quad , \quad Bu = 0 \quad \text{in} \quad \partial\Omega \quad .$$

This result substantially generalized some pioneering work of Klingelhöfer [10] and gives a sharp answer to a question raised in Kazdan-Warner [9]. See the above mentioned references for more details and generalizations.

References

1. S. Ahmad, A.C. Lazer and J.L. Paul, "Elementary critical point theory and perturbations of elliptic boundary value problems at resonance", Indiana Univ. Math. J. 25 (1976) 933-944.

2. A. Bahri and L. Sanchez, "Periodic solutions of a nonlinear telegraph equation in one dimension", Boll. Un. Mat. Ital. (5) 18 B (1981) 709-720.

3. M.S. Berger and M. Schechter, "On the solvability of semi-linear gradient operator equations", Adv. in Math. 25 (1977) 97-132.

4. H. Brézis, "Periodic solutions of nonlinear vibrating strings and duality principles", Bull. Amer. Math. Soc. (NS) 8 (1983) 409-426.

5. F. Clarke and I. Ekeland, "Nonlinear oscillations and boundary-value problems for Hamiltonian systems", Arch. Rat. Mech. Anal., to appear.

6. C.L. Dolph, "Nonlinear integral equations of the Hammerstein type", Trans. Amer. Math. Soc. 66 (1949) 289-307.

7. I. Ekeland, "Oscillations de systemes hamiltoniens non-linéaires, III", Bull. Soc. Math. France 109 (1981), 297-330.

8. I. Ekeland and R. Temam, Convex Analysis and Variational Problems, North Holland, Amsterdam, 1976.

9. J.L. Kazdan and F. Warner, "Remarks on some quasilinear elliptic equations", Comm. Pure Appl. Math. 28 (1975) 567-597.

10. K. Klingelhöfer, "Nonlinear boundary value problems with simple eigenvalues of the linear part", Arch. Rat. Mech. Anal. 37 (1970) 381-398.

11. J. Mawhin, "A Neumann boundary value problem with jumping monotone nonlinearity", Delft Progress Report 10 (1985) 44-52.

12. J. Mawhin, "The dual least action principle and nonlinear differential equations", in Qualitative Theory of Differential Equations, Edmonton, 1984, to appear.

13. J. Mawhin and M. Willem, "Critical points of convex perturbations of some indefinite quadratic forms and semi-linear boundary value problems at resonance", Ann. Inst. Poincaré, Analyse non-linéaire, to appear.

14. J. Mawhin and M. Willem, "Critical Point Theory and Hamiltonian Systems", in preparation.

15. J. Mawhin, J.R. Ward Jr. and M. Willem, "Variational methods and semi-linear elliptic equations", Arch. Rat. Mech. Anal., to appear.

16. J. Mawhin, J.R. Ward Jr. and M. Willem, "Necessary and sufficient conditions for the solvability of a nonlinear two-point boundary value problem", Proc. Amer. Math. Soc. 93 (1985) 667-674.

17. P. Rabinowitz, "Some minimax theorems and applications to nonlinear partial differential equations", in Nonlinear Analysis, a volume dedicated to E.H. Rothe, Academic Press, New York, 1978, 161-177.

18. M. Willem, "Remarks on the dual least action principle", Z. Anal. Anw. 1 (1982) 85-90.

A Survey of the Regular Weighted Sturm-Liouville Problem - The Non-Definite Case

Angelo B. Mingarelli [1,2]

1. *Introduction*

Let $p,q,r: [a,b] \to \mathbb{R}$ where $-\infty < a < b < +\infty$ and $p(x) > 0$ a.e., $p,r,$ $1/p \in L(a,b)$ and

$$\int_a^b |r(s)|ds > 0 \tag{1.1}$$

The weighted regular Sturm-Liouville problem consists in finding the values of a parameter λ (generally complex) for which the equation

$$-(p(x)y')' + q(x)y = \lambda r(x)y, \qquad a<x<b, \tag{1.2}$$

has a solution y (non-identically zero) satisfying a pair of homogeneous separated boundary conditions,

$$y(a)\cos\alpha - (py')(a)\sin\alpha = 0 \tag{1.3}$$

$$y(b)\cos\beta + (py')(b)\sin\beta = 0 \tag{1.4}$$

where $0\leq\alpha,\ \beta < \pi$.

Let $D = \{y:[a,b] \to \mathbb{C}\ |y,\ py' \in AC[a,b],\ r^{-1}\{(-py')' + qy\} \in L^2(a,b),$ y satisfies (1.3-4)$\}$ in the case $|r(x)| > 0$ a.e. on (a,b).

---

1. This research is supported, in part, by grant U0167 of the Natural Sciences and Engineering Research Council of Canada and a N.S.E.R.C. University Research Fellowship.

2. Respectfully dedicated to my wife, Leslie Jean.

Then associated with the problem (1.2-4) are the quadratic forms L and R, with domain D, where, for $y \in D$,

$$(Ly,y) = |y(a)|^2 \cot \alpha + (py')(a)y'(a) \cot \beta + \int_a^b \{p(x)|y'|^2 + q(x)|y|^2\}dx \tag{1.5}$$

and

$$(Ry,y) = \int_a^b r(x)|y|^2 dx \tag{1.6}$$

Here (,) denotes the usual L^2-inner-product. (Moreover we note, as is usual, that the cot α (resp. cot β) term in (1.5) is absent if $\alpha = 0$ (resp. $\beta = 0$) in (1.3-4)).

It was noticed at the turn of this century by Otto Haupt [29], [30] and Roland (George Dwight) Richardson, [56], [57], [61] (and it is very likely that David Hilbert was also aware, since Richardson went to Göttingen in 1908...) that the nature of the boundary problem (1.2-4) is dependent upon some "definiteness" conditions on the forms L and R.

Thus Hilbert, and his school, termed the problem (1.2-4) **polar** if the form (Ly,y) is definite on D, i.e., either $(Ly,y) > 0$ for each $y \not\equiv 0$ in D or $(Ly,y) < 0$ for each $y \not\equiv 0$ in D, [Modern terminology refers to this case as the **left-definite** case, cf., [22]].

The problem (1.2-4) was called **orthogonal** (or **right-definite** nowadays) if R is definite on D, (see above), whereas the general problem, i.e., when neither L nor R is definite on D, was dubbed **non-definite** by Richardson (see [61, p.285]). In this respect see also Haupt [30, p.91].

We retain Richardson's terminology "non-definite" in the sequel, in relation to the "general" Sturm-Liouville boundary problem (1.2-4). Thus, in the non-definite case, there exists functions y,z in D for which $(Ly,y) > 0$ and $(Lz,z) < 0$ and also, for a possibly different set of y,z, $(Ry,y) < 0$ and $(Rz,z) > 0$.

2. *The Early Theory of the Non-Definite Case*

The first published results pertaining albeit indirectly, to the non-definite Sturm-Liouville problem (1.2-4) appear to be due to Emil Hilb [32], (see also [33]), who considered the single equation

$$y'' + (A\phi(x) + B)y = 0, \qquad 0<x<1 \tag{2.1}$$

in the two parameters A, B and extended Klein's oscillation theorem to this case. In (2.1) ϕ is real and continuous and may change its sign in (0,1). Note that (2.1) is a special case of (1.2) with $\lambda \equiv A$, $p \equiv 1$, $r \equiv \phi$, $q \equiv -B$. In [32] Hilb noted the existence of parameter values (A,B) (corresponding to the Dirichlet problem) which gave rise to a "minimum oscillation number" for corresponding eigenfunctions, this, being one of the characteristics of non-definite problems, (however this may also occur in orthogonal problems, see [61, p.294]). Both Haupt [30, p.69 and p.88] and Richardson [59, p.289] cite Hilb's paper [32] although it is to be noted that Maxime Bôcher [11, p.173] only cites Richardson, in this context, (and the citation is by way of an oral communication of Richardson to Bôcher [11, p.173, footnote]), in his important survey of Sturm-Liouville theory up to about the year 1912. For a further up-date on the early developments of Sturmian Theory see Richardson's review article [62, pp.110-111] of Bôcher's classic book [13], Lichtenstein's paper [39] and the classic book by Ince [34, §§.10-11]. We note however, that in a later paper [12], Bôcher discovered Hilb's article [32], (see [12, p.7 footnote]).

Now it appears as if the first published theoretical investigations of the general non-definite problem (i.e., (1.2) with q not identically a constant function on (a,b)) are due to Haupt [29] in his dissertation and Richardson [58]. In each one of these works one finds the basis for a beautiful extension of the Sturm Oscillation Theorem in relation to a non-definite problem in which the coefficients are assumed, **a priori**, continuous over the finite interval under consideration. A revised version of Haupt's dissertation (along with a corrected version of the said oscillation theorem) appeared much later, in 1915, (see [30]). Richardson's version of the "general Sturm oscillation theorem" first appeared in [59] but the correct statement is to be found in [59, errata] after a comment by George D. Birkhoff.

In Haupt's paper [30] one finds many interesting results on the nature of non-definite problems, results which are, unfortunately, not mentioned explicitly in Richardson [61] a paper which deals essentially with "new" oscillation theorems for (1.2) in the non-definite case [61, § 4] and for the difficult case when q, in (1.2), is allowed to vary **non-linearly** with λ, (see also McCrea-Newing [42]). However Richardson does refer to Haupt [30] insofar as the oscillation theorem is concerned, as Haupt's theorem appeared one year earlier (cf., above). Thus the first version of an oscillation theorem (in the case $q \not\equiv$ constant) is due to Haupt [30]. On the other hand, a non-trivial sharpening, and the final beautiful form of the oscillation theorem, was formulated by Richardson in [59, errata], (see also Richardson [61, p.285]). Curiously enough, Richardson forgot to cite Hilb [32] in his paper [61], (although he had done so earlier as we have remarked above), and so it appeared to the author that Richardson was indeed the "father of the non-definite case" as one can infer from his remarks in [61, p.285], the claim being that the first investigations of a non-definite problem appeared in Richardson [58].

It is remarkable that Haupt's work [30] has remained obscure, even today. For example Ince [34, p.248] refers to Richardson but nowhere in the book does one find a reference to Haupt! One reason for this may be the following: As was mentioned above Hilbert and his school referred to left-definite problems as "polar" problems and to right-definite problems as "orthogonal" problems. However Haupt, in [30], decided to include "non-definite" problems (in Richardson's terminology) under the heading of "Polar" ones, Haupt [30; p.84, § 5], whereas the "orthogonal" case was included in the section entitled "non-polar case", Haupt [30, p.76, § 3]. It may then have appeared, to those influenced by Hilbert and his school, that, at first sight, Haupt was not doing anything radically new except for extensions, to equations in which the parameter λ appears non-linearly, of the Sturm oscillation theorem, Haupt [30, § 4]. Haupt's failure to emphasize the "**non-definite polar case**", Haupt [30, p.91], as a new and distinct case may have led to a low readership of his paper [30].

Theorem 2.1 (Haupt [30, pp. 84-85]).

In (1.2) let p, q, r be continuous in [a,b] along with p differentiable and $p(x) > 0$ there.

Then the eigenvalues of (1.2-4) are the zeros of an entire transcendental function which is not identically zero.

Remark 2.1 For results pertaining to the **order** of the entire transcendental function alluded to above by Haupt, see Halvorsen [27], Mingarelli [48, 50, 47] and Atkinson-Mingarelli [4].

Of course, it is necessary to assume in theorem 2.1 that there hold

(1.1) or else the entire function mentioned therein may, in fact, vanish identically on $\mathbb{C}$ as an easy example will show. This hypothesis is not explicit in Haupt [30] but is, however, necessary in the proof.

Corollary 2.1 The eigenvalues of (1.2-4) form a discrete subset of the complex plane (i.e., having no finite point of accumulation.)

In Haupt [30] a "normalization" is assumed for (1.2), i.e., the equation is transformed into a "normal form" in many places (see Haupt [30, pp.85-86] for the transformation and Mingarelli [44] for an assumption in the same vein).

The existence of possibly non-real eigenvalues was alluded to by Haupt [30, p.94] and by Richardson [61, p.289 and footnote] however neither author gave an instance of such an occurence (cf., also Mingarelli [46, Chapter 4]). Concrete examples of non-definite problems with non-real eigenvalues were obtained in Mingarelli [44, pp. 519-520], [47, p.376] in which some very early ideas of Hilb [33] were used to show that $\lambda = i$ may be turned into an eigenvalue for a Dirichlet problem associated with (1.2), (see [47] for details).

3. *Terminology and Notation*

The theory of linear operators in spaces with an indefinite metric, Azizov and Iohvidov [5], Bognár [14], together with their applications to quantum field theory, Nagy [53], suggests the following terminology (much of which was due to Werner Heisenberg).

An eigenfunction y of (1.2-4) corresponding to a non-real eigenvalue will be called a **complex ghost state** or complex ghost. (These necessarily satisfy $\int_a^b r(x)|y|^2 = 0$).

If the non-real eigenvalue λ is non-simple its corresponding eigenfunction y will be said to be **degenerate.** It is **non-degenerate** otherwise, i.e., $\int_a^b r(x)\{y(x)\}^2dx \neq 0$, (see §5).

An eigenfunction corresponding to a real non-simple eigenvalue will be called a **degenerate real ghost state** (or a **dipole ghost,** see [53]) whereas a real eigenfunction y corresponding to the (real) eigenvalue λ is a **non-degenerate real ghost state provided**

$$\text{sign}\ \{\lambda \int_a^b r(x)|y|^2dx\} < 0$$

The term **positive eigenfunction** (or **ground state**) will refer to a (real) eigenfunction whose values are strictly positive in the interior of the interval under consideration.

If y is any eigenfunction the quantity

$$\int_a^b r(x)|y|^2dx$$

will be called the **r-Kreĭn norm** of y. (This terminology is motivated by the formalism whereby one may cast the setting for a spectral theory of non-definite problems in a space with an indefinite metric, e.g., Kreĭn space, Pontrjagin space. [see, e.g., Azizov and Iohvidov [14], Bognár [5] and applications in Langer [37], Daho and Langer [16,17], Čurgus and Langer [15], Daho [18,19], and Mingarelli [46]]. Note that when $r(x) > 0$ a.e. on (a,b) the Krein-norm of y is the usual norm in the weighted Hilbert space $L_r^2(a,b)$.

(As is customary we will denote by $r_\pm$ the positive (negative) part of r, i.e., $r_{\pm}(x) = \max\{\pm r(x), 0\}$.

4. Some Motivation

Consider the boundary problem associated with Mathieu's equation

$$y'' + (-\alpha+\beta p(x))y = 0, \qquad -\infty < x < +\infty$$

where $(\alpha,\beta) \in \mathbb{R}^2$ are parameters and p is periodic or, more generally, locally Lebesgue integrable over $\mathbb{R}$. In the usual problem β is fixed and values of α are sought... However if we fix $\alpha << 0$ and seek β's the problem is generally non-definite. (We recall that Hilb [32] considered the same problem, but on a finite interval).

More generally, the equation

$$y'' + (-\alpha A(x) + \beta B(x))y = 0, \qquad -\infty < x < +\infty,$$

in the two parameters α,β where $A,B \in L_{loc}(\mathbb{R})$ also leads to a non-definite problem once one of the parameters α, β is held fixed and the other is treated as an eigenvalue parameter. The question of the existence of positive solutions of the latter equation has received some interest lately (e.g., the monograph of Halvorsen-Mingarelli [28]). For applications of these to laser theory see Heading [31], also McCrea-Newing [42] and the references therein.

Techniques from the theory of non-definite problems were recently used by Deift-Hempel [20] with applications to the theory of colour in crystals. Various other related problems are treated, for instance, in Barkovskii-Yudovich [6,7] in relation to Taylor vortex formation arising from rotating cylinders.

5. *On the Existence of Eigenvalues*

The problem of the actual existence of eigenvalues for the non-definite problem (1.2-4) is treated implicitly in Haupt [30] and somewhat more elaborately in Richardson [61], at least for the Dirichlet problem. Prüfer angle methods have been used in this respect in Halvorsen [27] to, at least, settle the existence of an infinite sequence of **real** eigen-values under the more general set of boundary conditions (1.3-4). A modified Prüfer angle method has been used in Atkinson-Mingarelli [4] which also settles the existence and, at the same time, yields their asymptotics.

Theorem 5.1 (Haupt [30], Richardson [61], Halvorsen [27], Atkinson-Mingarelli [4])

Whenever

$$\int_a^b r_+(s)ds > 0 \quad \text{and} \quad \int_a^b r_-(s)ds > 0$$

the problem (1.2-4) has two infinite sequences of real eigenvalues, one positive and one negative, and each one of which has $+\infty$ and $-\infty$ for its only point of accumulation.

Remark In Haupt [30], Richardson [61] the authors deal with the case in which the coefficients p,q,r are continuous in [a,b] whereas in Halvorsen [27] and Atkinson-Mingarelli [4] the more general case referred to in the introduction is considered.

For a non-definite problem (1.2-4), non-real eigenvalues may or may not occur! Both these cases are possible - With regards to the former see the comments in Haupt [30], Richardson [61], Daho and Langer [16,17] and the specific example in Mingarelli [44, p.520] and [51]. With regards the latter question see Mingarelli [44, p.521] or Atkinson-Jabon [3, Appendix] or Marziali [41].

The question of the possible existence of non-real eigenvalues for a non-definite problem (1.2-4) seems to have first been formulated by Haupt [30] and then by Richardson [61, p.286 and p.289].

The finiteness of the non-real point spectrum of (1.2-4) appears to be due to Haupt [30, p.100] for the case of continuous coefficients under the assumption that the parameter is "normalized" (see Haupt [30, pp.85-86]). A result in the same vein was obtained independently by the author in Mingarelli [44, theorem 2] the proof of which also includes the case of Lebesgue integrable coefficients.

Theorem 5.2 (Haupt [30], Mingarelli [44])

In the non-definite case and under the assumptions of §1, the problem (1.2-4) has at most finitely many non-real eigenvalues, their total number being even.

A particular case of a non-definite (singular) problem was explored in Daho-Langer [16,17], (see also Čurgus - Langer [15] and Mingarelli [44, chapters 3,4]).

There is indeed an intimate connection between the theory of symmetric linear operators in a Pontrjagin space (Krein space) and the theory of non-definite Sturm-Liouville problems. We will not delve into this matter here, for the sake of brevity, although we will refer the reader to the relevant literature herewith: For general notions on Pontrjagin spaces see the monograph by Bognár [14] and the excellent survey paper by Azizov-Iohvidov [5]. For applications of this theory to the problem at hand see Langer [37], Daho-Langer [16,17], Čurgus-Langer [15], while for interesting physico-theoretical applications of Pontrjagin spaces we refer to the monograph by Nagy [53]. We need only mention at this point that theorem 5.2 and other similar theorems, some of which we will present below, may be proved with the help of the theory of symmetric linear operators in a Pontrjagin space.

Now for λ real let $n(\lambda)$ denote the number of negative eigenvalues (counting multiplicities) of the problem

$$-(p(x)y')' + (q(x)-\lambda r(x))y = \nu y$$

where y is required to satisfy (1.3-4). Then the function $n(\lambda)$ has an absolute minimum which will be denoted by $n_\circ$, (This construction is due to the Haupt [30, p.85]).

Theorem 5.3 (Haupt [30], Mingarelli [44])

a) Let p,q,r be continuous on [a,b] and (1.2-4) non-definite. Then the number of pairs (i.e., an eigenvalue and its complex conjugate) of non-real eigenvalues does not exceed $n_\circ$, whenever the parameter is normalized, [30, p.100].

b) In (a) above we may replace $n_\circ$ by the number of negative eigenvalues of the problem

$$-(p(x)y')' + q(x)y = \lambda y \tag{5.1}$$

subject to (1.3-4) provided zero is not an eigenvalue of the said problem, [44]. Cases of equality here may be exhibited, see [44] or Marziali [41].

We now note that, using a Green's function argument, in the regular case of the non-definite problem (1.2-4) the spectrum is purely discrete (i.e., it consists only of eigenvalues of finite multiplicity, Mingarelli [43].)

We now turn our attention to the question of "simplicity" of the eigenvalues. For basic results regarding the nature of "simple" and "non-simple" eigenvalues we refer to Ince [34, § 10.72]. Thus let $y(x,\lambda)$ be a non-trivial solution of (1.2) which satisfies (1.3). Then the function

$$F(\lambda) \equiv y(b,\lambda)\cos\beta + (py')(b,\lambda)\sin\beta$$

is an entire function of $\lambda \in \mathbb{C}$ whose order, generally, does not exceed one-half (see e.g., Atkinson [1]). Actually it is now known that its order is **precisely** one-half as was shown directly by Halvorsen [27] and, as a consequence of the asymptotics, by Mingarelli [47], (cf., Mingarelli [48]), and Atkinson-Mingarelli [4]. For some extensions of the above result on the order of F see Mingarelli [50].

Now the zeros of F are in a one-to-one correspondence with the eigenvalues of the problem (1.2-4). We say that an eigenvalue $\lambda \in \mathbb{C}$ is **simple** if it is a simple zero of F (i.e., $F'(\lambda) \neq 0$). It is said to be **non-simple** otherwise, (i.e., if $0 = F(\lambda) = F'(\lambda)$).

Note: It does **not** follow from the above definition that a non-simple eigenvalue necessarily has two linearly independent eigenfunctions associated with it, for it this were ever the case, their span would generate the space of **all** solutions of (1.2) and clearly one could find solutions which do not satisfy the first boundary condition (1.3), (cf., Ince [34, p.241]).

Thus **for the problem under consideration, namely (1.2-4), there is a one-to-one correspondence between any real eigenvalue** and **its corresponding eigenfunction** (if it is suitably normalized).

It is known that $\lambda \in \mathbb{C}$ **is non-simple if and only if it has a** corresponding (real or complex) eigenfunction which is a degenerate ghost state. The latter result was anticipated by Richardson [61, p.294 corollary] albeit without proof. For a proof and an extension of the said result to the case of measurable coefficients, see Mingarelli [49].

It turns out that real degenerate ghost states may exist for (1.2-4), see the example in Mingarelli [49]. Furthermore there may also exist non-degenerate real ghost states, see, once again, an example in Mingarelli [49], thus answering a question raised by Haupt [30, p.100 footnote]. For all of the examples now known it appears that the complex ghost states are all non-degenerate. There is no known example of a degenerate complex ghost state although the author feels that these very likely do exist in some cases.

Theorem 5.4 Mingarelli [49]

Let $\lambda > 0$ and assume that zero is not an eigenvalue of (1.2-4).

Then the total number of non-degenerate and degenerate real ghost states is always finite and bounded above by the number of negative eigenvalues of (5.1-1.3-1.4). An analogous result holds for $\lambda < 0$.

Remark Specific examples, Marziali [41], seem to indicate that the bound appearing in theorem 5.4 is sharp. Consolidating the above results we may formulate

Theorem 5.5 In the non-definite case of (1.2-4) the spectrum is discrete, always consists of a doubly infinite sequence of real eigenvalues, having no finite limit point, and has at most a finite and even number of non-real eigenvalues (necessarily occuring in complex conjugate pairs) along with at most finitely many real non-simple eigenvalues. The totality of all such eigenvalues comprise the spectrum of (1.2-4).

For an extension of theorem 5.3-4 to an abstract setting, see Mingarelli [51] wherein the more general operator equation $Ax = \lambda Bx$ is considered as a generalized eigenvalue problem in a complex Hilbert space, thus allowing for extensions of the said results to partial differential equations with indefinite weight-functions.

Note **Open Problems**

1. For a given non-definite problem (1.2-4) can one find an **a priori** bound on the modulus (or real/imaginary part) of the "largest" non-real eigenvalue which might appear?

2. For a given non-definite problem (1.2-4) can one find an **a priori** bound on that interval of the real axis which contains all the non-simple eigenvalues and those eigenvalues corresponding to non-degenerate real ghost states? (Note that this interval is finite on account of theorem 5.4).

In relation to this question see Atkinson-Jabon [3] wherein this is done for a specific example.

3. Find a **sufficient** condition which will guarantee the existence of a non-real eigenvalue for a non-definite problem (1.2-4).

(**Necessary** conditions are widespread: See e.g. Mingarelli [44, p.521, theorem 1].)

4. According to Richardson [61, p.289] all sufficiently large real eigenvalues may be regarded as furnishing a minimum of a calculus of variations problem (since they all have positive r-Krein-norm?); however, no proof of this claim is given.

(Note, however, one of Richardson's last papers on this subject, Richardson [63]).

6. *A Generalized Sturm-Oscillation Theorem*

The classical Sturm-oscillation theorem for a left-definite Dirichlet problem associated with (1.2) states that, if r satisfies the hypothesis of theorem 5.1, the nth positive (negative) eigenvalue has an eigenfunction which has precisely n zeros in (a,b), (see e.g. Ince [34, p.235, theorem 3]).

Theorem 6.1 (Haupt [30], Richardson [61])

In the non-definite case of (1.2-4) there exists an integer $n_R \geq 0$ such that for each $n \geq n_R$ there are **at least** two solutions of (1.2-4) having exactly n zeros in (a,b) while for $n < n_R$ there are **no real solutions** having n zeros in (a,b).

Furthermore there exists a, possibly different, integer $n_H \geq n_R$ such that for each $n \geq n_H$ there are **precisely** two solutions of (1.2-4) having exactly n zeros in (a,b).

Remark We will call n_R the **Richardson index** (or number) and label n_H the **Haupt index** (or number) of the problem (1.2-4) for historical reasons - The existence of n_H appears to have been first established in the literature by Haupt [30, p.86] while the existence of n_R, in general, is almost certainly due to Richardson [58] and [61], (modulo Hilb's special case [32]).

It now follows from the **Haupt-Richardson oscillation theorem** (theorem 6.1) that, generally speaking, a non-definite problem will tend **not** to have a real ground state (positive eigenfunction)!

The existence of n_H in the case of measurable coefficients may be found in Atkinson-Jabon [3], (S.G. Halvorsen has also shown the author an independent proof).

The existence of n_R in the case of measurable coefficients was obtained by the author, Mingarelli [52], via a Prüfer transformation and use of the results in Mingarelli [49].

It was shown in Mingarelli [45] that **each one of the cases** $n_H = n_R$ **and** $n_H > n_R$ **may occur** : Indeed the problem (1.2) with $p \equiv 1$, $q \equiv -9\pi^2/16$, $r(x) = +1$ on $[0,1]$ and $r(x) = -1$ on $(1,2]$, $\alpha = \beta = 0$ on $[a,b] = [0,2]$ has precisely one pair of non-real eigenvalues (actually pure imaginary) situated at around $\pm 4.3i$ while the remaining real eigenvalues have eigenfunctions with at least one zero in (a,b), so that $n_R \geq 1$ and in fact $n_R = n_H = 1$, in this case.

On the other hand if we set $q \equiv -22.206$ (a lower approximation to $-9\pi^2/4$) and define the other quantities as in the above example, an explicit calculation shows that $n_H = 3$ while $n_R = 2$. More recently the calculations reported in Atkinson-Jabon [3] and Marziali [41] yield essentially unlimited examples of such pathological behavior (see also Richardson [61, p.298]).

Armed with the Haupt-Richardson oscillation theorem one may proceed to prove asymptotic estimates for the real eigenvalues of non-definite problems as was done in Mingarelli [47]- This is closely related to a conjecture of the late Konrad Jörgens [36]. Under assumptions which, in the aftermath, are very similar to those used by M.H. Stone [64], (e.g., if r is continuous, then r(x) changes sign finitely many times in (a,b)), the author showed

Theorem 6.2, (Mingarelli [47])

If the positive eigenvalues λ_n^+ of a given non-definite problem (1.2-4) are labeled according to the Haupt-Richardson oscillation theorem so that λ_n^+ has an eigenfunction with precisely n zeros in (a,b), then

$$\lambda_n^+ \sim n^2\pi^2 / \left\{ \int_a^b \sqrt{\left(\frac{r}{p}\right)_+}\, ds \right\}^2$$

as $n \to \infty$. An analogous formula holds for the negative eigenvalues as well.

Let $N(\lambda)$ denote the number of zeros of a non-trivial solution of (1.2) satisfying, say, (1.3). If we label our positive eigenvalues so that $N(\lambda_n^+) = n$ for $n > n_H$, then it will follow from (6.1) that

$$N(\lambda) \sim \sqrt{\lambda}\, \pi^{-1} \int_a^b \sqrt{\left(\frac{r}{p}\right)_+}\, ds \tag{6.2}$$

as $\lambda \to +\infty$. This last relationship entails what the author calls **Jörgens' conjecture** [36, p.5.16]. A proof of this conjecture, in general, is given, among other results, in Atkinson-Mingarelli [4] thus settling the question of the validity of (6.2), at least in the case when $p(x) > 0$ a.e. on (a,b). Now let $n_+(\lambda)$ **denote the number of positive eigenvalues of (1.2-4) which are in** $(0,\lambda)$. Then it follows from (6.1) once again that

$$n_+(\lambda) \sim \sqrt{\lambda}\, \pi^{-1} \int_a^b \sqrt{\left(\frac{r}{p}\right)_+}\, ds \tag{6.3}$$

as $\lambda \to +\infty$, (for a detailed calculation see Mampitiya [40, p.28, lemma 4.4). We mention, in passing, that (6.3) has been recently extended to incorporate polar **vector** Sturm-Liouville problems on a finite interval, see Mampitiya [40].

One of the truly remarkable results that one may extract from Richardson [61] is the following

Theorem 6.3 (Richardson [61, p.302, theorem 10])

Let r be continuous and not vanish identically in any right-neighborhood of $x = a$.

If $r(x)$ changes its sign precisely once in (a,b) then the roots of the real and imaginary parts u,v, of any non-real eigenfunction $y = u+iv$ corresponding to a non-real eigenvalue, separate one another.

Corollary 6.4 Under the conditions of theorem 6.3 it follows that any non-real eigenfunction y of (1.2-4) cannot have a zero for x in (a,b).

Note The initial assumption on r in theorem 6.3 is not explicitly mentioned in the proof given by Richardson [61] however it appears to be necessary for the validity of his proof.

For an extension of theorem 6.3 to the case when r(x) changes its sign finitely many times, see Mingarelli [44, p.525 theorem 3].

Finally we note that some interesting comparison theorems for non-definite problems are derived in Atkinson-Jabon [3, p.35, proposition 3,4].

7. *Further Results and Extensions*

With regards to an eigenfunction expansion in the regular non-definite case see Faierman [23] where the actual uniform convergence of such an expansion is also treated. Another approach to the same problem is to be found in Binding-Seddighi [9] with an aim towards an abstract expansion theorem which includes the non-definite case as a particular case, but with additional boundedness assumptions on the coefficients.

The problem of finding asymptotic expressions for the solutions of equations of the form (1.2) with additional smoothness on the coefficients, is an old one and much has been done in this direction - The book by Olver [54] is a good introduction to the subject as is the book by Erdelyi [21], especially chapter 4. Further results may be found in recent papers by Fedoryuk [24,25] and a paper by Olver [55], see also the references therein. Interesting results in this connection are also buried in Birkhoff-Langer [10].

Results in the case of **singular** non-definite Sturm-Liouville problems (dealing almost exclusively with a closed half-line) are more or less scattered, although the object of recent interest. In this direction see the pioneering papers by Atkinson-Everitt-Ong [2], and Daho-Langer [16,17] wherein an expansion theorem is formulated. The paper by Langer [37] sheds much insight into the role that Kreĭn spaces play in the study of non-definite problems. Further expansion results (full and half-range expansions) for higher-order scalar ordinary differential equations are given in Čurgus-Langer [15], see also Daho [18,19] for results dealing with the existence of a Titchmarsh-Weyl matrix function in a higher-order singular case.

For the most up-to-date results regarding left-and right-definite problems, which include (1.2), we refer to the paper by Everitt [22] and Bennewitz-Everitt [8] and the references therein.

The basis for the extension of the foregoing results to partial differential equations may be found in Fleckinger-Mingarelli [26], see also Mingarelli [51].

At this point many questions, even in the regular case, remain unanswered: For example, questions relating to the existence/non-existence of real ground states for (1.2-4); extensions of the preceding results to systems of second order differential equations and higher-dimensional analogs; an efficient method for calculating the non-real eigenvalues of non-definite problems (1.2-4), (current techniques, Marziali [41], Atkinson-Jabon [3], rely upon explicit computation of the zeros of $F(\lambda)$, (see §5), which is generally inefficient; and to find **a priori** estimates on the Haupt and Richardson indices for a given non-definite problem.

Acknowledgments

I am grateful to Professor F.V. Atkinson who, in 1977 introduced the author to this fascinating field. The main reference to Richardson [61] was only discovered in 1978 whereas the importance of Haupt's work was only recognized in 1984. All the other references to the early history of the non-definite problems are basically recent finds. It is uncanny that so much work had been done in connection with this problem at the turn of this century, work which seemingly came to a halt around 1920. Renewed theoretical interest in this area appears to be very recent, only after a gap of almost 60 years!

I am also grateful to Dr. Philip Hartman (formerly of Johns Hopkins University) who, in 1978, remarked to me that Hilb may have had something to do with this problem. This hunch proved correct as we have seen. I am grateful to Phil Hartman for also raising the question, in 1980, of the "simplicity" of the real eigenvalues of non-definite problems, a question which later led to my paper [49].

I wish to thank the following mathematicians for supplying me with preprints of their work - C. Bennewitz, Paul Binding, Karim Daho, Percy Deift, Mel Faierman, and David Jabon. I also wish to thank Derick Atkinson, Paul Binding, Patrick Browne, Mel Faierman and Gotskalk Halvorsen for interesting discussions. My gratitude also goes to Hans Kaper, for providing the reference [21], and Hubert Kalf for rekindling my interest in Lichtenstein's paper [39].

Bibliography

[1] F.V. Atkinson, Discrete and Continuous Boundary Problems, Academic Press, New York, 1964, xiv, 570p.

[2] F.V. Atkinson, W.N. Everitt, K.S. Ong, On the m-coefficient of Weyl for a differential equation with an indefinite weight-function, Proc. London Math. Soc., 29, (1974), 368-384.

[3] F.V. Atkinson, D. Jabon, Indefinite Sturm-Liouville problems, in Proceedings of the 1984 Workshop "Spectral Theory of Sturm-Liouville Differential Operators", Argonne National Laboratory, Argonne, Il. reprint ANL-84-73, 31-45.

[4] F.V. Atkinson, A.B. Mingarelli, Asymptotics of the eigenvalues of general Sturm-Liouville problems, preprint.

[5] T.Ya. Azizov, I.Ts. Iohvidov, Linear operators in spaces with an indefinite metric and their applications (Russian) in Matematicheskii Analiz, Tom 17, Itogi Nauki i Tekhniki, Moskva, (1979), 113-205

[6] Yu.S. Barkovskii, V.I. Yudovich, Taylor vortex formation in the case of variously rotating cylinders and the spectral properties of one class of boundary-value problems. Sov. Phys. Doklady., 23, (1978), 716-717.

[7] ————————, Spectral properties of a class of boundary value problems, (Russian) Matem. Sbornik. 114 (156), (1981), 438-450 (English translation: Math U.S.S.R. Sbornik, 42 (1982), 387-398).

[8] C. Bennewitz, W.N. Everitt, On second order left-definite boundary value problems, Uppsala University, Department of Mathematics Report 1982:7, (1982), 38p.

[9] P. Binding, K. Seddighi, Multiparameter spectral theory in the elliptic case, preprint.

[10] G.D. Birkholf, R.E. Langer, The boundary problems and developments associated with a system of ordinary linear differential equations of the first order, Proc. Amer. Acad. Arts and Sciences, 58, (1923), 51-128.

[11] M. Bôcher, Boundary problems in one dimension, 5th International Congress of Mathematicians, Proceedings, Cambridge University Press, 1912, Vol. 1, 163-195.

[12] ——————, The smallest characteristic numbers in a certain exceptional case, Bull. Amer. Math. Soc., 21, (1914), 6-9.

[13] ——————, Leçons sur les méthodes de Sturm, Gauthier-Villars, Paris, 1917.

[14] J. Bognár, Indefinite Inner Product Spaces, Springer-Verlag, Berlin; New York, 1974, ix, 223p.

[15] B. Čurgus, H. Langer, Spectral properties of selfadjoint ordinary differential operators with an indefinite weight-function in Proceedings of the 1984 Workshop "Spectral Theory of Sturm-Liouville Differential Operators" Argonne National Laboratory, Argonne, Il., reprint ANL-84-73; 73-80.

[16] K. Daho, H. Langer, Some remarks on a paper by W.N. Everitt, Proc. Royal Soc. Edinburgh, 78A, (1977), 71-79.

[17] ——————————, Sturm-Liouville operators with an indefinite weight-function, Proc. Royal Soc. Edinburgh, 78A, (1977), 161-191.

[18] K. Daho, On Titchmarsh-Weyl matrix-function, Linköping University Report LITH.-MAT-R-82-30, preprint.

[19] ——————————, Titchmarsh-Weyl matrix-functions and operators in spaces with an indefinite metric, Linköping University Report LITH.-MAT-R-82-31, preprint.

[20] P.A. Deift, R. Hempel, On the existence of eigenvalues of the Schrödinger operator H-λW in a gap of σ(H), preprint.

[21] A. Erdelyi, Asymptotic Expansions, Dover Publications, New York, 1956, vi, 108p.

[22] W.N. Everitt, On certain regular ordinary differential expressions and related differential operators, in 'Spectral Theory of Differential Operators' I.W. Knowles and R.T. Lewis eds., North-Holland, New York, 1981, 115-165.

[23] M. Faierman, Eigenfunction expansions associated with a left-definite two-parameter system of differential equations I (and II) preprints.

[24] M.V. Fedoryuk, A Sturm-Liouville problem with regular singularities, I, (Russian) Diff. Urav., 18, (1982), 2166-2173 [English translation: Differential Equations, 18, (1982), 1550-1557].

[25] ——————————, A Sturm-Liouville problem with regular singularities, II (Russian) Diff. Urav., 19, (1983),

278-286 [English translation: Differential Equations, 19, (1983), 208-215]

[26] J. Fleckinger, A.B. Mingarelli, On the eigenvalues of non-definite elliptic operators, in Differential Equations, I.W. Knowles and R.T. Lewis eds., North-Holland, New York, 1984, 219-227.

[27] S.G. Halvorsen, Oral communication, 5th Dundee Conference in Differential Equations, University of Dundee, Dundee, Scotland, March-April 1980, (unpublished).

[28] S.G. Halvorsen, A.B. Mingarelli, The Large-Scale Structure of the Domains of Non-Oscillation of Second-Order Differential Equations with Two Parameters, monograph, preprint.

[29] O. Haupt, Untersuchungen über Oszillationstheoreme, Teubner, Leipzig, 1911.

[30] ———————, Uber eine Methode zum Beweise von Oszillations-theoreme Mat. Ann., 76, (1915), 67-104.

[31] J. Heading, Polynomial-type eigenfunctions, J. Phys. A; Math. Gen. 15, (1982), 2355-2367.

[32] E. Hilb, Eine Erweiterung des Kleinschen Oszillationstheorems Jahresbericht d.d. Math. Ver., 16, (1907), 279-285.

[33] ———————, Uber Reihentwicklungen nach den Eigenfunktionen linear differential gleichungen 2ter Ordnung Mat. Ann., 71, (1912), 76-87.

[34] E.L. Ince, Ordinary Differential Equations, Dover Publications, New York, 1956, viii, 558p.

[35] D. Jabon, Indefinite Sturm-Liouville problems, preprint.

[36] K. Jörgens, Spectral Theory of Second Order Ordinary Differential Equations, Aarhus Universitet Lecture Notes, 1964.

[37] H. Langer, Sturm-Liouville problems with indefinite weight-function and operators in spaces with an indefinite metric, in 'Uppsala Conference on Differential Equations 1977', Almqvist and Wiksell, (1977), 114-124.

[38] R.E. Langer, The boundary problem associated with a differential equation in which the coefficient of the parameter changes sign, Trans. Amer. Math. Soc., 31, (1929), 1-24.

[39] L. Lichtenstein, Zur analysis der unendlichvielen Variabeln, I, Rendiconti Circolo Mat. Palermo., 38, (1914), 113-166.

[40] M.A.U. Mampitiya, Spectral asymptotics for polar vector Sturm-Liouville problems, M.Sc. thesis, Department of Mathematics, University of Ottawa, 1985, v, 44p.

[41] G. Marziali, A study of some symmetric and non-symmetric non-definite Sturm-Liouville problems, University of Ottawa Summer Research Project, manuscript, 1984.

[42] W.H. McCrea, R.A. Newing, Boundary conditions for the wave equation, Proc. London Math. Soc., 37, (1933), 520-534.

[43] A.B. Mingarelli, Boundary problems in a space with an indefinite metric, Oral communication, University of Alabama, Tuscaloosa, March 1980, (unpublished).

[44] ————————, Indefinite Sturm-Liouville problems, in 'Ordinary and Partial Differential Equations, W.N. Everitt ed., Lecture Notes in Mathematics 964, Springer-Verlag, Berlin-New York, (1982), 519-528.

[45] ————————, Non-definite Sturm-Liouville problems - remarks and examples, Oral communication, 64th Ontario Mathematics Meeting, Carleton University, Nov. 1982 (unpublished).

[46] ————————, Volterra-Stieltjes Integral Equations and Generalized Ordinary Differential Expressions, Lecture Notes in Mathematics 989, Springer-Verlag, Berlin-New York, 1983, XIV, 318 p.

[47] ————————, Asymptotic distribution of the eigenvalues of non-definite Sturm-Liouville problems, in 'Ordinary Differential Equations and Operators - A tribute to F.V. Atkinson', W.N. Everitt and R.T. Lewis eds., Lecture Notes in Mathematics 1032, Springer-Verlag, Berlin-New York, 1983, 375-383.

[48] ————————, Some remarks on the order of an entire function associated with a second order differential equation, in 'Ordinary Differential Equations and Operators - A tribute to F.V. Atkinson', W.N. Everitt and R.T. Lewis eds., Lecture Notes in Mathematics 1032, Springer-Verlag, Berlin-New York, 1983, 384-389.

[49] ————————, On the existence of non-simple real eigenvalues for general Sturm-Liouville problems, Proc. Amer. Math. Soc., 89, (1983), 457-460.

[50] ————————, Some remarks on the order of an entire function associated with a second order differential equation II, Comptes Rendus, Math. Rep. Acad. Sci. Canada, 6, (1984), 79-83.

[51] ————————, The non-real point spectrum of generalized eigenvalue problems, Comptes Rendus, Math. Rep. Acad. Sci. Canada, 6, (1984), 117-121.

[52] ————————, Non-definite Sturm-Liouville problems, Oral communication, University of Toronto, December 1984, (unpublished).

[53] K.L. Nagy, State Vector Spaces with Indefinite Metric in Quantum Field Theory, Akademiai Kiado and Noordhoff, Budapest, 1966, vii, 131p.

[54] F.W.J. Olver, Asymptotics and Special Functions, Academic Press, New York, 1974, 572p.

[55] ————————, General connexion formulae for Liouville-Green approximations in the complex plane, Phil. Trans. Royal Soc. London, 289, (1978), 501-548.

[56] R.G.D. Richardson, Das Jacobische Kriterium der Variationsrechnung und die oszillationseigenschaften linearer Differential gleichungen 2. Ordnung, Mat. Ann. 68, (1910), 279-304.

[57] ————————, Das Jacobische Kriterium der Variationsrechnung und die oszillationseigenschaften linearer Differential gleichungen 2. Ordnung (Part II), Mat. Ann., 71, (1912), 214-232.

[58] ————————, Theorems of oscillation for two linear differential equations of the second order with two parameters, Trans. Amer. Math. Soc., 13, (1912), 22-34.

[59] ————————, Ubér die notwendigen und hindreichenden Bedingungen für das Bestehen eines Kleinschen Oszillationstheorems, Mat. Ann., 73, (1913), 289-304 (errata, Mat. Ann., 74, (1913), 312).

[60] ————————, A new method in boundary problems for differential equations, Trans. Amer. Math. Soc., 18, (1917), 489-518.

[61] ————————, Contributions to the study of oscillation properties of the solutions of linear differential equations of the second order, Amer. J. Math., 40, (1918), 283-316.

[62] ————————, Bôcher's boundary problems for differential equations, (Book review), Bull. Amer. Math. Soc., 26, (1919), 108-124.

[63] ————————, A problem in the calculus of variations with an infinite number of auxiliary conditions, Trans. Amer. Math. Soc., 30, (1928), 155-189.

[64] M.H. Stone, An unusual expansion problem Trans. Amer. Math. Soc., 26, (1924), 335-355.

[65] A.P. Wheeler, Linear ordinary self-adjoint differential equations of the second order, Amer. J. Math., 49, (1927), 309-320.

Department of Mathematics
University of Ottawa,
Ottawa, Ontario
Canada, K1N 6N5

Pointwise and $L^p(\mathbb{R}^n)$ Convergence of Elliptic Eigenfunction Expansions

Louise A. Raphael*

Department of Mathematics

Howard University

Washington, D.C. 20059

USA

It is a great honor and privilege to have participated in the first International Workshop on Applied Differential Equations at Tsinghua University. This workshop has established a network for productive mathematical exchanges between Chinese and western mathematicians whose benefits we will enjoy for many years.

Abstract. In this two part paper we present some results on equisummability and regularization of eigenfunction expansions associated with a general class of elliptic operators on $\mathbb{R}^n$. The prototypical analogue of equisummability is equiconvergence. Equiconvergence asks: When does the pointwise convergence of an eigenfunction expansion associated with an unperturbed operator (say the Laplacian) imply the pointwise convergence of the perturbed operator (say the Laplacian plus non-zero potential)? The problem of equisummability is essentially the question of convergence to zero of the difference of the two summability means of the unperturbed and perturbed eigenfunction expansions. We apply the main equisummability

*Research supported by USARO grant number DAAG-29-84-G0004.

result to singular Sturm-Liouville theory and the heat equation.

Regularization of the ill-posed problem for partial differential equations where the solutions do not depend continuously on the data is presented in the latter part of this paper. That is, we develop a stable method for correct summation of expansions of data perturbed by error. Our method recovers from the perturbed expansion a good approximation to the correct data provided the data are sufficiently regular. Application to the study of time evolution of the Laplacian is mentioned.

§1. Equisummability of Closed Operators: Classical work in equiconvergence relating perturbed to unperturbed expansions began early in this century with the work of Haar [H] and Walsh [W], who showed that the pointwise difference of Sturm-Liouville eigenfunction expansions and the ordinary Fourier series tends to zero everywhere on finite intervals. Stone [ST1] and Tamarkin [Ta] studied analogous questions for expansions in eigenfunctions of higher order differential operators on intervals. Equiconvergence of singular Sturm-Liouville expansions on $[0,\infty)$ and Fourier transforms can be found in [T] and [LS]. Recently Komornik [K] has generalized these results to the one-dimensional Schrödinger operator with complex potentials.

When equiconvergence fails, Stone [St2] showed that equisummability is its appropriate replacement. Expansions with respect to two orthogonal systems are pointwise equisummable if, under a given summability method, the termwise difference of the two expansions converge to zero. Equisummability results can also be found in [LS] and [T]. Recently Benzinger [Be1, Be2] has shown equisummability for Riesz typical means of eigenfunction expansions of ordinary differential operators, and applied it to the heat equation to show the convergence of the solution to the initial value as time approaches zero.

In higher dimensions, Il'in [Il] studied Riesz equisummability of eigenfunction expansions of the Laplacian with Fourier integral expansions on $\mathbb{R}^n$. A survey of equisummability results in the Russian literature up to 1982 is given in [Go].

Gurarie and Kon [GK2] studied the question of equisummability via the Dunford-operator calculus formulation of summability. Their result gives conditions for which an L^p-function in eigenfunctions of an elliptic operator is equisummable with the corresponding expansion obtained from its leading term, within the class of analytic summation techniques.

We first present abstract criteria to establish equisummability of two closed linear operators on $\mathbb{R}^n$. Theorem 2 states that the analysis of $L^p(\mathbb{R}^n)$ equisummability for $1 \leq p \leq \infty$ of two closed linear operators is reduced to showing that the difference of the modified resolvent operators is uniformly bounded. This result can then be used to give a simple proof of Gurarie and Kon's equisummability result (Theorem 4). We begin by presenting basic definitions.

A linear operator A on $L^p(\mathbb{R}^n)$, $1 \leq p < \infty$, is <u>ϕ-summable</u> on $L^q(\mathbb{R}^n)$ $1 \leq q \leq \infty$, if $||\phi(\varepsilon A)f - f||_q \to 0$ as $|\varepsilon| \to 0$ for all f in $L^p(\mathbb{R}^n)$ and for complex ε in some domain D.

As a generic example, let A be a differential operator on $L^p(\mathbb{R}^n)$, $1 \leq p < \infty$. Let $\{u(x,\lambda)\}$ be the set of generalized eigenfunctions associated with the eigenvalue λ belonging to the spectrum $\sigma(A)$. Let $f \in L^p(\mathbb{R}^n)$ and assume that the eigenfunction expansion

$$f(x) \sim \int_{\sigma(A)} F(\lambda)\ u(x,\lambda) d\rho(\lambda) \tag{1}$$

and its generalized Fourier coefficient

$$f(\lambda) \sim \int_{\mathbb{R}^n} f(x)\ u(x,\lambda) dx$$

exist. Here ~ denotes L^p convergence as the limits of integration become infinite, and ρ is a combination of spectral functions. Let $\phi(\lambda)$ be continuous and $\phi(0) = 1$. We say that the eigenfunction expansion (1) is ϕ summable in an appropriate topology if the summability means

$$\phi(\varepsilon A)f = \int_{\sigma(A)} F(\lambda)\,\phi(\varepsilon\lambda)u(x,\lambda)d\rho(\lambda) \to f \tag{2}$$

as $|\varepsilon| \to 0$, ε again belonging to some domain D in $\mathbb{C}$. When the summator function $\phi(\varepsilon\lambda) = \frac{1}{1 - \varepsilon\lambda} = \frac{z}{z - \lambda}$, for $\varepsilon = \frac{1}{z}$ and $\frac{z}{z - A} f \to f$ as $|z| \to \infty$, z in D', $D' = \{z|\ -1/z \in D\}$, the eigenfunction expansion is resolvent summable to f.

Now let A, B be linear operators on $L^p(\mathbb{R}^n)$ for $1 \leq p < \infty$. Let $f \in L^p(\mathbb{R}^n)$ and ϕ be a function analytic on the spectrum of A and B with $\phi(0) = 1$. Let $\phi(\varepsilon A)f$ represent the ϕ-summability means of the expansion of f with respect to eigenfunctions of A. The question of equisummability is: When does convergence (as $\varepsilon \to 0$) of $\phi(\varepsilon A)f$ imply that of $\phi(\varepsilon B)f$? Specifically, we say that eigenfunction expansions with respect to A and B are ϕ-equisummable from L^p to L^q $1 \leq q \leq \infty$ if for $f \in L^p$

$$\lim_{\varepsilon\to 0} ||(\phi(\varepsilon A)f - \phi(\varepsilon B))f||_q \underset{\varepsilon\to 0}{\to} 0.$$

If $q = \infty$, the expansions are pointwise equisummable.

In particular we say that eigenfunction expansions in A and B are resolvent equisummable from $L^p(\mathbb{R}^n)$ to $L^q(\mathbb{R}^n)$, $1 \leq q \leq \infty$, if for $f \in L^p(\mathbb{R}^n)$

$$||z(zI - A)^{-1}f - z(zI - B)^{-1}f||_q \to 0 \quad \text{as} \quad |z| \to \infty, \text{ z complex.}$$

We remark that the advantage of the operator calculus approach to equisummability is its insensitivity to detailed spectral considerations, or even to the existence of a complete spectrum. For example, the approach addresses non-self-adjoint problems as readily as self-adjoint ones. The method is a strict generalization of the classical one.

Our first theorem gives sufficient conditions for two closed operators to be resolvent equisummable.

Theorem 1:

Let A, B be closed operators on $L^p(\mathbb{R}^n)$ into $L^q(\mathbb{R}^n)$, $1 \leq p < \infty$, $1 \leq q \leq \infty$. Let Ω be dense in $L^p(\mathbb{R}^n)$ and also contained in $L^q(\mathbb{R}^n)$ such that A, B map Ω into $L^p(\mathbb{R}^n) \cap L^q(\mathbb{R}^n)$. Assume $z(z - A)^{-1} - z(z - B)^{-1}$ is uniformly bounded from $L^p(\mathbb{R}^n)$ to $L^q(\mathbb{R}^n)$ for complex z in D. If $z(z - A)^{-1}$ and $z(z - B)^{-1}$ are uniformly bounded from $L^q(\mathbb{R}^n)$ to $L^q(\mathbb{R}^n)$, then A and B are resolvent equisummable from $L^p(\mathbb{R}^n) \to L^q(\mathbb{R}^n)$ in D.

The proof follows easily from an algebraic identity and a Banach space theorem. In applications, Ω is $C^\infty(\mathbb{R}^n)$.

The setting for our abstract criteria to establish equisummability follows. Let K denote a set in the complex plane $\mathbb{C}$ which is the union of a disc and wedge. Here ∂K denotes the boundary of K while K^c denotes the complement of K in $\mathbb{C}$. The regions K_i, (i = 1,2), defined in Theorem 2 can be informally thought of as concentric (skeleton) keyholes the intersection of one of which with the exterior of the other contains the contour Γ.

Theorem 2:

Let A, B be closed operators on $L^p(\mathbb{R}^n)$ for $1 \leq p < \infty$. Assume $R_z f = z(z - A)^{-1} f - z(z - B)^{-1} f$ is uniformly bounded from $L^p(\mathbb{R}^n) \to L^q(\mathbb{R}^n)$ where $1 \leq q \leq \infty$. Let $D_{r_i} = \{z \in \mathbb{C} \mid |z| \leq r_i\}$, and $W_{\theta_i} = \{z \in \mathbb{C} \mid |\arg z| \leq \theta_i\}$, and $K_i = D_{r_i} \cup W_{\theta_i}$ for i = 1,2 be such that $K_2 \subset K_1$.

Let A and B be resolvent equisummable on the complement of K_2. If ϕ is analytic on K_1 such that $\phi(0) = 1$, $\phi(z) = O(z^{-\delta})$, $(\delta > 0)$ z in $K_1 \cap K_2^c$, then A and B are ϕ-equisummable in ε, ε in $D = \{z \mid |\arg z| < \theta_1 - \theta_2\}$, from $L^p(\mathbb{R}^n) \to L^q(\mathbb{R}^n)$ for $1 \leq p < \infty$ and $1 \leq q \leq \infty$.

We now define our class of operators. Let $\alpha = (\alpha_1,\ldots,\alpha_n)$ be a multi-index, and $D^\alpha \equiv i^{-|\alpha|} \frac{\partial^{\alpha_1}}{\partial x_1^{\alpha_1}} \cdots \frac{\partial^{\alpha_n}}{\partial x_n^{\alpha_n}}$, $|\alpha| \equiv \alpha_1 + \ldots + \alpha_n$. Consider the differential operator

$$A = \sum_{|\alpha|\le m} b_\alpha(x) D^\alpha = A_o + B \tag{3}$$

where A_o contains the leading terms and B is the remainder. We assume that A_o is constant coefficient positive elliptic, while the coefficients of B can be expressed as sums of functions in certain L^p spaces: $b_\alpha(x) \in L^{r_\alpha} + L^\infty$ $(|\alpha| < m)$ where $d \equiv \sup_{|\alpha|<m}\{\frac{n}{r_\alpha} + |\alpha|\} < m$. We choose for the domain of A the L^p-Sobolev space $L^p_m = (1 - \Delta)^{-m/2} L^p$, $1 \le p \le \min r_\alpha$; if p is outside this range, then the domain must be smaller and A may not be densely defined. The resolvent $(z - A)^{-1}$ of A, if it exists, may be expressed as an integral operator with kernel denoted by $L_z(x,y)$, $x,y \in \mathbb{R}^n$. Sharp local bounds on the behavior of L_z were obtained in [GK2]. Specifically, let

$$h_{s,t}(x) = \begin{cases} |x|^{-s} \quad \text{if} \quad s > 0, \quad -\ln|x| \quad \text{if} \quad s = 0; & |x| \le 1 \\ |x|^{-t}; & |x| > 1. \end{cases} \tag{4}$$

The following is proved in [GK2].

Theorem 3: (Gurarie-Kon).

Let $1 \le p \le \min r_\alpha$. There is a constant $C > 0$ such that the L^p-spectrum of A is contained in the parabolic domain $\Omega = \{z = \rho e^{i\theta}\colon\ C\rho^{(d/m)-1} \ge |\sin \frac{\theta}{2}|^{n+2}\}$ containing $\mathbb{R}^+$. For $z \notin \Omega$ the kernel of $R_z = (z - A)^{-1}$ is estimated by

$$|L_z(x,y)| \le F(\rho,\theta)\rho^{(n/m)-1} h_{s,t}(\rho^{1/m}(x - y)), \tag{5}$$

where $t > n$, $s = \max(0,n - m)$, and

$$F(z) = F(\rho,\theta) = \frac{C}{|\sin \frac{\theta}{2}|^{n+2}}\left(1 - \frac{C\rho^{(d/m)-1}}{|\sin \frac{\theta}{2}|^{n+2}}\right)^{-1}. \tag{6}$$

Furthermore, the kernel $L_z^{(1)}$ of $R_z - R_z^o$, where $R_z^o = (z - A_o)^{-1}$, satisfies

$$|L_z^{(1)}(x,y)| \leq \frac{C}{|\sin \frac{\theta}{2}|^{n+2}} F(\rho,\theta)\rho^{((d+n)/m)-2}\, h_{s',t}(\rho^{1/m}(x - y)), \tag{7}$$

with $s' = \max(n - 2m + d, 0)$.

Consequences [GK2] of this theorem are that the resolvent of A maps a dense domain of $L^p(\mathbb{R}^n)$ into $L^p(\mathbb{R}^n) \cap L^\infty(\mathbb{R}^n)$ for $p > \frac{n}{m - d}$; and the spectrum of A, $\sigma(A)$, is contained in a parabolic region about $\mathbb{R}^+$. So the spectrum can be included in a K domain which is a union of a disc and a wedge.

Theorem 4: (Gurarie-Kon).

Let $A = A_o + C$ and $B = A_o + D$ be closed elliptic operators defined as above on the Sobolev space L_m^p, $1 \leq p \leq \min r_a$, where A_0 is constant coefficient positive elliptic containing the leading terms of order m. Assume the coefficients $b_\alpha(x)$ of A, B are singular on a nowhere dense set. Let K_1 and K_2 be as in Theorem 2 and ϕ be analytic on K_1, $\phi(0) = 1$, $\phi(z) = O(z^{-\delta})$, $\delta > 0$ and z in $K_1 \cap K_2^c$. Then

(a) A and B are ϕ-equisummable from $L^p(\mathbb{R}^n)$ into $L^\infty(\mathbb{R}^n)$ for $p > n/(m - d)$, and ε in $D = \{z \mid |\arg z| < \theta_1 - \theta_2\}$.

(b) $\phi(\varepsilon A)f \to f$ at $x \in \mathbb{R}^n$ if and only if $\phi(\varepsilon B)f \to f$ at x.

A simple proof follows from Theorems 2 and 3, the Minkowski integral and Hausdorff inequalities. The proofs of theorems 2 and 4 will appear in the Journal of Mathematical Analysis and Application [Ra].

As our first application of Theorem 4, consider the Sturm-Liouville operator $A = -\frac{d^2}{dx^2} + q(x)$, where $q \in (L^1 + L^\infty)(\mathbb{R})$ is real and continuous. For real λ, let u_1, u_2 solve $Au_i = \lambda u_i$, with $(u_1(0), u_1'(0)) = (0,1)$ and $(u_2(0), u_2'(0)) = (1,0)$. Then we can expand $f(x) \in L^2(\mathbb{R}^n)$ in the Sturm-Liouville expansion (see [LS]. It now follows that if ϕ is an analytic multiplier satisfying the conditions of Theorem 4, then the generalized Fourier transform of f is ϕ-summable to f(x) pointwise if and only if the Fourier transform of f is ϕ-summable to f(x) pointwise.

For our second equisummability application, consider the heat equation

$$Au \equiv (-a(x)\Delta + \vec{b}(x) \cdot \vec{\nabla} + q(x))u = -\frac{\partial u}{\partial t}, \tag{8a}$$

$$u(x,0) = u_0(x) \in (L^p + L^\infty)(\mathbb{R}^n) \tag{8b}$$

where $a(x) > 0$ represents a position dependent on heat conductivity, and $\vec{b}(x)$ and $q(x)$ are drift and dissipation terms. We assume $a(x) \in C^\infty(\mathbb{R}^n)$ such that all their partial derivatives are bounded, $b(x) \in L^{n+\varepsilon} + L^\infty$, and $q(x) \in L^{(n/2)+\varepsilon} + L^\infty$, for some $\varepsilon > 0$. Theorem 5 extends some results of Benzinger [Be2].

<u>Theorem 5</u>:

If $u_0 \in L^{(n/2)+\varepsilon} + L^\infty$, then $u(x,t) \underset{t\to 0}{\longrightarrow} u_0(x)$ at exactly the same set of points as the solution of the unperturbed problem $-a(x)\Delta u = -\frac{\partial u}{\partial t}$.

The theorem follows from the fact that $e^{-tA}u_0$ solves (8), and the semigroup falls into the class of multipliers ϕ in Theorem 4.

We close this first part by remarking that Professors Mark Kon of Columbia University, Alexander Ramm of Kansas State University and the author have proved equisummability for linear operators in Banach spaces by a different approach. Presently Professor Kon and the author are extending Theorem 4 to uniformly elliptic operators, as these are useful in physics.

2. Regularization:

The notion of a "well-posed" partial differential equation includes the requirement that solutions depend continuously on the data. However, in many physically motivated cases, partial differential equations or problems associated with them have solutions which lack such continuous dependence. The Cauchy problem for the Laplacian and the heat flow equation for negative time are such equations. One method of resolving such ill-posedness is the introduction of additional a priori assumptions on the solution. Tikhonov [Ti1], in a fundamental paper, applied the idea to the solution of ill-posed inverse problems. This idea was successfully used on the Cauchy problem for elliptic operators in pioneering work by Fritz John [Jo1], [Jo2].

The summation of eigenfunction expansions is an important ill-posed problem associated with partial differential equations. It is ill-posed in the sense that small changes in the coefficient vector can produce large and even unbounded pointwise changes in the function being expanded.

Mathematically, our regularization method is a family of linear operators $\{\phi_\varepsilon\}_{\varepsilon>0}$ on $L^p(\mathbb{R}^n)$ along with a scaling function $\varepsilon(\gamma)$, $\gamma > 0$, such that if $\{f_\gamma\}$, $\gamma > 0$ is a net of $L^p(\mathbb{R}^n)$ perturbations of f with $||f_\gamma - f||_p < \gamma$, then $\phi_\varepsilon(f_\gamma)$ approaches f in L^p and pointwise when $\varepsilon = \varepsilon(\gamma)$ and $\gamma \to 0$. The results we obtain are applicable to the class of analytic multipliers $\phi_\varepsilon(f) = \phi(\varepsilon A)f$, see where A is an elliptic operator on $\mathbb{R}^n$. Such regularization methods for expansions in eigenfunctions of ordinary differential operators have been widely studied. Results for partial differential operators are fewer, owing partly to the lack of a comprehensive theory of eigenfunction expansions. Our results apply to the class of elliptic operators on $\mathbb{R}^n$ defined in part one equation (3), and can be used for both discrete and continuous spectra. We use extensively the results and approach of [GK] and [KR].

We now define our class of regularization operators, which are expressed in terms of an analytic multiplier $\phi(z)$ with $\phi(0) = 1$. Let

$$W_\theta = \{z: \ |\arg z| \leqslant \theta\}, \ D_r = \{z: \ |z| \leqslant r\}. \tag{9}$$

Recall by Theorem 3, the spectrum $\sigma(A)$ is contained in $W_\theta \cup D_r$ for r sufficiently large. Let $\phi(z)$ be analytic in a domain $D \subset \mathbb{C}$ containing $W_\theta \cup D_r$, and Γ be the positively oriented contour consisting of the boundary $\partial(D_r \cup W_\theta)$. Using the analytic operator calculus we can define the operator $\phi(\varepsilon A)$: $L^p(\mathbb{R}^n) \to L^p(\mathbb{R}^n)$ by

$$\phi(\varepsilon A)f(x) = \frac{1}{2\pi i}\int_\Gamma \phi(z)(z - \varepsilon A)^{-1}f(x)dz,$$

where ε is small enough that the spectrum of εA lies within Γ.

To guarantee $\phi(\varepsilon A)$ is well defined we suppose

(a) $\displaystyle\int_{R_{\theta_1}} \left|\frac{\phi(z)}{z}\right| |dz| < \infty$ for any ray $R_{\theta_1} = \{z: \arg z = \theta_1\}$, $|\theta_1| \leqslant \theta\}$ (10a)

(b) $\displaystyle\int_{G_r} \left|\frac{\phi'(z)}{z}\right| |dz| \underset{r\to\infty}{\longrightarrow} 0$ where $G_r = \{z: \ |z| = r, \ |\arg z| \leqslant \theta\}$ (10b)

For our regularization methods, we must impose the following stronger integrability condition:

$$\int_\Gamma |\phi(z)|\,|z|^{-\delta}|dz| < \infty \tag{10c}$$

for all $\delta > 0$.

It is interesting to note the connection of the regularization operators $\phi(\varepsilon A)$ with the application of multipliers for summing expansions. Formally for integrals.

$$\phi(\varepsilon A)\int_{\sigma(A)} F(\lambda)u(x,\lambda)d\rho(\lambda) = \int_{\sigma(A)} F(\lambda)\phi(\varepsilon\lambda)u(x,\lambda)d\rho(\lambda)$$

with $u(x,\lambda)$ a generalized eigenfunction of A and λ its associated eigenvalue. The operator formulation on the left side is more general, in that it can be applied independently of detailed spectral considerations.

We can now formulate our problem. Let $f \in L^p(\mathbb{R}^n)$, $1 \leq p < \infty$, and $\{f_\gamma\}$, $\gamma > 0$, denote a net of functions in $L^p(\mathbb{R}^n)$ with $||f - f_\gamma||_p < \gamma$. We wish to determine a sharp scaling of ε and γ which guarantees that as $\gamma \to 0$, $\phi(\varepsilon A)f_\gamma \to f$ in $L^p(\mathbb{R}^n)$, $1 \leq p < \infty$, and pointwise on the Lebesgue set of f.

The following theorem provides a large class of L^p-regularizing operators of the form $\phi(\varepsilon A)$ and a sharp condition on the associated scaling functions.

Theorem 6:

Let A be an elliptic operator of order m on $\mathbb{R}^n$. satisfying the conditions above, and $\phi(z)$ be analytic on the spectrum of A, satisfying the boundedness conditions (10a,b,c). Let $p > \frac{n}{m}$, $f \in L^p(\mathbb{R}^n)$, and $\{f_\gamma\}$, $\gamma > 0$, be a net of $L^p(\mathbb{R}^n)$ functions, with $||f - f_\gamma||_p < \gamma$. If

$$\gamma\varepsilon^{-n/mp} \underset{\gamma\to 0}{\longrightarrow} 0$$

then $\phi(\varepsilon A)f_\gamma \to f$ in $L^p(\mathbb{R}^n)$ $(1 \leqslant p < \infty)$ and pointwise on the Lebesgue set of f. Furthermore, this scaling of ε with γ is sharp in that if (10) fails, then for each A there are f, $\{f_\gamma\}$ such that $\phi(\varepsilon A)f_\gamma \not\to f$ on the Lebesgue set of f.

The proof uses the fact that for f in L^p, $1 \leqslant p < \infty$, $\phi(\varepsilon A)f(x)$ converges as $\varepsilon \to 0$ to $f(x)$ in L^p norm $(1 \leqslant p < \infty)$ and pointwise on the Lebesgue set of f $(1 \leqslant p \leqslant \infty)$; it also depends on Theorem 3, the Minkowski integral inequality and Young's inequality. The proof is omitted for brevity, and will appear in the Bulletin of the London Mathematical Society.

As an application, consider a function f(t) in $L^2(\mathbb{R}^n)$ in a multiparameter time domain $(t = (t_1,\ldots,t_n))$ associated with the Laplacian, and so

$$f(\vec{t}) = \frac{1}{(2\pi)^{n/2}} \int_{R^n} \hat{f}(\vec{\omega}) e^{i\vec{t}\cdot\vec{\omega}} d\vec{\omega}. \tag{11}$$

where $\hat{f}$ is the Fourier transform. The basic question is how do frequency domain errors (in $\hat{f}(\vec{\omega})$) propagate into time domain $(f(\vec{t}))$ errors? In general, small function space or pointwise errors in $\hat{f}$ will lead to unbounded errors in pointwise estimation of f.

This type of problem has applications in many types of situations; one example is the problem of waves radiating from an antenna with fixed frequency ω, and a given aperture illumination f(x). Small perturbations in f(x) can lead to unbounded errors in the local radiation field intensity.

The following provides a regularization procedure for ameliorating the large pointwise errors in $f(\vec{t})$ above; we choose to work with L^2 for simplicity.

Corollary 7. Let f_γ be an L^2-perturbation of f, in $L^2(\mathbb{R}^n)$, of size γ, i.e., $||\hat{f} - \hat{f}_\gamma||_2 \leq \gamma$,

$$f_{\varepsilon,\gamma}(\vec{t}) \equiv \frac{1}{(2\pi)^{n/2}} \int_{R^n} \hat{f}\,(\vec{\omega})\phi(\varepsilon\vec{\omega}^2)e^{i\vec{\omega}\cdot\vec{t}}d\vec{\omega}, \tag{12}$$

where ϕ is a function analytic in a region $W_\theta \cup D_r$ (θ, r > 0; see (9)), and satisfies (10). Let $f(\vec{t})$ be the inverse Fourier transform of $\hat{f}$. If the scaling

$$\gamma\varepsilon^{-n/4} \underset{\gamma\to 0}{\to} 0 \tag{13}$$

holds, then the pointwise error in $f_{\varepsilon,\gamma}(\vec{t})$ vanishes as $\varepsilon \to 0$.

That is, if the summation parameter ε is scaled correctly with L^2-error γ, pointwise error in the time domain vanishes with function space error in the frequency domain. We remark that the exponent $-\frac{n}{4}$ in (13) arises from the fact that the expansion in (11) is in fact in eigenfunctions of the Laplacian. That is,

$$f_{\varepsilon,\gamma} = \phi(-\varepsilon\Delta)f_\gamma .$$

Acknowledgement: In closing, I wish to once again thank Professor Tung Chin-Chu, Professor Xiao Shutie and Professor Pu Fu-Quan for their kind invitation, friendship, wonderful hospitality and for providing this unique opportunity to exchange mathematical ideas with Chinese mathematicians.

Bibliography

[AT] Arsenin, V.Y. and A.N. Tikhonov, Solutions of ill-posed problems, Winston, 1977.

[Be1] Benzinger, H., Green's function for ordinary differential operators, J. Differential Equations 7, (1970), 478.

[Be2] Benzinger, H.E., Perturbation of the heat equation, J. Differential Equations 32 (1979), 398.

[DS] Dunford, N. and J.T. Schwartz, Linear Operators, Interscience, New York, 1959.

[Go] Golubov, B.I., Multiple series and Fourier integrals. Mathematical Analysis 19, Akad. Nauk SSSR, (1982), 3.

[GK1] Gurarie, D., and Kon, M., Radial bounds for perturbations of elliptic operators, J. Functional Analysis 56 (1984), 99.

[GK2] Gurarie, D., and Kon, M., Resolvents and regularity properties of elliptic operators. Operator Theory: Advances and Applications, 151-162, Birkhauser, Basel, 1983.

[H] Haar, A., Zur theorie der orthogonalen Funktionensysteme, Math. Annalen 69 (1910), 331.

[I1] Il'in, V.A., The Riesz equisummability of expansions in eigenfunctions and in the n-dimensional Fourier integral. Trudy Mat. Inst. Steklov. 128 (1972), 151.

[Jo1] John, F., A note on "improper" problems in partial differential equations Comm. Pure and Appl. Math., 8 (1955), 591.

[Jo2] John, F., Continuous dependence on data for solutions with a prescribed bound, Comm. Pure and Appl. Math., (4) 13 (1960), 551.

[K] Komornik, V., An equiconvergence theorem for the Schrödinger operator, Acta Math. Hungar. 44 (1984), 101.

[KR] Kon, M. and L. Raphael, New multiplier methods for summing classical eigenfunction expansions, J. Differential Equations, 50 (1983), 391.

[LS] Levitan, B., and Sargsjan, I., Introduction to Spectral Theory; Self-Adjoint Differential Operators, A.M.S. Translations, Providence, 1975.

[Ra] Raphael, L.A., Equisummability of eigenfunction expansions under analytic multipliers, J. Math. Anal. and Appl. (to appear).

[St1] Stone, M.H., A comparison of the series of Fourier and Birkhoff, Trans. A.M.S. 28 (1926), 695.

[St2] Stone, M.H., Irregular differential systems of order two and related expansion problems. Trans. A.M.S. 29 (1927), 23.

[Ta] Tamarkin, J.D., Sur quelques points de la theorie des equations differentielles lineaires ordinaires et sur la generalisation de la serie de Fourier, Rendiconti del Circolo Mathematico di Palermo 34 (1912), 345.

[Ti1] Tikhonov, A.N., The stability of inverse problems, Doklady Akad. Nauk SSSR (5) 39 (1943).

[Ti2] Tikhonov, A.N., Stable methods for the summation of Fourier series, Soviet Math. Dokl. 5 (1964), 641-644.

[T] Titchmarsh, E.C., Eigenfunction Expansions, Vol. I and II, Oxford, 1962.

[W] Walsh, J.L., On the convergence of the Sturm-Liouville series, Ann. Math 24 (1922-23), 109.

APPROXIMATIONS FOR NAVIER-STOKES PROBLEMS

R. Rautmann
Department of Mathematics
University of Paderborn, Warburger Str. 100
D-4790 Paderborn, West-Germany

This note gives a review of some recent theoretical and numerical approaches to the Navier-Stokes equations. Special attention has been given to the approxiation of the solution's spatial derivatives of first and second order even on the boundary of the flow region. The result is of practical importance for the computation of flow separation [6].

1. In a smoothly bounded 3-dimensional domain Ω we consider the evolution equation

$$(*) \qquad \frac{\partial}{\partial t} u + \mu A u = f \text{ for } t > 0,\ u = u_o \text{ for } t = 0$$

containing the Stokes operator $A = -P\Delta$ multiplied by the viscosity constant $\mu > 0$ (P denotes H. Weyl's orthogonal projection, Δ the Laplacian).

Let f_k be the restriction of the function f to the time grid $t_k = k \cdot h$ with step length $h > 0$, $k = 0,1,\ldots$. Then the sequence of resolvent equations

$$\binom{*}{*} \qquad \left(\frac{1}{h} + \mu A\right) u_{k+1} = f_k + \frac{1}{h} u_k$$

yields a difference scheme approximating the semigroup solution of (*) on the time grid. Its convergence with $h \to 0$ has been investigated in the fractional order Sobolev spaces $H_s(\Omega)$, $0 \le s < 3$. The result depends on the regularity of the datas u_o and f, and requires a compatibility condition for u_o in the cases $s \ge 5/2$, [4,5,8].

2. An efficient numerical realization of these approximating schemes has been achieved by means of a boundary integral equation method for each time step, [2,8].

3. Taking for f a suitable approximation [3] of the nonlinear convective term $- P(u\nabla u)$ of the Navier-Stokes equations, we get an approximation scheme for

the nonlinear Navier-Stokes initial-boundary value problem on Ω. The compatibility condition for the initial value at each time step can be established in a constructive way by a proposal of Solonnikov. However, due to the lack of global bounds in sufficiently strong norms, the stability of this scheme can be proved only locally in time or globally under suitable smallness conditions. The method has successfully been tested by computations in special 3-dimensional cases [8].

[1] FUJITA, H.: On the Semi-Discrete Finite Element Approximation of the Evolution Equation $u_t + A(t)\,u = 0$ of Parabolic Type, Topics in Num. Analysis, John J.H. Miller (ed.), Acad. Press New York 1977.

[2] HEBEKER, F.K.: A Boundary Element Method for Stokes' Equations in 3-D Exterior Domains, in: Whiteman, J.R. (ed.), The Mathematics of Finite Elements and Applications, Vol. 5 (Mafelap 1984) London 1985.

[3] RAUTMANN, R.: On the Navier-Stokes Initial Value Problem (Sir G.I. Taylor Memorial Lecture), Proc. Int. Symp. "Problems of Nonlinear Continuum Mechanics" Kharagpur 1980, Indian Soc. Theor. Appl. Mech. Kharagpur, 1 - 16 (1982).

[4] RAUTMANN, R.: Zur Konvergenz des Rothe-Verfahrens für instationäre Stokes-Probleme in dreidimensionalen Gebieten, ZAMM 64 (1984) T 387-388.

[5] RAUTMANN, R.: On Optimum Regularity of Navier-Stokes Solutions at Time t = 0, Math. Z. 184 (1983) 141-149.

[6] RAUTMANN, R.: Eine Bemerkung zum Ablöseproblem zäher Strömungen, ZAMM 65 (1985), T 358-360.

[7] RAUTMANN. R., VARNHORN, W.: Die Navier-Stokessche Randwertaufgabe mit einer Differenzennäherung, ZAMM 65 (1985), T 360-362.

[8] VARNHORN, W.: Zur Numerik der Gleichungen von Navier-Stokes, Thesis Paderborn 1985.

Supported by "Deutsche Forschungsgemeinschaft".

THREE PHASE FLOW IN A POROUS MEDIUM AND THE CLASSIFICATION OF NON-STRICTLY HYPERBOLIC CONSERVATION LAWS

David G. Schaeffer
Department of Mathematics
Duke University
Durham, North Carolina 27706

Michael Shearer
Department of Mathematics
North Carolina State University
Raleigh, North Carolina 27695

1. INTRODUCTION. A system of conservation laws in one space dimension

$$U_t + F(U)_x = 0 \qquad -\infty < x < \infty, \quad t > 0 \tag{1.1}$$

is strictly hyperbolic at U if $dF(U)$ has distinct real eigenvalues. Many hyperbolic systems of physical interest fail to be strictly hyperbolic at every point. In this paper, we summarize recent work on 2×2 hyperbolic systems that are strictly hyperbolic except at a single point called an umbilic point.

Equations with umbilic points arise as models of three phase flow in a porous medium. In this context, the Riemann initial value problem assumes especial importance, since it is central to numerical front tracking methods based on Glimm's scheme. The Riemann problem for (1.1) has jump initial data

$$U(x,0) = \begin{cases} U_L & \text{if} \quad x < 0 \\ U_R & \text{if} \quad x > 0 \end{cases}. \tag{1.2}$$

Solutions of (1.1), (1.2) involve combinations of centered shock and rarefaction waves. Each wave in a specific characteristic family may be characterized by the state U_- on the left of the wave in (x,t)-space and the wave strength. For fixed U_-, the strength of the wave parameterizes a one-parameter family of states U_+ on the right of the wave. These one-parameter families (one for each characteristic family) are called wave curves. The umbilic point has a profound effect on the structure of wave curves. An overall goal of this research is to classify the various effects produced by the presence of such a point. The classification of 2×2 systems in [3] identifies four different regimes that we describe in §2. In §3 we illustrate the effect of the umbilic point by describing the solution of a prototype Riemann problem. Of particular note are two new types of shock wave violating the Lax entropy condition. These overcompressive and undercompressive shocks are associated with the presence of the umbilic point, which confuses the labelling of the characteristic speeds. Details of the results and properties described here, together with new techniques for studying Riemann problems, are discussed fully in [3,4].

2. CLASSIFICATION OF EQUATIONS WITH UMBILIC POINTS. To motivate the analysis, we begin by describing equations modelling three phase flow in a porous medium. Let u,v,w denote volume fractions of the phases, with corresponding relative permeabilities f,g,h. We assume the flow is one-dimensional, is not subject to external forces, and that the pressure in each of the phases is the same. We make the constitutive, or modelling, assumption that f,g,h depend only on u,v,w *respectively*: $f = f(u)$, etc. As usual, the conservation of momentum is approximated by Darcy's law, which for 1-d flows enables us to express the velocities in terms of the volume fractions. This puts the conservation of mass in the following form:

$$\left.\begin{aligned} u_t + \left(\frac{f(u)}{D}\right)_x &= 0 \\ v_t + \left(\frac{g(v)}{D}\right)_x &= 0, \end{aligned}\right] \qquad (2.1)$$

with $D = f(u) + g(v) + h(w)$ and $u + v + w = 1$. The physical range for u,v and $w = 1 - u - v$ is $[0,1]$. Note that the corresponding equation for w is redundant and thus excluded from (2.1), and that f,g,h are taken to include the corresponding viscosities.

Here are properties of f,g,h that reflect experimental results for 2-phase flow:

(A) $$f(0) = f'(0) = 0,\ f''(u) > 0,\ 0 \le u \le 1,$$

and similarly for g,h.

Let us return to terminology for general 2×2 systems in order to state a result for system (2.1):

$$U_t + F(U)_x = 0\ , \qquad (2.2)$$

where $U = U(x,t) \in \mathbb{R}^2$, $F\colon \mathbb{R}^2 \to \mathbb{R}^2$. System (2.2) is hyperbolic if $dF(U)$ has real eigenvalues (*characteristic speeds*), which we label $\lambda_1(U) \le \lambda_2(U)$. If $\lambda_1(U) = \lambda_2(U)$, we call U an *umbilic point*. An umbilic point U^* is *essential* if the set of matrices $\mathcal{T} = \{dF(U)\colon U \text{ near } U^*\}$ is a smooth 2-dimensional manifold, in which U^* is the only umbilic point, and $dF(U^*)$ is diagonal. The term "essential" reflects the property that such umbilic points cannot be removed by perturbations of F.

Theorem 2.1. Under assumption (A), system (2.1) has a unique essential umbilic point in the triangle $0 < u, v, 1 - u - v < 1$.

We next turn to the structure of wave curves for system (2.2) near an essential umbilic point U^*. Consider the Taylor series for $F(U)$ about U^*

$$F(U) = F(U^*) + dF(U^*)(U - U^*) + Q(U - U^*) + \text{h.o.t.},$$

where $Q = \frac{1}{2} d^2F(U^*)$ and h.o.t. indicates the remainder term. Since $F(U^*)$ is constant and $dF(U^*)$ is a multiple of the identity (and may be removed by changing to a moving system of coordinates), the first term to affect wave curves is $Q(U - U^*)$. Assuming Q is nondegenerate in a sense specified in [3], the higher order terms do not change the qualitative properties of wave curves near the umbilic point. We shall ignore the higher order terms in what follows. Taking $U^* = 0$ without loss of generality, we are left with the system

$$U_t + Q(U)_x = 0 \qquad -\infty < x < \infty, \quad t > 0 \tag{2.3}$$

Now $U = 0$ is automatically an umbilic point for (2.3), and is essential if $dQ(U)$ has distinct (real) eigenvalues for every $U \neq 0$.

Now linear changes of dependent variable U in (2.3) do not affect the characteristic speeds, or other features such as the admissible shock waves and the rarefaction waves. Accordingly, it is appropriate in a classification of equations (2.3) to call two equations equivalent if a linear change of variable converts one to the other. That is, quadratic maps $Q_k\colon \mathbb{R}^2 \to \mathbb{R}^2$ $(k = 1,2)$ are equivalent if there exists a (constant) matrix S such that $Q_1(SU) = SQ_2(U)$ for all $U \in \mathbb{R}^2$. If $Q(U) = dC(U)$ for a homogeneous cubic scalar $C\colon \mathbb{R}^3 \to \mathbb{R}$, then (2.3) is symmetric, and automatically hyperbolic. The following result says that up to equivalence, all equations (2.3) with an essential umbilic point are symmetric.

Theorem 2.2. If $U = 0$ is an essential umbilic point for (2.3), then Q is equivalent to dC, where

$$C(u,v) = au^3/3 + bu^2v + uv^2 \tag{2.4}$$

and $a \neq 1 + b^2$.

Note that (2.4) is not the most general cubic scalar. We have rotated coordinates to eliminate the v^3 term and scaled to let the coefficient of uv^2 be unity. Both of these transformations preserve equivalence. Theorem 2.2 reduces the study of equations (2.3) to an investigation of a 2-parameter family of equations:

$$U_t + dC(U)_x = 0 \tag{2.5}$$

Recall that shock waves for a 2×2 system are compressive if they satisfy the Lax entropy condition [1], which requires characteristics of one family to enter the shock from both sides and characteristics of the other family to pass through the shock. There are also what we call overcompressive shocks for which both families of characteristics enter the shock, and noncompressive shocks for which both families of characteristics pass through the shock. If we fix U_L, all possible shock waves with U_L on the left are identified by the set of U_R lying on a shock wave curve.

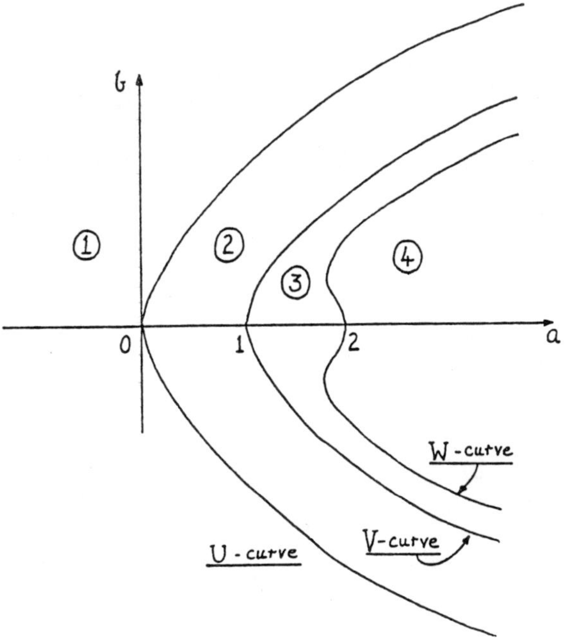

Figure 1. Regions in the (a,b)-plane.

These U_R satisfy the Rankine-Hugoniot condition and correspond to compressive and overcompressive shocks. (Certain noncompressive shocks also need to be selected in general. This is discussed briefly in §3.) We also have <u>rarefaction curves</u> which give the values of U through a centered rarefaction wave. These values lie on an integral curve of one of the two right eigenvectors of $dF(U)$. In Figure 2, we show these wave curves for (2.5) with $U_L = 0$ on the left of the wave. There are four regions in which the structure of the wave curves changes; these are shown in Figure 1.

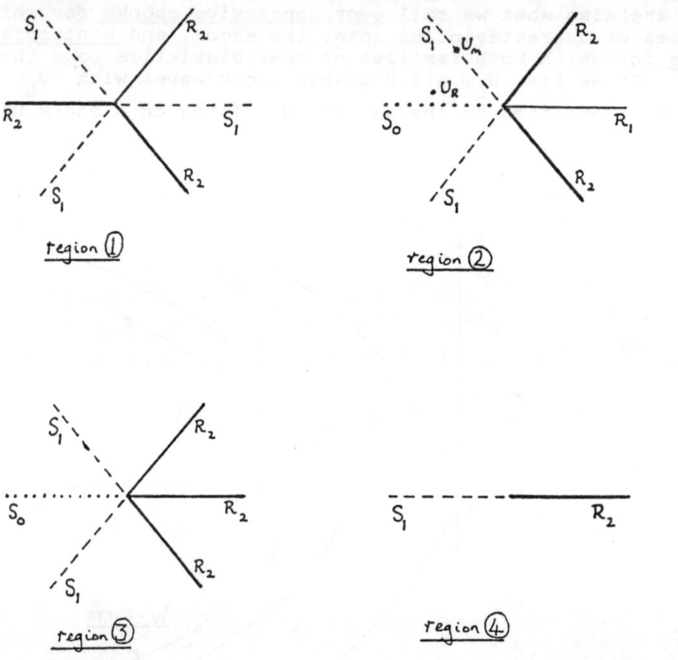

S_1: slow shocks $\qquad$ S_0: overcompressive shocks

R_1: slow rarefactions $\qquad$ R_2: fast rarefactions

Figure 2. Wave curves originating at the umbilic point.

To understand the overcompressive shocks, consider the Riemann problem (1.2), (2.5) in region 2, when $U_L = 0$ and U_R is close to the curve S_0 representing overcompressive shocks in Figure 2. The solution involves two compressive shock waves of nearly equal speed separated by a state U_m on the curve S_1. This is illustrated in Figure 4. Thus, an overcompressive shock wave is a superposition of fast and slow compressive shock waves travelling at the same speed.

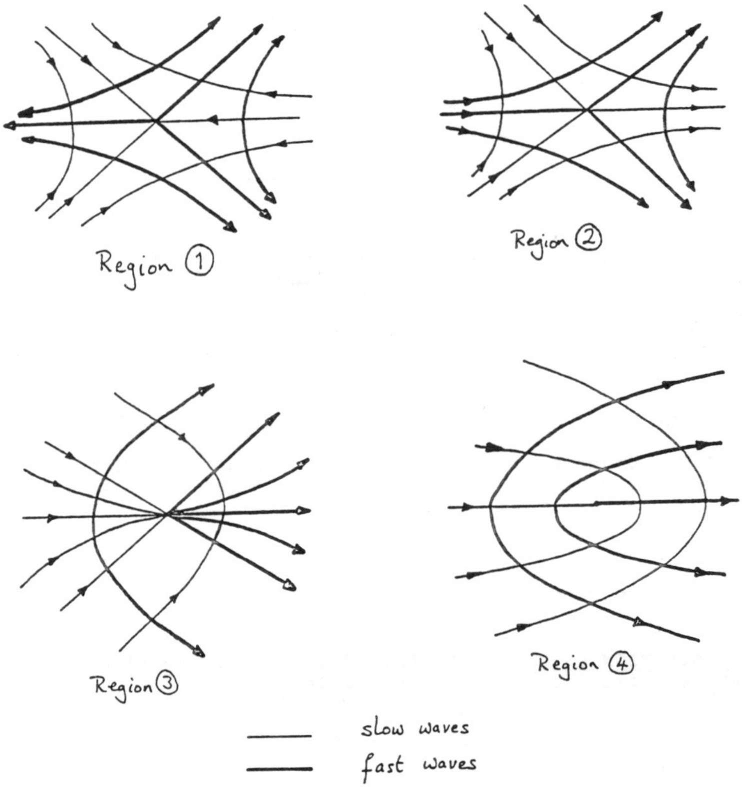

Figure 3. Rarefaction curves.

In Figure 3, we show the patterns of rarefaction curves for (2.5) in the four cases. Note that the arrows indicate the direction of increasing characteristic speed.

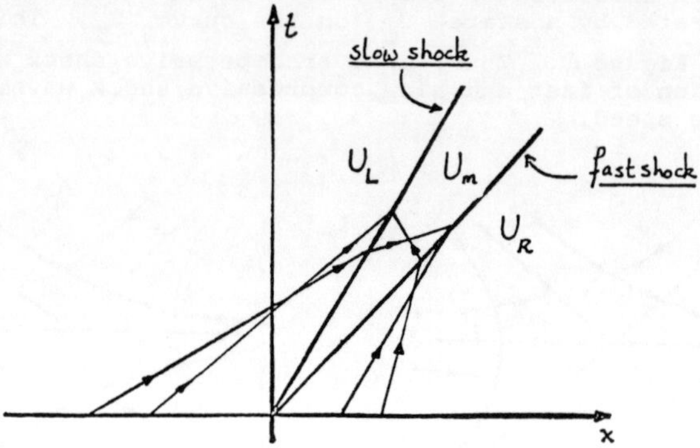

Figure 4. Solution of Riemann problem.
$U_L = 0$, U_R near S_0.

The following result shows how the 3-phase flow model (2.1) fits into the classification. By Theorem 2.1, system (2.1) has an umbilic point (u^*, v^*). Let Q denote the quadratic terms of the Taylor expansion of the nonlinearity of (2.1) about (u^*, v^*).

<u>Theorem 2.3</u>. Under assumption (A), the quadratic mapping Q for system (2.1) is equivalent to dC, for C given by (2.4) with $a < 1 + b^2$.

In other words, system (2.1) has properties near (u^*, v^*) corresponding only to quadratic nonlinearities in regions 1 and 2 of Figure 1.

3. SOLUTION OF A RIEMANN PROBLEM.

In this section, we explain the solution of the Riemann problem

$$z_t - (\bar{z}^2)_x = 0 \tag{3.1}$$

$$z(x,0) = \begin{cases} z_L & x < 0 \\ z_R & x > 0 \end{cases}, \tag{3.2}$$

where $z = u + iv$. Note that equation (3.1) has the symmetry property that $e^{2\pi i/3}z$ and $\bar{z}$ are solutions whenever z is a solution. For each z_L, z_R, there is a combination of centered shock and rarefaction waves that join z_L to z_R. For fixed z_L, each combination of successively faster waves determines a value of z_R, and as the strengths of the waves change, so does the value of z_R. In Figure 5, each possible combination of shock waves (S) and rarefaction waves (R) is indicated and associated with a region.

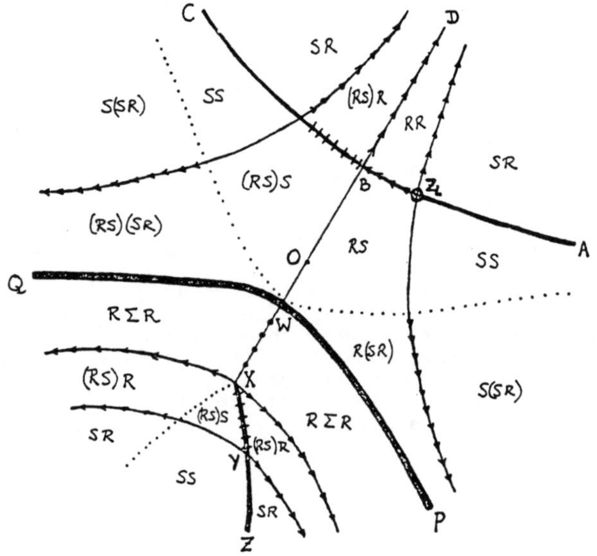

Figure 5. Solution of the Riemann problem (3.1), (3.2) for a typical z_L.

For example, if z_R lies in a region labelled (RS)S, then the solution of the Riemann problem involes a slow rarefaction-shock, and a fast shock wave. The label Σ indicates an <u>undercompressive shock</u>. These are noncompressive shocks that have viscous profiles associated with the Burgers' system

$$z_t - (\bar{z}^2)_x = \epsilon\, z_{xx} . \tag{3.3}$$

Now as z_L changes, the z_R regions get distorted. The regions have the shapes indicated in Figure 5 only for z_L between the lines $M_1^+ = \{z > 0\}$ and $M_2^- = \{z = \alpha e^{i\pi/3} : \alpha > 0\}$, which are lines of symmetry for equation (3.1). As z_L crosses M_2^- or M_1^+, several regions coalesce and collapse, and then reform in a different arrangement that is easily obtained by symmetry.

One interesting feature of Figure 5 emphasizes the role of the umbilic point $z = 0$. The curve PQ separates two different constructions of solutions of the Riemann problem. If z_R lies above PQ, then the standard construction of Liu, for strictly hyperbolic systems, applies. This says that there is a point z_m on the curve ABC (representing slow waves) such that z_R lies on the curve through z_m representing fast waves. Below PQ, the construction is similar except that the slower waves (including the undercompressive shocks) are represented by the curve WXYZ, which does not include z_L. The "coordinate system" that was used above PQ is, roughly speaking, rotated through 90° below PQ.

It should be emphasized that the solution of (3.1), (3.2) has relied upon the symmetry property of (3.3) to distinguish the undercompressive shocks. Without the symmetry, or if the diffusion matrix fails to be a multiple of the identity, then characterizing shocks possessing viscous profiles is significantly more difficult.

Reference

1. P. D. Lax. Hyperbolic systems of conservation laws II. Comm. Pure Appl. Math. 10 (1957), 537-566.

2. T.-P. Liu. The Riemann problem for general 2 × 2 conservation laws. Trans. Amer. Math. Soc. 199 (1974), 89-112.

3. D. G. Schaeffer and M. Shearer. The classification of non-strictly hyperbolic conservation laws, with application to oil recovery. Submitted to Comm. Pure Appl. Math.

4. M. Shearer, D. G. Schaeffer, D. Marchesin and P. L. Paes-Leme. Solution of the Riemann problem for a prototype 2 × 2 system of non-strictly hyperbolic conservation laws. Submitted to Arch. Rat. Mech. Anal.

WEAK SOLUTIONS OF PROBLEMS FOR SYSTEMS OF FERRO-MAGNETIC CHAIN

Zhou Yu-lin (周毓麟) & Guo Bo-ling (郭柏灵)

Institute of Applied Physics and Computational Mathematics

1. In the classical study of one-dimensional motion of ferro-magnetic chain, the so-called Landau-Lipschitz equation of isotropic Heisenberg chain is of the form

$$\vec{S}_t = \vec{S} \times \vec{S}_{xx} + \vec{S} \times \vec{h} \tag{1}$$

where $\vec{S} = (S_1, S_2, S_3)$ and $\vec{h} = (0, 0, h(t))$ are three-dimensional vector functions, $h(t)$ is a constant or a function of t and "$\times$" denotes the cross-product operator of two three-dimensional vectors.
The system with small diffusion term

$$\vec{S}_t = \vec{S} \times \vec{S}_{xx} + \varepsilon \vec{S}_{xx} \tag{2}$$

is called the spin system. In recent years there are many works concerning the soliton solutions, interactions of soliton waves, conservative laws and other properties for the Landau-Lipschitz equation [1-4]. These systems also appear in the study of problems in physics of condensed state medium.

We are interesting to study the weak solutions of various problems for systems of ferro-magnetic chain in somewhat more general form

$$Z_t = Z \times Z_{xx} + f(x,t,Z) \tag{3}$$

by different method [5-11], where $Z = (u, v, w)$ and $f(x,t,Z)$ are three-dimensional vector functions. The coefficient matrix of the terms of second order derivatives

$$A(Z) = \begin{pmatrix} 0, & -w, & v \\ w, & 0, & -u \\ -v, & u, & 0 \end{pmatrix} \tag{4}$$

is singular and zero-definite, i.e., the determinant $|A(z)| \equiv 0$ and for any three-dimensional vector $\xi \in \mathbb{R}^3$, $\xi \cdot A(z)\xi \equiv 0$.

The corresponding spin system with small diffusion term

$$z_t = \varepsilon z_{xx} + z \times z_{xx} + f(x,t,z) \tag{5}$$

is a quasilinesr non-degenerate parabolic system, where $\varepsilon > 0$.

In [12] for the homogeneous boundary problem of system with several independent variables

$$z_t = z \times \Delta z + f(x,t,z) \tag{6}$$

and the corresponding spin system

$$z_t = \varepsilon \Delta z + z \times \Delta z + f(x,t,z) \tag{7}$$

the weak solutions are constructed by the Galerkin method, where $x = (x_1, \cdots, x_m)$, $m \geq 2$.

2. Let us consider the boundary problem with periodic boundary condition

$$z(x+2D,t) = z(x,t) \tag{8}$$

and the initial value condition

$$z(x,0) = \varphi(x) \tag{9}$$

for the above mentioned systems (3) and (5), where $\varphi(x)$ is three-dimensional initial vector function.

Suppose that the following conditions are satisfied.

(I) $f(x,t,z)$ is continuously differentiable with respect to $x \in [-D,D]$ and $z \in \mathbb{R}^3$. The 3×3 Jacobi derivative matrix $f_z(x,t,z)$ is semibounded, i.e., there is a constant b, such that

$$\xi \cdot f_z(x,t,z)\xi \leq b|\xi|^2 \tag{10}$$

for any $(x,t) \in Q_T^{(D)} = \{-D \leq x \leq D, 0 \leq t \leq T\}$ and $\xi, z \in \mathbb{R}^3$.

(II) $\varphi(x) \in H^1(-D,D)$ is periodic with period $2D$.

<u>THEOREM</u> 1. Under the conditions (I) and (II), the periodic boundary problem (8) and (9) for the quasilinear parabolic spin system (5) with $\varepsilon > 0$ has a unique three-dimensional generalized vector solution $z_\varepsilon(x,t) \in W_2^{(2,1)}(Q_T^{(D)})$, which satisfies the system (5) in generalized sense and the boundary condition and the initial condition in classical sense.

By energy method, we have the following estimations.

LEMMA 1. Under the conditions (I) and (II) for the solutions $z_\varepsilon(x,t) \in W_2^{(2,1)}(Q_T^{(D)})$ of periodic boundary problem (8) and (9) for the spin system (5) ($\varepsilon > 0$), there are estimations

$$\sup_{0\le t\le T} \|z_\varepsilon(\cdot,t)\|_{H^1(-D,D)} + \sup_{0\le t\le T} \|z_{\varepsilon t}(\cdot,t)\|_{H^{-1}(-D,D)} \le K_1 \qquad (11)$$

where K_1 is a constant independent of $\varepsilon > 0$.

DEFINITION 1. The three-dimensional vector function $z(x,t) \in L_2(0,T;H^1(-D,D))$ is called the weak solution of periodic boundary problem (8) and (9) for the system (3) of ferro-magnetic chain, if the integral relation

$$\iint_{Q_T^{(D)}} [\xi_t z - \xi_x (z \times z_x) + \xi f(x,t,z)]\,dx\,dt + \int_{-D}^{D} \xi(x,0)\varphi(x)\,dx = 0 \qquad (12)$$

holds for any periodic test function $\xi(x,t) \in \Phi \equiv \{\xi \mid \xi \in C^{(1)}(Q_T^{(D)}),\ \xi(x,T)=0,\ \xi(x+2D,t) = \xi(x,t)\}$.

Using the estimations of Lemma 1, from the set of solutions $\{z_\varepsilon(x,t)\}$ of periodic boundary problem (8) and (9) for the spin system(5) ($\varepsilon > 0$), we can select a sequence $\{z_{\varepsilon_i}(x,t)\}$, such that as $\varepsilon_i \to 0$, $\{z_{\varepsilon_i}(x,t)\}$ converges uniformly in $Q_T^{(D)}$ to $z(x,t)$ and $\{z_{\varepsilon_i x}(x,t)\}$ converges weakly to $z_x(x,t)$. And $z(x,t)$ is a weak solution of periodic boundary problem (8) and (9) for the strongly degenerate system (3) of ferro-magnetic chain.

THEOREM 2. Under the conditions (I) and (II) the periodic boundary problem (8) and (9) for the system (3) of ferro-magnetic chain has at least one three-dimensional weak solution $z(x,t) \in L_\infty(0,T;H^1(-D,D)) \cap C^{(\frac{1}{2},\frac{1}{4})}(Q_T^{(D)})$, which satisfies the boundary condition (8) and the initial condition (9) in classical sense.

3. Now we turn to consider the initial value problem in infinite domain $Q_T^* = \{x \in \mathbb{R}, 0 \le t \le T\}$ with initial condition

$$z(x,0) = \varphi(x), \qquad x \in \mathbb{R} \qquad (13)$$

for the systems (3) and (5).

Let us assume that the following conditions are satistied.

(I*) $f(x,t,z)$ is continuously differentiable with respect to $x \in \mathbb{R}$ and $z \in \mathbb{R}^3$. The Jacobi derivative matrix $f_z(x,t,z)$ is semibounded, i.e., there is a constant b, such that (10) holds for any $(x,t) \in Q_T^*$ and $\xi, z \in \mathbb{R}^3$.

(II*) $\varphi(x) \in H^1(\mathbb{R})$.

(III*) For any $(x,t,z) \in Q_T^* \times \mathbb{R}^3$, there is

$$|f(x,t,z)|, |f_x(x,t,z)| \leq a(x,t) F(z) + b(x,t) \tag{14}$$

or

$$|f(x,t,z)|, |f_x(x,t,z)| \leq c(x,t) |z|^k + d(x,t), \tag{15}$$

where $k \geq 1$ is a constant, $a(x,t)$, $b(x,t)$, $d(x,t) \in L_\infty(0,T; L_2(\mathbb{R}))$, $c(x,t) \in L_\infty(Q_T^*)$ and $F(z)$ is a continuous function of $z \in \mathbb{R}^3$.

DEFINITION 2. The three-dimensional vector function $z(x,t) \in L_2(0,T; H^1(\mathbb{R}))$ is called the weak solution of initial value problem (13) for the system (3) of ferro-magnetic chain, if the integral equality

$$\iint_{Q_T^*} [\xi_t z - \xi_x(z \times z_x) + \xi f(x,t,z)]\,dx\,dt + \int_{-\infty}^{\infty} \xi(x,0)\varphi(x)\,dx = 0 \tag{16}$$

holds, for any test function of finite support $\xi(x,t) \in \Phi^* \equiv \{\xi \mid \xi \in C^{(1)}(\bar{Q}_T^*),\ \xi(x,T) = 0,\ \operatorname{supp} \xi < \infty\}$.

LEMMA 2. Suppose that the conditions (I*), (II*) and (III*) are fulfilled. For the vector solution $z_\varepsilon(x,t)$ of periodic boundary problem (8) and (9) for the spin system (5) ($\varepsilon > 0$), there is

$$\sup_{0 \leq t \leq T} \|z_\varepsilon(\cdot,t)\|_{H^1(-D,D)} + \sup_{0 \leq t \leq T} \|z_{\varepsilon t}(\cdot,t)\|_{H^{-1}(-D,D)} \leq K_2 \tag{17}$$

where K_2 is a constant independent of $\varepsilon > 0$ and $D > 0$.

THEOREM 3. When $D \to \infty$, the unique three-dimensional generalized vector solution $z_\varepsilon^{(D)}(x,t)$ of periodic boundary problem (8) and (9) for the spin system (5) in rectangular domain $Q_T^{(D)}$ tends to the unique three-dimensional generalized vector solution $z_\varepsilon(x,t)$ of initial value problem (13) for the spin system (5) in domain Q_T^*, in the sense that $\{z_\varepsilon^{(D)}(x,t)\}$ and $\{z_{\varepsilon x}^{(D)}(x,t)\}$ converge uniformly in any finite domain of Q_T^* to $z_\varepsilon(x,t)$ and $z_{\varepsilon x}(x,t)$ respectively, $\{z_{\varepsilon xx}^{(D)}(x,t)\}$ and $\{z_{\varepsilon t}^{(D)}(x,t)\}$ weakly converge to $z_{\varepsilon xx}(x,t)$ and $z_{\varepsilon t}(x,t)$ respectively in $L_2(Q_T^*)$.

THEOREM 4. Under the conditions (I*), (II*) and (III*), the initial value problem (13) for the system (3) of ferro-magnetic chain has at least one three-dimensional weak solution $z(x,t) \in L_\infty(0,T;H^1(R)) \cap C^{(\frac{1}{2},\frac{1}{4})}_{loc}(Q_T^*)$, which satisfies the initial condition (13) in classical sense.

4. It is naturally to arise a question, how about the boundary problems for there strongly degenerate systems of ferro-magnetic chain ?

By similar method of approach, for the boundary problems with any one set of boundary conditions

$$\begin{aligned} z(0,t) &= z(l,t) = 0 \ ; \\ z_x(0,t) &= z_x(l,t) = 0 \ ; \\ z(0,t) &= z_x(l,t) = 0 \ ; \\ z_x(0,t) &= z(l,t) = 0 \end{aligned} \tag{18}$$

and the initial condition (9), we can prove that :

<1> the unique three-dimensional generalized vector solution $z_\varepsilon^{(l)}(x,t) \in W_2^{(2,1)}(Q_T^{(l)})$ ($\varepsilon > 0$) for the spin system (5) exists in rectangular domain $Q_T^{(l)} = \{0 \le x \le l, 0 \le t \le T\}$;

<2> the problems for the system (3) of ferro-magnetic chain have at least one three-dimensional weak vector solution $z^{(l)}(x,t) \in L_\infty(0,T;H^1(0,l)) \cap C^{(\frac{1}{2},\frac{1}{4})}(Q_T^{(l)}\}$ and

<3> the weak vector solution $z^{(l)}(x,t)$ for the system (3) is a limit of the set of generalized vector solutions $\{z_\varepsilon^{(l)}(x,t)\}$ for the spin system (5) as $\varepsilon \to 0$.

For the boundary problems in infinite domain $\overset{+}{Q}{}_T^* = \{x \in R^+, 0 \le t \le T\}$ with boundary conditions

$$z(0,t) = 0 \qquad \text{or} \qquad z_x(0,t) = 0 \ , \tag{19}$$

we can also prove that :

<4> as $l \to \infty$, the unique three-dimensional generalized vector solution $z_\varepsilon^{(l)}(x,t) \in W_2^{(2,1)}(Q_T^{(l)})$ of boundary problem (18) and (9) for the spin system (5) tends to the unique three-dimensional generalized vector solution $z_\varepsilon(x,t) \in W_2^{(2,1)}(\overset{+}{Q}{}_T^*)$ of the corresponding boundary problem (19) and (9) for the spin system (5) in $\overset{+}{Q}{}_T^*$ and

⟨5⟩ there exist sequences $l_i \to 0$ and $\varepsilon_i \to 0$ as $i \to \infty$, such that $z_{\varepsilon_i}^{(l_i)}(x,t)$ tends to a three-dimensional weak vector solution $z(x,t) \in L_\infty(0,T; H^1(\mathbb{R}^+)) \cap C_{loc}^{(t,\frac{1}{2})}(\bar{Q}_T^*)$ of boundary problem (19) and (9) for system (3) of ferro-magnetic chain. Hence

⟨6⟩ the boundary problem (19) and (9) for spin system (5) have unique three-dimensional generalized vector solution $z_\varepsilon(x,t)$ and for system (3) of ferro-magnetic chain have at least one three-dimensional weak vector solution $z(x,t)$.

5. On the lateral sides of rectangular domain $Q_T^{(l)}$, let us give the nonlinear boundary conditions

$$\begin{aligned} z_x(0,t) &= \operatorname{grad} \psi_0(t, z(0,t)), \\ -z_x(l,t) &= \operatorname{grad} \psi_1(t, z(l,t)), \end{aligned} \tag{20}$$

where $\psi_0(t,z)$ and $\psi_1(t,z)$ are two scalar functions of $t \in [0,T]$ and $z \in \mathbb{R}^3$ and " grad " denotes the gradient operator with respect to $z \in \mathbb{R}^3$.

Suppose that the following conditions are satisfied.

(I_N) $f(x,t,z)$ is continuously differentiable in $Q_T^{(l)} \times \mathbb{R}^3$. The Jacobi derivative matrix $f_z(x,t,z)$ is semibounded.

(II_N) $\varphi(x) \in H^{(2)}(0,l)$ is a three-dimensional initial vector function, satisfying the nonlinear boundary conditions (20) at the ends of the interval $[0,l]$.

(IV_N) $\psi_0(t,z)$ and $\psi_1(t,z)$ are two scalar functions, which have the continuous derivatives with respect to $t \in [0,T]$, the continuous mixed derivatives with respect to $t \in [0,T]$ and $z \in \mathbb{R}^3$ and the continuous derivatives of second order with respect to $z \in \mathbb{R}^3$. The 3×3 Hessian matrices $H_0(t,z) = \operatorname{grad}^2 \psi_0(t,z)$ and $H_1(t,z) = \operatorname{grad}^2 \psi_1(t,z)$ for $\psi_0(t,z)$ and $\psi_1(t,z)$ with respect to z respectively are nonnegatively definite. And $\operatorname{grad} \psi_0(t,0) \equiv \operatorname{grad} \psi_1(t,0) \equiv 0$

The finite interval $[0,l]$ is divided into small grid by the points $x_j = jh$ ($j = 0,1,\cdots,J$), where $l = Jh$. Let us now construct the system of ordinary differential equations

$$z_j'(t) = \varepsilon \frac{\Delta_+\Delta_- z_j(t)}{h^2} + z_j(t) \times \frac{\Delta_+\Delta_- z_j(t)}{h^2} + f(x_j,t,z_j(t)), \quad (j = 1,2,\cdots,J-1) \tag{21}$$

with nonlinear boundary conditions

$$\frac{\Delta_+ z_0(t)}{h} = \operatorname{grad} \psi_0(t, z_0(t)),$$
$$-\frac{\Delta_- z_J(t)}{h} = \operatorname{grad} \psi_1(t, z_0(t)) \tag{22}$$

and the initial conditions

$$z_j(0) = \varphi_j = \varphi(x_j), \quad j = 1, 2, \cdots, J-1 \tag{23}$$

and $z_0(0)$ and $z_J(0)$ satisfying

$$\varphi_1 = z_0(0) + h \operatorname{grad} \psi_0(0, z_0(0)),$$
$$\varphi_{J-1} = z_J(0) + h \operatorname{grad} \psi_1(0, z_J(0)) \tag{24}$$

respectively

The solution $z_h(t) = \{z_j(t) \mid j = 0, 1, \cdots, J\}$ of the above system (21)-(24) can be regarded as the approximate solution of the nonlinear boundary problem (20) and (9) for the spin system (5) obtained by discretization of space variable. Under the above mentioned conditions, the system (21)-(24) has a unique global solution $z_h(t)$ ($0 \le t \le T$), such that $z_j(t) \in C^{(2)}([0,T])$($j = 1, \cdots, J-1$) and $z_0(t)$, $z_J(t) \in C^{(1)}([0,T])$.

THEOREM 5. Under the conditions (I_N), (II_N) and (IV_N), when $h \longrightarrow 0$, the three-dimensional discrete vector solution $z_h(t)$ of the system (21)-(24) of ordinary differential equations tends to the unique three-dimensional generalized vector solution $z(x,t) \in W_2^{(2,1)}(Q_T^{(\ell)})$ of the nonlinear boundary problem (20) and (9) for the spin system (5) ($\varepsilon > 0$), in the sense that $\{z_h(t)\}$ and $\left\{\frac{\Delta_+ z_h(t)}{h}\right\}$ are uniformaly convergent to $z_\varepsilon(x,t)$ and $z_{\varepsilon x}(x,t)$ in Q_T respectively and $\left\{\frac{\Delta_+\Delta_- z_h(t)}{h^2}\right\}$ and $\{z_h'(t)\}$ are weakly convergent to $z_{\varepsilon xx}(x,t)$ and $z_{\varepsilon t}(x,t)$ respectively.

THEOREM 6. Under the conditions (I_N), (II_N) and (IV_N), the nonlinear boundary problem (20) and (9) for the spin system (5) has a unique three-dimensional global generalized vector solution $z_\varepsilon(x,t)$, having the continuous derivatives $z_x(x,t)$, the generalized derivatives $z_{xx}(x,t)$, $z_t(x,t) \in L_\infty(0,T; L_2(0,\ell))$ and $z_{xt}(x,t)$, $z_{xxx}(x,t) \in L_2(Q_T^{(\ell)})$ and satisfying the spin system (5) in generalized sense and the nonlinear boundary conditions (20) and the initial condition (9) in classical sense.

Similarly we can get some uniform estimations independent of $\varepsilon > 0$ for the generalized vector solutions $Z_\varepsilon(x,t)$ of the nonlinear boundary problem (20) and (9) for spin system (5).

DEFINITION 3. The three-dimensional vector function $Z(x,t) \in L_2(0,T;H^1(0,\ell)) \cap C(Q_T^{(\ell)})$ is called the weak solution of the nonlinear boundary problem (20) and (9) for the system (3) of ferro-magnetic chain, if the integral relation

$$\iint_{Q_T^{(\ell)}} [\xi_t Z - \xi_x (Z \times Z_x) + \xi f(x,t,Z)]\,dx\,dt + \int_0^\ell \xi(x,0)\varphi(x)\,dx \\ - \int_0^T \xi(\ell,t)(Z(\ell,t) \times \operatorname{grad} \psi_1(t, Z(\ell,t)))\,dt - \int_0^T \xi(0,t)(Z(0,t) \times \operatorname{grad} \psi_0(t, Z(0,t)))\,dt = 0 \tag{25}$$

holds for any test function $\xi(x,t) \in H^1(Q_T^{(\ell)})$ with $\xi(x,T) \equiv 0$.

THEOREM 7. Under the conditions (I_N), (II_N) and (IV_N), the nonlinear boundary problem (20) and(9) for the system (3) of ferro-magnetic chain has at least one three-dimensional global weak vector solution $Z(x,t) \in L_\infty(0,T;H^1(0,\ell)) \cap C^{(\frac{1}{2},\frac{1}{4})}(Q_T^{(\ell)})$.

6. For the boundary problem in $Q_T^{(\ell)}$ with mixed nonlinear boundary conditions

$$Z_x(0,t) = \operatorname{grad} \psi_0(t, Z(0,t)), \\ Z(\ell,t) = 0 \tag{26}$$

and the initial condition (9), we assume the condition (I_N) and

(II_{0N}) $\varphi(x) \in H^{(2)}(0,\ell)$ satisfies the condition (26).

(IV_{0N}) $\psi_0(t,Z)$ has the same continuous derivatives and properties as in (IV_N).

The approximate solution $Z_h(t)$ can also be construct similarly by the process of discretization of space variable. By analogous way of approach, we get the following theorems.

THEOREM 8. Under the conditions (I_N), (II_{0N}), (IV_{0N}) and the condition $f(\ell,t,0) \equiv 0$, the mixed nonlinear boundary problem (26) and (9) for spin system (5) has a unique three-dimensional generalized vector solution $Z_\varepsilon(x,t) \in L_\infty(0,T;H^2(0,\ell)) \cap W_\infty^{(1)}(0,T;L_2(0,\ell)) \cap L_2(0,T;H^3(0,\ell))$.

THEOREM 9. Suppose that the conditions (I_N), (II_{0N}) and (IV_{0N}) are satisfied and suppose that $f(\ell,t,0) \equiv 0$. Then the mixed nonlinear boundary problem (26) and (9) for the system (3) of ferro-magnetic chain has at least one three-dimensional weak vector solution $z(x,t) \in L_\infty(0,T; H^1(0,l)) \cap C^{(\frac{1}{2},\frac{1}{4})}(Q_T^{(l)})$.

In the infinite domain $\overset{+}{Q}{}_T^*$, let us consider the boundary problem with nonlinear boundary condition

$$z_x(0,t) = \operatorname{grad} \psi_0(t, z(0,t)) \tag{27}$$

and the initial condition (9) for systems (3) and (5).

<u>THEOREM</u> 10. Assume that the conditions (I_N), (II_{0N}) and two following conditions are fulfilled :

(II_N^*) $\varphi(x) \in H^1(R^+)$ satisfies (27) at $x=0$.

(III_N^*) for $(x,t,z) \in \overset{+}{Q}{}_T^* \times R^3$, there is

$$|f(x,t,z)|, |f_x(x,t,z)|, |f_t(x,t,z)| \leq a(x,t) F(z) + b(x,t) \tag{28}$$

or

$$|f(x,t,z)|, |f_x(x,t,z)|, |f_t(x,t,z)| \leq c(x,t) |z|^k + d(x,t) \tag{29}$$

where $k \geq 1$ is a constant ; $a(x,t)$, $b(x,t)$, $d(x,t) \in L_\infty(0,T; L_2(R^+))$, $c(x,t) \in L_\infty(\overset{+}{Q}{}_T^*)$ and $F(z)$ is a continuous function of $z \in R^3$. Then the nonlinear boundary problem (27) and (9) for spin system (5) ($\varepsilon > 0$) has a unique three-dimensional generalized vector solution $z_\varepsilon(x,t) \in L_\infty(0,T; H^2(R^+)) \cap W_\infty^{(1)}(0,T; L_2(R^+)) \cap L_2(0,T; H^3(R^+))$.

<u>THEOREM</u> 11. Under the conditions of Theorem 10, when $l \to \infty$, the unique generalized vector solution $z_\varepsilon^{(l)}(x,t)$ of nonlinear boundary problem (26) and (9) for spin system (5) in $Q_T^{(l)}$ tends to the unique generalized vector solution $z_\varepsilon(x,t)$ of the nonlinear boundary problem (27) and (9) for the spin system (5) in $\overset{+}{Q}{}_T^*$.

<u>THEOREM</u> 12. Suppose that the conditions of Theorem 10 are satisfied. The nonlinear problem (27) and (9) for the system (3) of ferro-magnetic chain in the infinite domain $\overset{+}{Q}{}_T^*$ has at least one three-dimensional weak vector solution $z(x,t) \in L_\infty(0,T; H^1(R^+)) \cap C_{loc}^{(\frac{1}{2},\frac{1}{4})}(\overset{+}{Q}{}_T^*)$.

7. Let us consider in this section the more general boundary problem with the nonlinear mutual boundary conditions

$$\begin{aligned} Z_x(0,t) &= \operatorname{grad}_0 \psi(Z(0,t), Z(\ell,t)), \\ -Z_x(\ell,t) &= \operatorname{grad}_1 \psi(Z(0,t), Z(\ell,t)) \end{aligned} \tag{30}$$

and the initial condition (9) for the system (3) of ferro-magnetic chain directly by the finite difference method, where $\psi(Z_0, Z_1)$ is a scalar function of $Z_0, Z_1 \in \mathbb{R}^3$ and " grad_0 " and " grad_1 " denote the gradient operators with respect to Z_0 and Z_1 respectively.

We take the finite difference approximation of the nonlinear mutual boundary problem (30) and (9) for the system (3) of ferro-magnetic chain as follows

$$\frac{Z_j^n - Z_j^{n-1}}{\Delta t} = Z_j^n \times \frac{\Delta_+\Delta_- Z_j^n}{h^2} + f(x_j, t^n, Z_j^n), \tag{31}$$

$$j = 1, 2, \cdots, J-1 \; ; \; n = 1, 2, \cdots, N$$

with finite difference boundary conditions corresponding to non-linear mutual boundary conditions (30)

$$\begin{aligned} \frac{Z_1^n - Z_0^n}{h} &= \int_0^1 \operatorname{grad}_0 \psi(\tau Z_1^n + (1-\tau) Z_1^{n-1}, \tau Z_{J-1}^n + (1-\tau) Z_{J-1}^{n-1})\, d\tau, \\ \frac{Z_J^n - Z_{J-1}^n}{h} &= \int_0^1 \operatorname{grad}_1 \psi(\tau Z_1^n + (1-\tau) Z_1^{n-1}, \tau Z_{J-1}^n + (1-\tau) Z_{J-1}^{n-1})\, d\tau \end{aligned} \tag{32}$$

and the finite difference initial condition

$$Z_j(0) = \varphi_j \qquad (j = 0, 1, \cdots, J) \tag{33}$$

The existence of the solution Z_j^n $(j = 0, 1, \cdots, J \; ; n = 0, 1, \cdots, N)$ of the finite difference system (31)-(33) can be proved by the fixed-point theorem.

<u>THEOREM</u> 13. Suppose that the following conditions are fulfilled :

(I_M) $f(x,t,z)$ and $f_x(x,t,z)$ are two three-dimensional continuous vector functions for $(x,t,z) \in Q_T^{(\ell)} \times \mathbb{R}^3$. There is a constant b , such that

$$(z-y) \cdot (f(x,t,z) - f(x,t,y)) \le b|z-y|^2 \tag{34}$$

for any $(x,t) \in Q_T^{(\ell)}$ and $z, y \in \mathbb{R}^3$.

(II_M) $\varphi(x) \in H^1(0,\ell)$.

(IV_M) $\psi(z_0, z_1)$ is a continuously differentiable scalar function of $z_0, z_1 \in R^3$.

Then the nonlinear mutual boundaryproblem (30) and (9) for the system (3) of ferro-magnetic chain has at least one thre-dimensional weak vector solution $z(x,t) \in L_\infty(0,T; H^1(0,\ell)) \cap C^{(\frac{1}{2},\frac{1}{4})}(Q_T^{(\ell)})$.

8. For the case of systems with several space variables, let us consider the homogeneous boundary problem

$$z(x,t) = 0 , \qquad x \in \partial\Omega , \ 0 \le t \le T \tag{35}$$

with initial condition

$$z(x,0) = \varphi(x) , \qquad x \in \bar{\Omega} \tag{36}$$

for the systems (6) and (7), where Ω is a finite domain in $\mathbb{R}^m$ ($m \geq 2$) and $\partial\Omega$ is twice continuously differentiable.

DEFINITION 4. The three-dimensional vector function $z(x,t)$ $L_2(0,T; H_0^1(\Omega))$ is the weak solution of the homogeneous boundary problem (35) and (36) for the spin system (7) ($\varepsilon > 0$) or the system (6) ($\varepsilon = 0$) of ferro-magnetic chain with several space variables, if the integral relation

$$\int_0^T\int_\Omega [\xi_t z - \nabla\xi * (z \times \nabla z) - \varepsilon \nabla\xi * \nabla z + \xi f(x,t,z)]\,dx\,dt + \int_\Omega \xi(x,0)\varphi(x)\,dx = 0 \tag{37}$$

holds, for any test function $\xi(x,t) \in C^{(1)}(\bar{\Omega}\times[0,T])$ with $\xi(x,t) = 0$ on $\{x \in \bar{\Omega}, t=T\} \cup \{x \in \partial\Omega, 0 \le t \le T\}$, where "$*$" and "∇" denote the scalar product operator and the gradient operator in space R^m respectively and "$\times$" denotes the cross-product operator in $\mathbb{R}^3$.

Suppose that the following conditions are satisfied :

(I') The Jacobi derivative matrix $f_z(x,t,z)$ is semibounded.

(II') $\varphi(x) \in H_0^1(\Omega)$.

(III') For $(x,t,z) \in \bar{\Omega} \times [0,T] \times \mathbb{R}^3$,

$$\begin{aligned} &|f(x,t,z)| \le A(T)|z|^\ell + B(T), \\ &|\nabla f(x,t,z)| \le A(T)|z|^{1+\frac{2}{m}} + B(T), \\ &f(x,t,0) \equiv 0 , \end{aligned} \tag{38}$$

where $A(T)$ and $B(T)$ are constants dependent on T and $2 \leq l \leq 2+\frac{4}{m-2}$ ($m \geq 2$).

THEOREM 14. Under the conditions (I'), (II') and (III'), the homogeneous boundary problem (35) and (36) for the spin system (7) ($\varepsilon > 0$) or the system (6) ($\varepsilon = 0$) of ferro-magnetic chain with several space variables has at least one three-dimensional weak vector solution $z(x,t) \in L_\infty(0,T; H_0^1(\Omega)) \cap C^{(0,\frac{1}{3+[\frac{m}{2}]})}(0,T; L_2(\Omega))$, which satisfies the homogeneous boundary condition (35) almost everywhere on the lateral manifold $\partial\Omega \times [0,T]$ of the cylindrical domain $\bar{\Omega} \times [0,T]$.

The Galerkin approximation method is used to study the mentioned problem. The solution of the system (6) ($\varepsilon = 0$) can be obtained by the limiting process of the solutions of system (7) ($\varepsilon > 0$), as $\varepsilon \to 0$.

9. In the discussions of previous sections, the constant T is arbitrary. Hence all results obtained in previous sections are also valid for $T = \infty$. So we have generalized solutions and weak solutions for $0 \leq t < \infty$.

Suppose that the constant b appeared in the condition of semiboundedness of the Jacobi derivative matrix $f_z(x,t,z)$ for the vector function $f(x,t,z)$ is negative. Then we have

$$\frac{d}{dt} \| z(\cdot,t) \|^2_{L_2(\Omega)} \leq 2b \| z(\cdot,t) \|^2_{L_2(\Omega)}$$

THEOREM 15. Suppose that $b < 0$. For the weak solution $z(x,t)$ of homogeneous boundary problem (35) and (36) for the spin system (7) ($\varepsilon > 0$) and the system (6) ($\varepsilon = 0$) of ferro-magnetic chain, $\lim_{t \to \infty} \| z(\cdot,t) \|_{L_2(\Omega)} = 0.$

The generalized global solution of various problems for the spin system (5) and the weak global solution of various problems for the system (3) of ferro-magnetic chain have the similar asymptotic behavior for $t \to \infty$, provided that the constant b is negative.

10. Suppose that the part $f(x,t,z)$ of the system (3) or (6) of ferro-magnetic chain satisfies the inequality

$$z \cdot f(x,t,z) \geqslant c_0 |z|^{2+\delta}$$

for any $(x,t,z) \in \bar{\Omega} \times [0,T] \times \mathbb{R}^3$, where $c_0 > 0$ and $\delta > 0$. Then the generalized solution of the system of ferro-magnetic chain blows up.

THEOREM 16. Suppose that $f(x,t,z)$ satisfies the condition (39) and suppose that $\|\varphi\|_{L_2(\Omega)} > 0$, $2 \leq p < \infty$. Then the three-dimensional generalized vector solution $z(x,t) \in W_2^{(2,1)}(\Omega \times \mathbb{R}^+)$ of the system (6) of ferro-magnetic chain blows up in a finite interval of time t, in the sense that $\|z(\cdot,t)\|_{L_p(\Omega)}$ becomes infinite in certain finite value of t for $2 \leq p < \infty$.

The "blow up" behavior are also available for the three-dimensional generalized and weak global vector solution $z(x,t) \in W_2^{(2,1)}(Q_T)$ of various problems for the system (3) of ferro-magnetic ahain discussed in previous sections.

11. Of course the uniqueness of the weak solutions of various problems for the strongly degenerate system (3) or (6) of ferro-magnetic chain is in general not to be expected. But for the system (3) or (6) of ferro-magnetic chain, the smooth classical vector solution $z(x,t) \in C^{(3,1)}$ is unique, where $f(x,t,z)$ is twice continuously differentiable with respect to $x \in \mathbb{R}^m$ ($m > 1$) and $z \in \mathbb{R}^3$.

REFERENCE

[1] L.A. Takhtajan, Integration the Continuous Heisenberg Spin Chain Through the Inverse Scattering Method, Phys. Lett., **64** A (1977), 235.

[2] J. Tjon & J. Wright, Solitons in the Continuous Heisenberg Spin Chain, Phys. Rev., B **15** (1977), 3470-3476.

[3] M. Lakshmanan, Continum Spin System as an Exactly Solvable Dynamical System, Phys. Lett., **61** A (1977), 53.

[4] H. C. Fogedby, Theoretical Aspects of Mainly Low Dimensional Magnetic Systems, Lecture Notes in Physics, **131**, Springer-Verlag, New York, 1980.

[5] Zhou Yu-lin & Guo Bo-ling, The Solvability of the Initial Value Problem for the Quasilinear Degenerate Parabolic System (Proceedings of DD-3 Symposium, 1982).

[6] Zhou Yu-lin & Guo Bo-ling, On the Boundary Problems for the Systems of Ferro-Magnetic Chain (Proceedings of DD-3 Symposium, 1982).

[7] Zhou Yu-lin & Guo Bo-ling, The Existence of Weak Solutions of Boundary Problems for the Systems of Ferro-Magnetic Chain, Scientia Sinica, A **27** (1984), 799-811.

[8] Zhou Yu-lin & Guo Bo-ling, Finite Difference Solutions of the Boundary Problems for the Systems of Ferro-Magnetic Chain, J. Comp. Math., **1** (1983), 294-302.

[9] Zhou Yu-lin & Xu Guo-rong, Finite Difference Solutions of the Nonlinear Mutual Boundary Problems for the System of Ferro-Magnetic Chain, J. Comp. Math., 2(1984), 263-271.

[10] Zhou Yu-lin & Guo Bo-ling, Some Boundary Problems of the Spin Systems and the Systems of Ferro-Magnetic Chain, I. Nonlinear Boundary Problem (to appear).

[11] Zhou Yu-lin & Guo Bo-ling, Some Boundary Problems of the Spin Systems and the Systems of Ferro-Magnetic Chain, II. Mixed Problems and Others (to appear).

[12] Zhou Yu-lin & Guo Bo-ling, The Weak Solution of Homogeneous Boundary Value Problem for the System of Ferro-Magnetic Chain with Several Variables (to appear).

THE INFILTRATION PROBLEM WITH LARGE CONSTANT SURFACE FLUX IN PARTIALLY SATURATED POROUS MEDIA

An Lianjun

1. Introduction

Consider a filtration process of a homogeneous, isotropic porous medium. For one-dimensional case, the continuity equation and Darcy's law yield the equation

$$\theta(\varphi)_t = (k(\varphi)\varphi_x)_x - k(\varphi)_x \tag{1.1}$$

where θ is the volumetric moisture content, φ the capillary piezometric head and k the hydraulic conductivity. The equation contains a gravitational term, the downward direction being positive. We assume that θ and k are given functions depending only on φ and $\theta(\varphi) = \theta_s$, $k(\varphi) = k_s$ for $\varphi \geq 0$ and $0 \leq \varphi(\varphi) < \theta_s$, $0 \leq k(\varphi) < k_s$ for $\varphi < 0$, where θ_s is the saturated moisture and k_s the saturated hydraulic conductivity. Equation (1.1) is parabolic when $-\infty < \varphi < 0$, but $\frac{\theta'(\varphi)}{k(\varphi)} \to \infty$ as $\varphi \to -\infty$ and $\frac{\theta'(\varphi)}{k(\varphi)} \to 0$ as $\varphi \to 0^+$, i.e. Eq. (1.1) is degenerated at the interface between the dry and wet regions and at the interface between the saturated and unsaturated regions.

We make the transformation

$$u(x, t) = \int_{-\infty}^{\varphi(x, t)} k(s)ds \tag{1.2}$$

and assume $1 = \int_{-\infty}^{0} k(s)ds$. The function $\varphi = \varphi(u)$ is well-defined and Eq. (1.1) becomes

$$c(u)_t = (u_x - k(u))_x \tag{1.3}$$

where $c(u) = \theta \cdot \varphi(u)$ (Fig. 1.1). Equation (1.3) degenerate at $u = 0$ and $u = 1$.

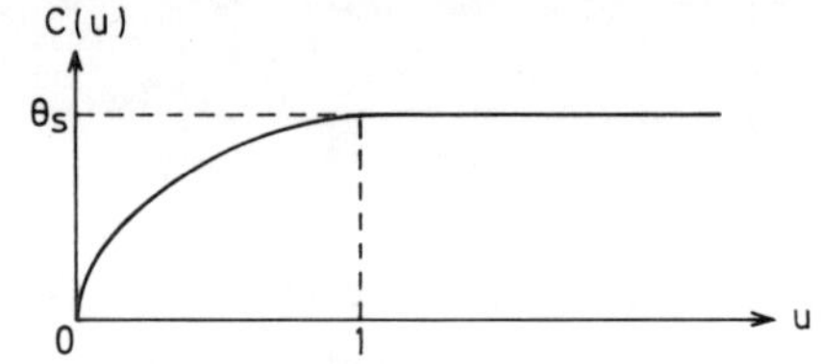

Fig. 1.1

Equation (1.3) was first studied by van Duyn and Peletier [1, 2] in 1982. Subsequently the study of Eq. (1.3) was taken up by An [3], Zhou [4], Su and Li [5, 6], Kröner [7], Bertsch, Hulshof and Peletier [8, 9]. They proved existence, uniqueness and regularity for the solutions of Eq. (1.3), but they avoided the situation in which both dry and saturated states appear. Xiao, Huang and Zhou [10] studied this case first with small constant surface flux. They proved existence, uniqueness and asymptotic behavior of the weak solution.

In this paper, we continue to discuss the problem posed by Xiao, Huang and Zhou. Under mild assumption on c, k and v_0, we study the problem with general constant surface flux

$$\left.\begin{aligned} &c(u)_t = (u_x - k(u))_x \quad \text{in } H_T = (0,\infty)\times(0,T] ,\\ &(u_x - k(u))|_{x=0} = -R \quad \text{on } [0,T] \quad (R > k_s) ,\\ &c(u(x,0)) = v_0(x) \quad \text{on } [0,\infty) . \end{aligned}\right\} \tag{I}$$

According to the physical sense, we shall make the following assumptions:

H1a. $c(0)=0$, $c(s)=\theta_s$ when $s\geq 1$ and $c(s)$ is strictly increasing when $0<s<1$. $c(s)$ is Lipschitz continuous on $[\delta, \infty)$ for any $\delta>0$.

H2a. $k(0)=0$, $k(s)=k_s$ when $s\geq 1$ and $k(s)$ is nondecreasing on $[0, \infty)$. $k(s)\in C^{0+1}[0, \infty)$.

H3a. There exists $u_0(x)\in C^{0+1}[0, \infty)$ such that $u_0(x)=0$ when $x\geq x_1>0$ and $c(u_0(x))=v_0(x)$.

H4a. $(k(s_1)-k(s_2))^2\leq L(s_1-s_2)(c(s_1)-c(s_2))$ for any $s_1>s_2\geq 0$.

H4b. $\lambda c(s)-k(s)>0$ for $s\in(0,\infty)$ when $\lambda\in[\lambda_0-\delta, \lambda_0+\delta]$ for some $\delta>0$, where $\lambda_0=\dfrac{R}{\theta_s}$.

In order to obtain some properties, we make further hypotheses:

H1b. $c(s)$ is concave ($c''\leq 0$ if it exists).

H1c. $c(s)$ twice continuously differentiable in $(0, \delta)$ for some $\delta<1$. And there exists a constant $M>0$, such that

$$\left|\frac{cc''}{c'^2}\right|\leq M.$$

H1d. $c(s)$ is differentiable in $(0, 1)$ and satisfies $\dfrac{sc'(s)}{c(s)}\geq\sigma>0$ in the neighborhood of 0^+.

H2b. $k(s)$ is twice continuously differentiable in $(0, \delta)$ for some $\delta<1$. And there exists a constant $M>0$, such that $|k''|\leq M$.

H3b. $u_0\in C^{2+\alpha}[0, \infty)$ $(0<\alpha<1)$ and

$$u_{0xx}-k'(u_0)u_{0x}\geq -\chi c'(u_0)$$

for some $\chi\geq 0$, whenever $u_0(x)>0$.

H3c. $u_0(x)>1-(R-k_s)x$ for $x\in[0, \dfrac{1}{R-k_s}]$ and there exists $x_2<x_1$, such that $v_0(x)=\theta_s$ for $0\leq x\leq x_2$, $0<v_0(x)<\theta_s$ for $x_2<x<x_1$ and $v_0(x)=0$ for $x\geq x_1$.

H3d. u_0 satisfies $u_{0x}-k(u_0)+\lambda_0 c(u_0)\leq 0$ whenever $u_0(x)>0$.

H4c. $\lambda c(s)-k(s)$ is a nondecreasing function on $(0, \infty)$ for $\lambda\in[\lambda_0-\delta, \lambda_0+\delta]$.

where H3b is something like the concavity of u_0 and H3d is about the monotonicity of u_0.

Definition: A function u defined a.e. on H_T is called a weak solution of problem I, if it satisfies the following conditions:

1) u is real and non-negative on $\bar{H}_T$ and $u \in L^2(0, T; H^1_{loc}(0, \infty))$;
2) $c(u) \in C(\bar{H}_T)$ and $k(u) \in C^{0+1}(\bar{H}_T)$, u possibly redefined on a set of measure zero;
3) $u(x, t)$ satisfies the identity

$$\iint_{H_T} (\varphi_x(u_x - k(u)) - \varphi_t c(u))dxdt = \int_0^T R\varphi(0, t)dt + \int_0^\infty \varphi(x,0)V_0(x)dx$$

for all $\varphi \in C^1(\bar{H}_T)$, which vanishes for $t = T$ and large x.

In Section 2, we establish the uniqueness and the existence of weak solution of Problem I. In Section 3, we prove a regularity result with respect to t under the condition H3b and an important property that $c(u)$ has compact support if and only if $\int_{0^+} \frac{ds}{c(s)} < \infty$. Comparing the condition $\int_{0^+} \frac{D(\theta)}{\theta} d\theta < \infty$ which is a necessary and sufficient condition of θ having compact support for any $t > 0$ when $\theta(x, 0)$ has compact support for equation $\theta_t = (D(\theta)\theta_x)_x$ [16], $\int_{0^+} \frac{ds}{c(s)} < \infty$ is a version of $\int_{0^+} \frac{D(\theta)}{\theta} d\theta < \infty$ under the transformation $u = \int_{-\infty}^{\varphi} k(s)ds$. Since $D = k\frac{d\varphi}{d\theta}$, $\int_{0^+} \frac{D(\theta)}{\theta} d\theta < \infty$ implies $\int_{-\infty} \frac{K(\varphi)}{\theta(\varphi)} d\varphi < \infty$ and then $\int_{0^+} \frac{du}{\theta(\varphi(u))} < \infty$ i.e. $\int_{0^+} \frac{ds}{c(s)} < \infty$.

In Section 4, we obtain the existence of two interfaces, i.e. the dry-wet interface or wetting front $\xi(t)$ and the saturated-unsaturated interface or water table $\zeta(t)$, and prove $\xi(t)$ and $\zeta(t)$ are Lipschitz-continuous under the condition H3b-d.

In Section 5, the asymptotic behavior of the solution as $t\to\infty$ is studied by using the comb functions and intersection of the solution which were used by Stokes [11] and Huang [12].

2. Uniqueness and Existence

Since a weak solution of Problem I belong to $L^2(0, T; H^1_{loc}(0, \infty))$, the uniqueness of weak solutions can be proved by the same method as X.H.Z. [10]. The existence of solution is obtained by a standard approximation process. First, we choose smooth functions $\{c_n^\varepsilon\}$, $\{k_n\}$ and $\{u_{on}^\varepsilon\}$ satisfying the following conditions:

1) c_n^ε, k_n and u_{0n}^ε are defined and infinitely differentiable on R, R and $[0, \infty)$ respectively.
2) $\lim_{\substack{\varepsilon\to 0\\ n\to\infty}} c_n^\varepsilon(s) = c(s)$, $c_n^\varepsilon \to c_n$ as $\varepsilon\to 0$ and $c_n^\varepsilon \to c^\varepsilon$ as $n\to\infty$ uniformly on bounded subsets of $[0, \infty)$. $c_n^\varepsilon \searrow c^\varepsilon$ as $n\to\infty$ for $s>1$.
3) $\frac{1}{n} \le (c_n^\varepsilon)'(s) \le \frac{1}{\varepsilon}$ for all $n\ge 1$, $\varepsilon>0$ and $s\in R$.
4) $c_n^\varepsilon(1) = \theta_s$, $c_n^\varepsilon(0) = 0$ and that if c is concave, then $(c_n^\varepsilon)'' \le 0$ on R for all $n\ge 1$.
5) $k_n\to k$ as $n\to\infty$ uniformly on $[0, \infty)$, $k_n(0) = 0$, $k_n \le k_s$ and $k_n' \le M_{k'}$ for all $s\in[0, \infty)$ and $n\ge 1$.
6) $\lambda c_n^\varepsilon(s) > k_n(s)$ for all $s\in(0, \infty)$, $n\ge 1$ and $\varepsilon>0$ when $\lambda\in[\lambda_0 - \delta, \lambda_0 + \delta]$.
7) $(k_n(s_1) - k_n(s_2))^2 \le L(s_1 - s_2)(c_n(s_1) - c_n(s_2))$ for all $s_1 > s_2 \ge 0$ and $n\ge 1$.
8) $u_{0n}^\varepsilon \to u_0^\varepsilon = \max\{\varepsilon, u_0\}$ as $n\to\infty$ uniformly on $[0, \infty)$ and $|(u_{on}^\varepsilon)'| \le L$.
9) $(u_{on}^\varepsilon)'(0) - k_n(u_{0n}^\varepsilon(0)) = -R$ and $u_{0n}^\varepsilon = \varepsilon$ for $x\ge x_1$.

Consider, for each $n\ge x_1$, the problem

$$
(I_n^\varepsilon)\quad \begin{cases} (c_n^\varepsilon)'(u)u_t = u_{xx} - k_n(u) & \text{in } H_T^n = (0, n)\times(0, T] \\ (u_x - k_n(u))|_{x=0} = -R \quad u(n, t) = \varepsilon & \text{on } [0, T] \\ u(0, x) = u_{0n}^\varepsilon & \text{on } [0, h] \end{cases} \tag{2.1}
$$

The properties of c_n^ε, k_n and u_{0n}^ε guarantee that Problem I_n^ε has an unique solution $u_n^\varepsilon \in C^{2,1}([0, n] \times [0, T]) \cap C^\infty((0, n) \times (0, T])$ [13, 14]. Next, we shall make some uniform estimation on u_n^ε and u_{nx}^ε and then extract a subsequence $u_{n'}^{\varepsilon'}$ which converges to a weak solution of Problem I.

Lemma 1. Let u_n^ε be the solution of Problem I_n^ε, then there exists a constant $M_0 > 0$ such that

$$\varepsilon \le u_n \le M_0 \quad \text{on} \quad H_T^n \quad \text{and} \quad \iint_{H_T^n} (u_n - \varepsilon)^2 dxdt \le M_0 \quad .$$

It is easy to obtain $u_n^\varepsilon(x, t) \ge \varepsilon$ on H_T^n by the maximum principle and to get the bound from above by using the travelling wave solutions

$$x - \lambda_0 t - x_1 = \int_L^{V_n^\varepsilon} \frac{ds}{k_n(s) - \lambda_0 c_n^\varepsilon(s)}$$

as supersolutions. The proof of $\iint_{H_T^n} (u_n - \varepsilon)^2 dxdt \le M_0$ can be found in [10].

Lemma 2. Let u_n^ε be the solution of Problem I_n^ε, then there exists a constant $M_1 > 0$ such that

$$|u_{nx}^\varepsilon| \le M_1 \quad ,$$

for all $n \ge 1$ and $\varepsilon > 0$.

Proof. We prove $0 \ge u_{nx}^\varepsilon(n, t) \ge -M$ first. Since $u_n^\varepsilon \ge \varepsilon$ and $u_n^\varepsilon(n, t) = \varepsilon$, we have $u_{nx}^\varepsilon(n, t) \le 0$. Consider a function

$$w = MM_{k'}^2 (\frac{2}{M} (n - x) - (x - n)^2) - (u_n^\varepsilon - \varepsilon)$$

in the region $\left[n - \frac{1}{2M_{k'}}, n\right] \times (0, T]$. We can choose M such that

$w_x(x, 0) \le 0$ for $x \in [n - \frac{1}{2M_{k'}}, n]$ and $w(n - \frac{1}{2M_{k'}}, t) > 0$. From

$$(c_n^\varepsilon)'w_t - w_{xx} + k_n'w_x > 0 \quad ,$$

we have

$$w_x(n, t) = -2MM_{k'} - u_{nx}^\varepsilon(n, t) \le 0 \quad ,$$

that is

$$u_{nx}^\varepsilon(n, t) \ge -2MM_{k'} \quad .$$

The next step is to investigate the equation which the function $v = u_{nx}^\varepsilon - k_n(u_n^\varepsilon)$ satisfies. By maximum principle, we get $\max_{H_T^n}|v| \le \max_{\Gamma_T^n}|v|$ and then $|u_{nx}^\varepsilon| \le M_1$ for some constant $M_1 > 0$ which is independent of n and ε.

In addition to $(u_n^\varepsilon - \varepsilon) \in L^2(0, T; L^2(0, \infty)) \cap L^\infty(0, T; W^{1,\infty}(0, \infty))$ (assume that $u_n^\varepsilon = \varepsilon$ for $x \ge n$), we can prove that there exist constants M_ε and M [2] such that

$$|c_n^\varepsilon(u_n^\varepsilon(x_1, t_1)) - c_n^\varepsilon(u_n^\varepsilon(x_2, t_2))| \le M_\varepsilon(|x_1 - x_2|^2 + |t_1 - t_2|)^{\frac{1}{2}} \quad ,$$

$$|k_n(u_n^\varepsilon(x_1, t_1)) - k_n(u_n^\varepsilon(x_2, t_2))| \le M(|x_1 - x_2|^2 + |t_1 - t_2|)^{\frac{1}{2}} \quad .$$

These facts guarantee that there exist a function $(u^\varepsilon - \varepsilon) \in L^2(0,T; L^2(0, \infty)) \cap L^\infty(0, T; W^{1,\infty}(0, \infty))$ and a subsequence $\{u_{n'}^\varepsilon\}$ such that

$$u_{n'}^\varepsilon \to u^\varepsilon \qquad \text{in} \quad L^2(0, T; L^1_{loc}(0, \infty)) \ \text{weakly} \tag{2.2}$$

$$c_{n'}^\varepsilon(u_{n'}^\varepsilon) \to c^\varepsilon(u^\varepsilon) \qquad \text{in} \quad C^{0+\beta}(H_T) \quad \text{strongly}$$

$$k_{n'}(u_{n'}^\varepsilon) \to k(u^\varepsilon) \tag{2.3}$$

with $\beta \in (0, 1)$. So that u satisfies the following identity

$$\iint_{H_T} (\varphi_x(u_x - k(u^\varepsilon)) - \varphi_t c^\varepsilon(u^\varepsilon))dxdt$$

$$= \int_0^T R\varphi(0,t)dt + \int_0^\infty \varphi(x, 0)c^\varepsilon(u_0^\varepsilon(x))dx$$

when $\varepsilon > \varepsilon'$, we have $u_{0n}^{\varepsilon} \geq u_{0n'}^{\varepsilon'}$. This gives $c_n^{\varepsilon}(u_n^{\varepsilon}) \geq c_n^{\varepsilon'}(u_n^{\varepsilon'})$ and $k_n(u_n^{\varepsilon}) \geq k_n(u_n^{\varepsilon'})$. Recalling (2.2) and (2.3), we conclude that

$$c^{\varepsilon}(u^{\varepsilon}) \geq c^{\varepsilon'}(u^{\varepsilon'}) \quad , \quad k(u^{\varepsilon}) \geq k(u^{\varepsilon'}) \quad .$$

Together with $(u^{\varepsilon} - \varepsilon) \in L^2(0, T; H^1_{loc}(0, \infty))$, we can find a function $u \in L^2_{loc}(0, T; H^1(0, \infty))$ and a subsequence $\varepsilon_k \to 0$ as $k \to \infty$ such that

$$u^{\varepsilon_k} \to u \quad \text{in} \quad L^2(0, T: H^1_{loc}(0, \infty)) \quad \text{weakly} \quad ,$$

$$c^{\varepsilon}(u^{\varepsilon}) \to c(u) \quad , \quad k(u^{\varepsilon}) \to k(u) \quad \text{monotonically} \quad .$$

It is now easy to verify that $c(u) \in C(\bar{H}_T)$, $k(u) \in C^{0+1}(\bar{H}_T)$ and then u is a solution of Problem I.

<u>Theorem 1</u>. Suppose that the functions c, k and v_0 satisfy the hypotheses H1a, H2a, H3a and H4a. Then Problem I has a unique weak solution.

<u>Remark</u>. In fact, ε- and n-limiting process deal with the degenerations at $u = 0$ and $u = 1$ respectively and independently. If we consider $u\big|_{x=\frac{1}{\varepsilon}} = \varepsilon$ instead of $u|_{x=n} = \varepsilon$, the order of two limits can be changed. Let $\varepsilon_k \to 0$, we obtain $u_n^{\varepsilon_k} \to u_n$ weakly in L^2, $c_n^{\varepsilon_k}(u_n^{\varepsilon_k}) \to c_n(u_n)$ and $k_n(u_n^{\varepsilon_k}) - k_n(u_n)$ monotonically where the monotonicity holds for sufficiently large $\frac{1}{\varepsilon_k}$, while $c_n(u_n) \in C(\bar{H}_T) \cap W^{0,\infty}(\bar{H}_T)$, $k_n(u_n) \in C^{0+1}(\bar{H}_T)$. Next let $n' \to \infty$, then $u_{n'} \to \bar{u}$ in $L^2(0, T; H^1_{loc}(0, \infty))$ weakly, $c_{n'}(u_{n'}) \to c(\bar{u})$ in $L^2_{loc}(H_T)$ weakly and $k_{n'}(u_{n'}) \to k(\bar{u})$ in $C^{0+\beta}(\bar{H}_T)$ strongly with $\beta \in (0, 1)$. Using a similar argument as that in proving the uniqueness theorem, we may show that $u = \bar{u}$ a.e. in H_T and $c(u)$ and $k(u)$ are continuous.

3. General Properties of the Solution

We shall prove in this section some regularity results and an important property that $c(u)$ has compact support if and only if $\int_{0^+} \frac{ds}{c(s)} < \infty$. Moreover, we have an estimation on $\left(\int_0^u \frac{ds}{c(s)}\right)_x$.

Theorem 2. Suppose the hypotheses H1a-b, H2a, H3a-b and H4a are satisfied. Let u be the weak solution of Problem I. Then there exists a function $u^* : \bar{H}_T \to R^+$ such that $u^* = u$ a.e. in H_T and

$$u^*(x, t_1) - u^*(x, t_2) \geq -\chi(t_1 - t_2)$$

for all $(x, t_1), (x, t_2) \in H_T$ and $0 \leq t_2 \leq t_1 < T$.

Consider the equation and the initial-boundary conditions which a function $q = u^{\varepsilon}_{nt} e^{\lambda t}$ satisfies for large λ. Using the hypothesis H3b and maximum principle, we can obtain the conclusion of Theorem 3 [2].

Theorem 3. Let u be the weak solution of Problem I in which the data satisfies H1a, H2a, H3a, H3b with $\chi = 0$ and H4a. Then $\bar{u} \in L^2(0, T; H^2(0, \infty)) \cap L^\infty(0, T; W^{1,\infty}(0, \infty))$, where $\bar{u} = \max\{u, \sigma\}$, $(0 < \sigma < 1)$. And if $c(s)$ and $k(s)$ are twice continuously differentiable in $(0, 1)$, then the weak solution of Problem I is a classical solution of Eq. (1.3) in the region

$$D = \{(x, t) | 0 < c(u(x, t)) < 1\} \quad .$$

The proof of Theorem 2 can be found in [6].

Theorem 4. Suppose that the hypotheses H1a, H2a, H3a and H4a-b are satisfied and c is differentiable on $(0, 1)$. Let u be the weak solution of Problem I. Then $c(u(x, t))$ has compact support for any $t > 0$ if and only if

$$\int_{0^+} \frac{ds}{c(s)} < \infty \quad .$$

Proof. Constructing a travelling wave solution of Problem I

$$\begin{cases} x - \lambda_0 t - \alpha = \displaystyle\int_0^{v(x, t)} \frac{ds}{k(s) - \lambda_0 c(s)} & \text{for } x - \lambda_0 t - \alpha < 0 \\ v(x, t) = 0 \quad \text{for } x - \lambda_0 t - \alpha \geq 0 \end{cases} ,$$

we choose sufficiently large α such that

$$v(x, 0) \geq u_0(x) \quad \text{and} \quad v(0, t) > 1 \quad .$$

In virtue of the comparison theorem for weak solution [2], we have

$$c(v(x, t)) \geq c(u(x, t)) \quad \text{for} \quad (x, t) \in H_T \quad .$$

Hence $\int_{0^+} \frac{ds}{c(s)} < \infty$ is a sufficient condition for $c(u)$ having compact for any $t > 0$.

Next, we prove the necessity. It is enough to prove that, if $\int_{0^+} \frac{ds}{c(s)} = \infty$, then, for any $t > 0$, either (1) $c(u(x, t)) > 0$ for all $x \in [0, \infty)$ or (2) $c(u(x, t)) = 0$ for all $x \in [0, \infty)$.

Suppose that there exists $t_1 \in (0, T]$ and points $x', x'' \in (0, \infty)$ such that

$$u(x', t_1) > 0 \quad \text{and} \quad u(x'', t_1) = 0 \quad .$$

In virtue of the remark of Theorem 1, we have, for large $n \geq N$

$$u_n(x', t_1) > 0 \quad .$$

Without loss of generality, we take $x' = 0$, $t_1 = T$ and $x'' > 0$. Let $\{u_n^{\varepsilon_k}\}$ be a sequence of classical solutions from which u_n was constructed in the remark of Theorem 1. We also note, since $\{u_n^{\varepsilon_k}\}$ is decreasing

$$u_n^{\varepsilon_k}(0, t) \geq \mu$$

for all $t \in [0, T]$, all k and $n \geq N$.

It can be done that $c_n(s) = c(s)$ for $s \in [0, \frac{1}{2})$ when $c(s)$ is differentiable on $(0, 1)$. So we have $\int_{0^+} \frac{ds}{c_n(s)} = \infty$. Making the transformation $s = c_n^{-1}(t)$, we obtain $\int_{0^+} \frac{dt}{tc_n'(c_n^{-1}(t))} = \infty$. This implies that the problem

$$\begin{cases} \left(\frac{1}{c_n'(c_n^{-1}(f))} f'\right)' + \frac{1}{2}\eta f' = 0 \quad , \quad 0 < \eta < \infty \\ f(0) = \frac{1}{2}\mu \quad , \quad f(\infty) = 0 \end{cases}$$

has a unique positive solution $f(\eta)$ on $[0, \infty)$ [15]. Since $f'(\eta) < 0$ for all $\eta \in [0, \infty)$, so $0 < f(\eta) < \frac{1}{2}\mu < \frac{1}{2}$ and then $f(\eta)$ is independent of n. By a similar argument as that in [6], we find that

$$c_n^{\varepsilon_k}(u_n^{\varepsilon_k}(x, t)) \geq f((x + \nu t)(t + \tau_k)^{-\frac{1}{2}}) \quad \text{for all} \quad (x, t) \in \bar{H}_T^{\frac{1}{\varepsilon_k}} \quad .$$

Let $k \to \infty$, and using the monotonicity of f, we obtain

$$c_n(u_n(x, t)) \geq f((x + \nu t)t^{-\frac{1}{2}}) \quad \text{for all} \quad (x, t) \in H_T \quad .$$

Let $n \to \infty$, then

$$c(u(x, t)) \geq f((x + \nu t)t^{-\frac{1}{2}}) \quad \text{for all} \quad (x, t) \in H_T \quad .$$

Particularly this means

$$c(u(x'', t_1)) > 0$$

which contradicts the assumption.

From the identity

$$\int_0^\infty c(u(x, t))ds = Rt + \int_0^\infty v_0(x)dx \quad ,$$

it follows that $\int_{0^+} \frac{ds}{c(s)} < \infty$ is a necessary condition of the weak solution having compact support for any $t > 0$ when u_0 has compact support.

Let $v = \int_0^u \frac{ds}{c(s)}$. In the following theorem, we shall prove that v_x is uniformly bound, which will be used to prove the Lipschitz continuity of the dry-wet interface $\xi(t)$.

Theorem 5. Suppose that c, k and v_0 satisfy the hypotheses H1a-c, H2a-b, H3a and H4a-b and $u_0(x) > \delta$ for $x \in [0, \delta)$, $\int_{0^+} \frac{ds}{c(s)} < \infty$. Then there exists a constant $M > 0$ such that

$$|v_x| \le M \quad \text{for} \quad (x, t) \in [0, \infty) \times (\tau, T] \quad .$$

Moreover, if $\int_0^{u_0(x)} \frac{ds}{c(s)}$ is Lipschitz continuous, then

$$|v_x| \le M \quad \text{in} \quad H_T \quad .$$

We use the Bernstein technique. Let $w_n = \int_0^{u_n} \frac{ds}{c_n(s)\theta_n(s)}$, where

$$\theta_n(u_n) = \bar{M} + \frac{u_n}{c_n(u_n)} - \int_0^{u_n} \frac{ds}{c_n(s)} \quad .$$

Consider the function $z = \zeta^2(x, t) w_{nt}^2(x, t)$ in which $\zeta \in C^2([a,b] \times [0,T])$ is a cut-off function. From the condition z should satisfy at the point of its maximal value, the equation which $q = w_{nx}$ satisfies and the hypothesis H1c, we can deduce

$$|v_{nx}| \le M \quad \text{on} \quad [a_1, b_1] \times (\tau, T] \cap \{u_n(x, t) \le \delta\} \quad ,$$

where $a < a_1 < b_1 < b$. When $u_n > \delta$ at the positive maximum of z in $[a_1, b_1] \times (\tau, T]$, we get directly

$$|v_{nx}| = \left|\frac{1}{c_n(u_n)} u_{nx}\right| \le \left|\frac{1}{c_n(\delta)} u_{nx}\right| \le M \quad .$$

The condition $u_0(x) > \delta$ for $x \in (0, \delta)$ gives $|v_{nx}| \le M$ at the neighborhood of $x = 0^+$. So we have

$$|v_x| \leq M \quad \text{on} \quad [0, \infty) \times [\tau, T) \quad .$$

4. Interface

In previous section, we have introduced the unsaturated and wet region D. Now we define the saturated region

$$P = \{(x, t) \in H_T | c(u(x, t)) = 1\}$$

and the dry region

$$J = \{(x, t) \in H_T | c(u(x, t)) = 0\} \quad .$$

We shall discuss the interface between saturated and unsaturated regions and the interface between dry and wet regions. The behavior of interfaces could be very complicated for a general initial data. Thus, we impose the hypotheses H1d, H3c and H4c. Under the conditions H1d, H3a, c and H4c, it can be proved that there exist unique dry-wet interface $\xi(t)$ and unique saturated-unsaturated interface $\zeta(t)$ (as in [2, 17]).

In this section, we prove some regularity properties of the interfaces and the behavior of weak solution in the saturated region.

Theorem 6. Suppose that the hypotheses H1a-d, H2a-b, H3a, c and H4a-c are satisfied and that $\int_0^{u_0(x)} \frac{ds}{c(s)}$ is Lipschitz continuous. Then $\xi(t)$ is nondecreasingly Lipschitz continuous.

From $|v_x| \leq M$ on H_T of Theorem 5, the conclusion would be obtained as that in [16]. The proof has little change because we consider u_n^ε instead of u.

Lemma 3. Suppose that the hypothesis H3d holds. Then the function

$$v = u_x - k(u) + \lambda_0 c(u) \leq 0 \qquad \text{a.e. in } D \quad .$$

This conclusion follows from the maximum principle for the equation which $v_n^\varepsilon = u_n^\varepsilon - k_n(u_n^\varepsilon) + \lambda_0 c_n^\varepsilon(u_n^\varepsilon)$ satisfies.

Theorem 7. Suppose that the hypotheses H1a, H2a, H3a,c and H4a-c are satisfied. If H3b with $\chi = 0$ holds, then $\zeta(t)$ is an increasingly continuous function. Or, if H3d holds, then $\zeta(t)$ is a continuous function.

Proof. If H3b with $\chi = 0$ holds, from Theorem 3, $\zeta(t)$ is increasing and $\zeta(t)$ has only discontinuous points of the first kind. Set

$$\zeta(t_0^-) = \lim_{t \nearrow t_0^-} \zeta(t) \quad , \quad \zeta(t_0^+) = \lim_{t \searrow t_0^+} \zeta(t)$$

for any $t_0 \in (0, T)$.

From the continuity of $c(u(x, t))$, we have

$$c(u(\zeta(t), t)) = c(u(\zeta(t_0^+), t_0)) = c(u(\zeta(t_0^-), t_0)) = 1 \quad . \tag{4.1}$$

By the monotonicity of ζ and the uniqueness of the interface, we have

$$\zeta(t_0^-) \le \zeta(t_0) \le \zeta(t_0^+) \quad , \quad \zeta(t_0) = \zeta(t_0^+) \quad .$$

If $\zeta(t_0^-) < \zeta(t_0)$, then there would exist a $\alpha > 0$ and $b > \zeta(t_0)$ such that

$$1_\alpha = \{(x, t) \mid t = t_0 - \alpha m(x - \zeta(t_0^-)),\ x \le b\} \subset D \quad .$$

Consider the function $w = (2 - \exp(m(t - t_0 + \alpha(x - \zeta(t_0^-)))))$ which satisfies

$$c'w_t - w_{xx} + k'w_x > 0 \quad \text{for large} \quad m > 0 \quad .$$

Since $w = 1$ along 1_α and $w < 1$ on $\{x = b\} \cap \{t_0 - m(x - \zeta(t_0^-)) < t \le t_0\}$, by the maximum principle,

$$w(x,t) \ge u(x,t) \quad \text{on} \quad S_\alpha = \{(x,t) \mid t_0 \ge t \ge t_0 - \alpha m(x - \zeta(t_0^-)),\ \zeta(t_0^-) \le x \le b\}$$

That is

$$u(x,\ t) \le (2 - e^{m(t-t_0) + m(x - (t_0^-))}) \quad \text{on} \quad S_\alpha \quad .$$

In particular,

$$u(\zeta(t_0),\ t_0) \le (2 - e^{\alpha m(\zeta(t_0) - \zeta(t_0^-)}) < 1$$

which contradicts (4.1). Hence

$$\zeta(t_0^-) = \zeta(t_0) = \zeta(t_0^+) \quad \text{for} \quad t_0 \in (0,\ T) \quad .$$

If H3d holds, it follows from Lemma 3 that

$$u_x(x,\ t) < -\sigma \qquad \text{for some} \quad \sigma > 0$$

in the right neighborhood of the interface $\zeta(t)$. Thus as that in [2, 6], we are able to prove that $\zeta(t)$ is continuous.

Theorem 8. Suppose that the hypotheses H1a-d, H2a-b, H3a, c-d, H4a-c, $\int_{0^+} \frac{ds}{c(s)} < \infty$ and H3b with $\chi = 0$ are satisfied. Then $\zeta(t)$ is increasingly Lipschitz continuous.

Proof. From Lemma 2 of Section 2,

$$-M \le u_x \le M \qquad \text{for all} \quad (x,\ t) \in H_T \quad .$$

Hence

$$0 \le u(x,\ t_0) - 1 \le M(\zeta(t_0) - x) \qquad \text{for} \quad x \in (0,\ \zeta(t_0)) \quad .$$

We take $t_0 = 0$ and $\zeta(t_0) = b$ for simplicity. Consider a travelling wave solution

$$\begin{cases} x - \lambda_1 t - b = \displaystyle\int_1^{U_1} \frac{ds}{k(s) - \lambda_1 c(s)} & x - \lambda_1 t - b < \displaystyle\int_0^1 \frac{ds}{\lambda_1 c(s) - k(s)} \\ U_1(x,\ t) = 0 & x - \lambda_1 t - b \ge \displaystyle\int_0^1 \frac{ds}{\lambda_1 c(s) - k(s)} \end{cases} .$$

It can be found that

$$u_0(x) \le U_1(x, 0) \quad \text{for} \quad x \in (0, b)$$

for large $\lambda_1 \ge \lambda_0$. When $u_0(x)$ satisfies H3d, by Lemma 3,

$$u_x(x, t_0) - k(u(x, t_0)) + \lambda_0 c(u(x, t_0)) \le 0$$

$$\text{for} \quad x \in (\zeta(t_0) \quad , \quad \xi(t_0)) \quad .$$

It is not difficult to obtain that the travelling wave solution

$$\begin{cases} x - \lambda_2 t - b = \int_1^{U_2} \frac{ds}{K(s) - \lambda_2 c(s)} & x - \lambda_2 t - b < \int_0^1 \frac{ds}{\lambda_2 c(s) - K(s)} \\ U_2(x, t) = 0 & x - \lambda_2 t - b \ge \int_0^1 \frac{ds}{\lambda_2 c(s) - K(s)} \end{cases}$$

with $\lambda_2 < \lambda_0$ satisfies

$$u_0(x) \le U_2(x, 0) \quad \text{for} \quad x \in (b, \infty) \quad .$$

Consider the weak solution $\bar{U}$ of the problem

$$\begin{cases} c(u)_t = u_{xx} - k(u)_x \\ (u_x - k(u))|_{x=0} = -\lambda_1 \theta_s \\ u(x, 0) = \max\{U_1(x, 0), U_2(x, 0)\} \end{cases} \quad .$$

$U(x, t) = \max\{U_1(x, t), U_2(x, t)\}$ is also a solution of the problem above. By the uniqueness

$$\bar{U}(x, t) = \max\{U_1(x, t), U_2(x, t)\} \quad \text{a.e. in} \quad H_T \quad .$$

From the comparison principle

$$c(u(x, t) \le c(U(x, t)) \quad \text{a.e. in} \quad H_T \quad .$$

Since $\lambda_1 > \lambda_2$, we have

$$c(U(x, t)) < 1 \quad \text{for} \quad x > \lambda_1 t + b \quad \text{and} \quad t \geq 0 \quad .$$

Therefore

$$c(u(x, t)) < 1 \quad \text{for} \quad x > \lambda_1 t + b \quad \text{and} \quad t \geq 0 \quad .$$

Thus $\zeta(t) \leq \lambda_1 t + b$, that is

$$0 \leq \zeta(t) - \zeta(t_0) \leq \lambda_1 (t - t_0) \quad \text{for all} \quad t > t_0 \geq 0 \quad .$$

Theorem 9. Suppose that H1a, H2a, H3a, c, H3b with $\chi = 0$ and H4a-c hold and that $\zeta(t)$ is continuous. Let u be the weak solution of Problem I. Then

$$u(x, t) = (k_s - R) \cdot (x - \zeta(t)) + 1 \qquad 0 \leq x \leq \zeta(t) \tag{4.2}$$

almost everywhere on $(0, T)$. Moreover, if c is twice differentiable and H3d holds, then (4.2) holds for all $t \in (0, T)$.

It is easy to prove this theorem by using the facts that $\max\{u, \frac{1}{2}\} \in L^2(0, T; H^2(0, \infty))$ and the definition of the solution, see [2].

5. Asymptotic Behavior for Large Time

In this section, we consider the solution of Problem I in the region $H = \bigcup_{T>0} H_T$. We shall prove that the solution of Problem I tends to a travelling wave solution as $t \to \infty$.

Here we cite directly the definitions and properties established by Huang [12] and give a sketch of the process of the proof.

First, we make a hodograph transformation $P(u, t) = -u_x(x, t)$ and then a family of curves $u(x, \alpha)$ on (x, u)-plane corresponds to a trajectory $P = P(u)$ in the hodograph (u, p)-plane. Consider a travelling wave solution $U_n(x - \lambda t - \alpha)$

$$x - \lambda t - \alpha = \int_0^{U_n} \frac{ds}{k_n(s) - \lambda c_n(s) + \delta} \tag{5.1}$$

which corresponds to a stationary trajectory $P_n^{\lambda}(u)$ in the hodograph (u, p)-plane.

Secondly, we prove that $P_n^{\lambda_0}(u)$ is a limit trajectory. For any given $\varepsilon > 0$, we choose the following functions:

1). Shallow right-comb function $U_n^-(x - \lambda t - \alpha)$.

Take $\lambda_- < \lambda_0$, $\delta > 0$ in (5.1) such that the corresponding $P_n^-(u)$ has a zero C_1 with $0 < C_1 < \varepsilon/2$ and

$$0 < P_n^0(u) - P_n^-(u) < \varepsilon \qquad \text{for} \qquad u \in [\varepsilon, M_0] \quad .$$

Let

$$x - \lambda_- t - \alpha = \int_{\varepsilon}^{U_n^-} \frac{ds}{k_n(s) - \lambda_- c_n(s) + \delta}$$

where $C > C_1$ such that $U_n^-(x - \alpha) > u_{0n}(x)$ for all $\alpha \geq 0$. The functions $U_n^-(x - \lambda_- t - \alpha)$ have the property that the number of intersections of $U_n^-(x - \lambda_- t - \alpha)$ and $u_n(x, t)$ cannot be more than one and $U_{nx}^- > u_{nx}$ at the intersection point.

2). Underfunction $U_n^=$ associated to U_n^-

Define $U_n^=$ as follows

$$\begin{cases} x - \lambda_= t - \alpha = \displaystyle\int_0^{U_n^=} \frac{ds}{k_n(s) - \lambda_= c_n(s)} & x - \lambda_= t - \alpha < 0 \\ U_n^=(x, t) = 0 & x - \lambda_= t - \alpha \geq 0 \end{cases}$$

where $\lambda_= \in (\lambda_-, \lambda_0)$. It is easy to check

$$U_n^=(x - \lambda_= t) \leq u_n(x, t) \qquad \text{for all} \qquad (x, t) \in H_T \quad .$$

3). Steep left-comb function U_n^+

Take $\lambda_+ > \lambda_0$ in (5.1) such that the corresponding $P_n^+(u)$ satisfies

$$0 < P_n^+(u) - P_n^0(u) < \varepsilon \qquad \text{for} \qquad u \in [\varepsilon, M_0].$$

Define

$$\begin{cases} x - \lambda_+ t + \alpha = \displaystyle\int_0^{U_n^+} \frac{ds}{K_n(s) - \lambda_+ c_n(s)} & x - \lambda_+ t + \alpha < 0 \\ U_n^+(x, t) = 0 & x - \lambda_+ t + \alpha \geq 0 \end{cases}$$

for all $\alpha \geq 0$. The function $U_n^+(x - \lambda_+ t + \alpha)$ has the property that the number of intersections of $U_n^+(x - \lambda_+ t + \alpha)$ and $u_n(x, t)$ cannot be more than one and $U_{nx}^+ < u_{nx}$ at the intersection point.

4). Overfunction U_n^{++} associated to U_n^+

Define U_n^{++} as follows

$$x - \lambda_{++} t = \int_{d_2}^{U_n^{++}} \frac{ds}{k_n(s) - \lambda_{++} c_n(s) + \delta}$$

where $\lambda_{++} \in (\lambda_0, \lambda_+)$. Choosing δ such that the corresponding $P_n^{++}(u)$ has a zero $d_1 \in (0, \varepsilon/2)$ and d_2 such that $u_{0n}(x) \leq U_n^{++}(x, 0)$ and so $u_n(x, t) \leq U_n^{++}(x - \lambda_{++} t)$.

Using these functions and the comb-region and the argument which is easier than that in [12], we get the following conclusion.

<u>Theorem 10</u>. For any $\varepsilon > 0$, there exists $T_1 > 0$ such that

$$|P(u, t) - P^0(u)| < \varepsilon$$

for all $u \in [\varepsilon, M_0]$ and $t > T_1$.

Finally, we prove that $u(x, t) \to U_0(x - \lambda_0 t - \alpha)$ as $t \to \infty$ uniformly for some α. According to the definition of U_0, for any $\varepsilon > 0$, there exists α_t and $\alpha > 0$ which is independent of t such that

$$U_0(\zeta(t) - \lambda_0 t - \alpha_t) = 1 \quad \text{and} \quad U_0(\zeta(t) - \lambda_0 t - \alpha_t + \alpha) = \frac{3}{4}\varepsilon \quad .$$

By theorem 10, we can find a $\eta \in (0, \min_{\frac{\varepsilon}{2} < u < 1} P_0(u))$ such that

$$|P(u, T) - P_0(u)| < \eta \quad \text{for} \quad u \in [\eta, 1] \quad \text{and} \quad T > T_1 \quad .$$

And then

$$u(x, T) - U_0(x - \lambda_0 T - \alpha_T) < \varepsilon/4 \quad \text{for} \quad x \in [\zeta(T), \zeta(T) + \alpha] \quad . \tag{5.2}$$

By Theorem 9,

$$u(x, t) = U_0(x - \lambda_0 T - \alpha_T) \quad \text{for} \quad x \in [0, \zeta(T)] \quad .$$

By (5.2),

$$u(x_2, T) \leq U_0(x_2 - \lambda_0 T - \alpha_T) + \frac{\varepsilon}{4} = \varepsilon$$

where $x_2 = \zeta(T) + \alpha$. On the other hand, for any x satisfying $u(x, T) = \varepsilon$,

$$u_x(x, T) = -P(u, T) < \eta - P_n(u) = \eta - P_0(\varepsilon) < 0 \quad .$$

So for $x \geq x_2$, we have $0 \leq u(x, T) \leq \varepsilon$. Together with $0 \leq U_0(x - \lambda_0 T - \alpha_T) \leq \frac{3}{4}\varepsilon$ for $x \geq x_2$, we have

$$|u(x, T) - U_0(x - \lambda_0 T - \alpha_T)| < \varepsilon \quad \text{for all} \quad x \in R^+ \quad .$$

We proceed with a similar procedure as in Section 2. For any $t_1 > T$ we consider two sequences of classical solutions $\{u_n^\varepsilon\}$ and $\{v_n^\varepsilon\}$ in the region $(0, n) \times (T, t_1)$ which converge u and U_0, respectively, in $L^2(T, t_1; H^1_{loc}(0, \infty))$. By the maximum principle,

$$|u_n^\varepsilon(x, t) - v_n^\varepsilon(x, t)| \leq |u_n^\varepsilon(x, T) - v_n^\varepsilon(x, T)| \quad \text{for} \quad t > T \quad .$$

Hence

$$|u(x, t) - U_0(x - \lambda_0 t - \alpha_T)| \le M|u(x, T) - U_0(x - \lambda_0 T - \alpha_T)|$$

for $t > T$.

where M is independent of t. Then using the same argument as that in [12], we obtain the result.

Theorem 11. Suppose that the hypotheses H1a, H2a, H3a, c-d and H4a-c are satisfied. Let u be the weak solution of Problem I in the region $H = (0, \infty) \times (0, \infty)$. Then there exists constant α such that

$$\lim_{t \to \infty} \max_{0 \le x < \infty} |u(x, t) - U_0(x - \lambda_0 t - \alpha)| = 0 \quad .$$

The author gratefully acknowledges the guidance and help of Prof. Xiao Shutie during the writing of this paper.

References

[1] Van Duyn, C.J., "Nonstationary filtration in partialed media: Continuity of the free boundary", Arch. Rat. Mech. Anal., 79 (3), 261-265 (1982).

[2] Van Duyn, C.J. and Peletier, L.A., "Nonstationary filtration in partially saturated porous media", Arch. Rat. Mech. Anal., 78 (3), 173-198 (1982).

[3] An Lianjun, "Some properties of the solution of a filtration problem in partially saturated porous media", Acta Math. Appl. Sinica, 1 (1), 44-56 (1984).

[4] Zhou Chuanzhong, The Cauchy problem for an equation of partially saturated filtration, The dissertation of MA in Peking University (1983).

[5] Li Jianguo and Su Ning, The solution of an infiltration problem with a bonded surface flux condition, (to appear) (1984).

[6] Su Ning and Li Jianguo, "The boundary value problems for the filtration equation in partially saturated porous media", Acta Math. Appl. Sinica 1 (2), 180-192 (1984).

[7] Kröner, D., "Parabolic regularization and behaviour of the free boundary for unsaturated flow in a porous medium", Journal fur die reine und angewandte Mathmatik 348, 180-196 (1984).

[8] Bertshch, M. and Peletier, L.A., "Nonstationary filtration in partially saturated media, in: Analytical and numerical approaches to asymptotic problems in analysis", North-Holland Math. Studies 47, 205-212 (1980).

[9] Hurshof, J. and Peletier, L.A., An elliptic-parabolic free boundary problem, Report of Leiden University. The Netherlands (1984).

[10] Xiao Shutie, Huang Zhida and Zhou Chuanzhong, "The infiltration problem with constant rate in partially saturated porous media", Acta Math. Appl. Sinica, 1 (2), 108-126 (1984).

[11] Stokes, A.N., "Intersection of solution of nonlinear parabolic equations", J. Math. Anal. Appl., 60 (3), 721-727 (1977).

[12] Huang Zhida, "Asymptotic behavior of the generalized solution of infiltration problem with constant surface flux", Acta Math. Sinica, 26 (6), 676-698 (1983).

[13] Ladyzhenskaja, D.A., Solonnikov, V.A. and Ural'cera, N.N., Linear and quasilinear equations of parabolic type, Translations of Mathematical Society, Providence, R.I. (1968).

[14] Fokina, T.I., On a boundary problem for parabolic equations with strong nonlinearties, Vect. Moscow University, Math. Mech. 2, 22-28 (1975) (in Russian).

[15] Atkinson, F.V. and Peletier, L.A., "Similarity solutions of the nonlinear diffusion equation", Arch. Rat. Mech. Anal. 54, 373-392 (1974).

[16] Gilding, B.H., "A nonlinear degenerate parabolic equation, Annali della Seuola Normale Superiore di Pisa.", Serie IV, Vol. IV, n. 3, 393-432 (1977).

[17] Gilding, B.H. and Peletier, L.A., "The Cauchy problem for an equation in the theory of infiltration", Arch. Rat. Mech. Anal., 65 (3), 203-225 (1977).

AN ISOSPECTRAL CLASS FOR A GENERALIZED HILL'S EQUATION

Cao Cewen
Zhengzhou University

A classical result concerning the KdV equation states (see [1]): if $q(x,t)$ is a solution to

$$q_t - 6qq_x + q_{xxx} = 0 \quad , \qquad q(x+T, t) = q(x, t) \quad ,$$

then the periodic eigenvalues of the Hill's equation:

$$(-\partial^2 + q(x, t))\psi = \lambda\psi \quad , \qquad (\partial = \partial/\partial x)$$

are conserved integrals: $\lambda_t = 0$. In the present paper we are going to investigate a generalized Hill's equation:

$$L\psi \equiv (-\partial p \partial + q)\psi = \lambda + \psi \quad , \qquad (p > 0,\ r > 0) \quad , \tag{0.1}$$

where $p, q, r \in C_T^\infty$, which is the set of real-valued smooth functions with period T in regard to x. A class of nonlinear partial differential equations with isospectral properties is discovered, with some important equations like Harry-Dym's equation as its special cases.

The inner product in $L^2(0, T)$ is denoted by $(\cdot, \cdot)$, and the symbol a is used to denote the multiplicative operator multiplied by the function $a(x, t)$.

1. PRELIMINARIES

Our starting point is the following differential relations, which can be verified directly:

Proposition 1.1 Let ψ satisfy (0.1), then ψ^2, ψ_x^2 satisfy the following linear ordinary differential equations respectively:

$$(\partial p \partial p \partial)\psi^2 = 2(p(q-\lambda r)\partial + \partial p(q-\lambda r))\psi^2 \quad ,$$

$$(p^2\partial \frac{1}{q-\lambda r} \partial \frac{1}{q-\lambda r} \partial p^2)\psi_x^2 = 2(\frac{p^3}{q-\lambda r}\partial + \partial \frac{p^3}{q-\lambda r})\psi_x^2 \quad .$$

Corollary 1. If ψ satisfies (0.1), then

$$A\psi^2 = -4\lambda J\psi^2 \quad , \tag{1.1}$$

where A, J are

$$\left.\begin{aligned} A &= \partial p \partial p \partial - 2(pq\partial + \partial pq) \quad , \\ J &= \frac{1}{2}(pr\partial + \partial pr) = \sqrt{pr}\,\partial\,\sqrt{pr} \quad . \end{aligned}\right\} \tag{1.2}$$

Corollary 2. If $q \equiv 0$, and ψ satisfies (0.1), then

$$B\psi_x^2 = -4\lambda I\psi_x^2 \quad , \tag{1.3}$$

where B, I are

$$\left.\begin{aligned} B &= p^2\partial \frac{1}{r} \partial \frac{1}{r} \partial p^2 \quad , \\ I &= \frac{1}{2}(\frac{p^3}{r}\partial + \partial \frac{p^3}{r}) = \frac{p^2}{\sqrt{pr}}\,\partial\,\frac{p^2}{\sqrt{pr}} \end{aligned}\right\} \tag{1.4}$$

Proposition 1.2 Suppose $p, q, r \in C_T^\infty$, then the gradients of the periodic eigenvalue λ of (0.1) with regard to p, q, r are

$$\frac{\delta\lambda}{\delta p} = \psi_x^2 \quad , \quad \frac{\delta\lambda}{\delta q} = \psi^2 \quad , \quad \frac{\delta\lambda}{\delta r} = -\lambda\psi^2 \quad ,$$

ψ being the corresponding eigenfunction.

Proof. Let $p \to p + \varepsilon v$, $\forall v \in C_T^\infty$. Denote $' = \left.\frac{d}{d\varepsilon}\right|_{\varepsilon=0}$. Differentiate the Hill's equation with respect to ε and let $\varepsilon = 0$, we have:

$$L\psi' - (v\psi_x)_x = \lambda r\psi' + \lambda' r\psi \quad ,$$

$$(L\psi', \psi) - ((v\psi_x)_x, \psi) = (\lambda r\psi', \psi) + \lambda' \quad .$$

Since $(L\psi', \psi) = (\psi', L\psi) = (\psi', \lambda r\psi)$, so

$$\lambda' = -((v\psi_x)_x, \psi) = (v\psi_x, \psi_x) = (v, \psi_x^2) \quad .$$

This means that $\delta\lambda/\delta p = \psi_x^2$. The other two formulas can be proved similarly.

2. THE ISOSPECTRAL EVOLUTION EQUATIONS OF q AND r

Lemma 2.1 A differential operator with real coefficients is anti-symmetric if and only if it is a linear combination of $p_m(x)\partial^{2m+1} + \partial^{2m+1}p_m(x)$, (see [2]).

Example. A, J, B, I defined by (1.2) and (1.4) are anti-symmetric.

Lemma 2.2 Let M be an anti-symmetric differential operator, then $uMv + vMu$ is a total derivative (TD), in particular, $uMu = TD$.

Proposition 2.1 A sequence of functions G_m can be determined recursively by the following equations:

$$G_1 = \frac{1}{\sqrt{pr}} \quad , \qquad AG_m = JG_{m+1} \quad , \qquad m = 1,2,\ldots \qquad (2.1)$$

Proof. We use the method of induction. Suppose that G_m has already been expressed finitely by p, q, r and their derivatives, then we have

$$\partial\sqrt{pr}\, G_{m+1} = \frac{1}{\sqrt{pr}} AG_m = G_1 \cdot AG_m \quad .$$

Now a question: is the right-hand side a TD? The answer is affirmative. In fact, we have

$$\begin{aligned} G_1 \cdot AG_m &= TD - AG_1 \cdot G_m = TD - JG_2 \cdot G_m \\ &= TD + G_2 \cdot JG_m = TD + G_2 \cdot AG_{m-1} \\ &= \ldots = TD' + G_i \cdot AG_{m+1-i} = TD + G_i \cdot J_{m+2-i} \\ &= \begin{cases} TD + G_n \cdot AG_n \quad , & m = 2n+1 \quad ; \\ TD + G_{n+1} \cdot JG_{n+1} \quad , & m = 2n \quad ; \end{cases} \\ &= TD \quad . \end{aligned}$$

Examples. $G_1 = \frac{1}{\sqrt{pr}}$

$$G_2 = \frac{1}{r}\, \partial p \partial \frac{1}{\sqrt{pr}} - \frac{1}{2\sqrt{pr}} \left(p\partial \frac{1}{\sqrt{pr}}\right)^2 - \frac{2q}{r\sqrt{pr}} + \frac{\text{const}}{\sqrt{pr}} \quad .$$

Remark. It can be shown by induction that $(pr)^{3m+\frac{1}{2}} G_{m+1}$ is a polynomial of p, q, r and their derivatives; and G_m is homogeneous of order $-m+\frac{1}{2}$ with regard to r by properly choosing integral constants.

Proposition 2.2 Let ψ_k be the eigenfunction corresponding to the eignevalue λ_k, then $(m, n = 1,2,\ldots)$:

$$\text{i)} \quad (G_m, JG_n) = 0 \quad ;$$

$$\text{ii)} \quad (\psi_m^2, JG_n) = 0 \quad ;$$

$$\text{iii)} \quad (\psi_m^2, J\psi_n^2) = 0 \quad . \tag{2.2}$$

Proof. i) $G_m \cdot JG_n = TD$. ii) Since $A\psi^2 = -4\lambda J\psi^2$, $(\psi^2, JG_n) = (\psi^2, AG_{n-1}) = -(A\psi^2, G_{n-1}) = 4\lambda(J\psi^2, G_{n-1}) = -4\lambda(\psi^2, JG_{n-1}) = \ldots = (-4\lambda)^{n-1}(\psi^2, JG_1) = 0$. iii) Let ψ, φ correspond to the eigenvalues $\lambda \neq \mu$ respectively, then

$$-4\mu(\psi^2, J\varphi^2) = (\psi^2, A\varphi^2) = -(A\psi^2, \varphi^2)$$
$$= 4\lambda(J\psi^2, \varphi^2) = -4\lambda(\psi^2, J\varphi^2) \quad .$$

So we have $4(\lambda-\mu)(\psi^2, J\varphi^2) = 0$, hence $(\psi^2, J\varphi^2) = 0$.

Proposition 2.3 The periodic eigenvalues of (0.1) are conserved integrals of the following system of evolution equations:

$$p_t = 0 \quad , \quad q_t = JG_k \quad , \quad r_t = JG_\ell \quad .$$

Proof. $\lambda_t = (\frac{\delta\lambda}{\delta p}, p_t) + (\frac{\delta\lambda}{\delta q}, q_t) + (\frac{\delta\lambda}{\delta r}, r_t)$

$$= (\psi^2, JG_k) - \lambda(\psi^2, JG_\ell) = 0 \quad .$$

Example 1. $p = r = 1$, $G_2 = -2q$, $A = \partial^3 - 2(2q\partial + q_x)$, $k = 3$:

$$q_t = AG_2 = 12qq_x - 2q_{xxx} \quad . \quad \text{(KdV equation)}$$

Example 2. $p = 1$, $q = 0$, $G_1 = \frac{1}{\sqrt{r}}$, $A = \partial^3$, $\ell = 2$:

$$r_t = AG_1 = (\frac{1}{\sqrt{r}})_{xxx} \quad . \quad \text{(Harry-Dym's equation)}$$

3. HAMILTONIAN STRUCTURE OF $q_t = JG_k$

Theorem 3.1 There exists a functional F_m such that

$$\frac{\delta F_m}{\delta q} = G_m \quad .$$

What we have to solve is essentially an inverse variational problem. The method used here stems from [1]. First we have:

Lemma 3.1 $G(u)$ is a gradient of a functional if and only if the derivative operator $N(u)$ of $G(u)$ is symmetric. (see [1], p. 356).

So it is sufficient to prove the symmetry of the derivative operator N_m of G_m. In this section, p, r are independent of the time t. Let $q(\varepsilon)$ be a smooth path in C_T^∞. Set $' = d/d\varepsilon$. Then $G_m' = N_m q'$. It is easy to calculate the first two derivatives:

$$N_1 = 0 \quad , \quad N_2 = -\frac{2}{r\sqrt{pr}} \quad .$$

Both are multiplicative operators and thus are symmetric.

Now we shall use the method of induction to prove the symmetric property of N_m. Suppose N_{m-1}, N_m are symmetric. Differentiate the identity $AG_m = JG_{m+1}$ with respect to ε, we obtain:

$$A'G_m + AN_m q' = JN_{m+1} q' \quad .$$

Since:

$$A'G_m = -(4pG_{mx} + 2p_x G_m + 2pG_m \partial)\, q' \overset{\Delta}{=} K_m q' \quad .$$

we have:

$$K_m + AN_m = JN_{m+1} \quad ,$$

and

$$K_{m-1} + AN_{m-1} = JN_m \quad .$$

Multiply the first equation by J on the right, the second by A on the right, then substract the second from the first, we get

$$JN_{m+1}J = (K_m J - K_{m-1}A) - (AN_{m-1}A) + (AN_m J + JN_m A) \quad .$$

The second and the third terms are symmetric since A, J are anti-symmetric. Concerning the first term, we have

Lemma 3.2 $K_m J - K_{m-1} A$ is a symmetric operator.

Proof. Through direct calculations we get

$$(K_m J)^* - (K_m J) = (2pJG_m)\partial + \partial(2pJG_m) \quad ,$$

$$(K_{m-1}A)^* - (K_{m-1}A) = (2pAG_{m-1})\partial + \partial(2pAG_{m-1})$$

where the expressions in the round parentheses on the right are functions, and $(\cdot)$ on the right are multiplicative operators. The Lemma is proved since $JG_m = AG_{m-1}$.

Hence $JN_{m+1}J$ is symmetric on C_T^∞. Define a subset of C_T^∞:

$$E = JC_T^\infty = \{v \in C_T^\infty | v = Ju,\ u \in C_T^\infty\} \quad .$$

Then N_{m+1} is symmetric on E. It can be verified that

$$E = \left\{ v \in C_T^\infty \middle| \int_0^T \frac{v}{\sqrt{pr}}\, dx = 0 \right\} \quad .$$

So E contains arbitrarily many functional-independent elements, e.g. $\sqrt{p(x)r(x)}\cos(2n\pi x/T)$, $n = 1,2,\ldots$. Thus E is infinite-dimensional.

Since G_{m+1} is a polynomial of $q, q_x, \ldots,$ we can define

$$F_{m+1}(q) = \int_0^1 (G_{m+1}(\varepsilon q),\ q)d\varepsilon = \int_0^T P_{m+1}(q,\ q_x, \ldots)dx \quad ,$$

where P_{m+1} is a polynomial of $q, q_x, \ldots$. Hence we have

$$\left.\frac{d}{d\eta}\right|_{\eta=0} F_{m+1}(q + \eta v) = (\hat{G}_{m+1}(q),\ v) \quad , \tag{3.1}$$

and $\hat{G}_{m+1}$ is also a polynomial of $q, q_x, \ldots$. Next we shall omit the subscript $m+1$ for short.

Lemma 3.3 If the derivative operator $N(q)$ of $G(q)$ is symmetric on the underlying space S (S may be E or C_T^∞), then

$$\left.\frac{d}{d\eta}\right|_{\eta=0} F(q + \eta v) = (G(q),\ v) \quad , \quad \forall q,\ v \in S \quad . \tag{3.2}$$

Proof. The left-hand side is equal to

$$\text{LHS} = \left.\frac{d}{d\eta}\right|_{\eta=0} \int_0^1 (G(\varepsilon(q + \eta v)),\ q + \eta v)d\varepsilon$$

$$= \int_0^1 (N(\varepsilon q)\varepsilon v, q)d\varepsilon + \int_0^1 (G(\varepsilon q), v)d\varepsilon$$

$$= \int_0^1 (N(\varepsilon q)q, v)\varepsilon d\varepsilon + \int_0^1 (G(\varepsilon q), v)d\varepsilon \quad .$$

Since

$$\frac{d}{d\varepsilon}\{(G(\varepsilon q), v)\varepsilon\} = (N(\varepsilon q)q, v)\varepsilon + (G(\varepsilon q), v) \quad .$$

So

$$\text{LHS} = (G(\varepsilon q), v)\varepsilon\Big|_{\varepsilon=0}^{\varepsilon=1} = (G(q), v) \quad .$$

Proof of Theorem 3.1. Subtract (3.1) from (3.2) with $S = E$, we have:

$$(G(q) - \hat{G}(q), v) = 0 \quad , \quad \forall q, v \in E \quad .$$

Since v can be expressed as $v = Ju$, $u \in C_T^\infty$, we get

$$(J(G(q) - \hat{G}(q)), u) = 0 \quad , \quad \forall q \in E \quad , \quad u \in C_T^\infty \quad .$$

$$J(G(q) - \hat{G}(q)) = 0 \quad , \quad \forall q \in E \quad . \tag{3.3}$$

The left-hand side is a polynomial of $q, q_x, \ldots$:

$$\Sigma\, a_{k_1 \ldots k_s} q^{(k_1)}(x) \ldots q^{(k_s)}(x) = 0 \quad .$$

This is an ordinary differential equation of algebraic type, its solution manifold is finite dimensional. On the other hand, any element of the infinite-dimensional manifold E is a solution to (3.3), thus it must be an identity; hence $a_{k_1 \ldots k_s}(x) \equiv 0$. Therefore (3.3) is valid for any q in C_T^∞, from which we can solve

$$G(q) = \hat{G}(q) + \frac{\gamma}{\sqrt{pr}} \quad , \quad \gamma = \text{const.} \quad \forall q \in C_T^\infty \quad .$$

It can be easily verified that $G(q)$ is the gradient of the functional $F(q) + (\gamma/\sqrt{pr}, q)$ on the underlying space C_T^∞, so $N(q)$ is symmetric on the space C_T^∞. Let $S = C_T^\infty$ in the Lemma 3.3, we see $G(q)$ is the gradient of $F(q)$.

Theorem 3.2 The evolution equation $q_t = JG_k$ has the Hamiltonian form:

$$q_t = J\frac{\delta F_k}{\delta q} = A\frac{\delta F_{k-1}}{\delta q} \quad ,$$

which possesses two sets of conserved integrals: F_m, λ_n, $m, n = 1,2,\ldots,$ satisfying the following involutive relations:

$$[F_m, F_n] = [\lambda_m, F_n] = [\lambda_m, \lambda_n] = 0 \quad , \tag{3.4}$$

where $[\cdot, \cdot]$ is the Poisson's bracket with respect to q:

$$[F, K] \triangleq \left(\frac{\delta F}{\delta q}, J\frac{\delta K}{\delta q}\right) \quad .$$

Proof. (3.4) are equivalent to (2.2). F_m, λ_n are conserved since

$$\frac{d}{dt}F_m = \left(\frac{\delta F_m}{\delta q}, q_t\right) = (G_m, JG_k) = 0 \quad ;$$

$$\frac{d}{dt}\lambda_n = \left(\frac{\delta \lambda_n}{\delta q}, q_t\right) = (\psi_n^2, JG_k) = 0 \quad .$$

4. HAMILTONIAN STRUCTURE OF $r_t = JG_\ell$

Theorem 4.1 There exists a functional $\tilde{F}_m$ such that

$$\frac{\delta \tilde{F}_m}{\delta r} = G_m \quad .$$

Proof. In view of the Remark to the Proposition 2.1, we have $G_m(\varepsilon r) = \varepsilon^{-m+\frac{1}{2}} G_m(r)$. Differentiating with respect to ε and then letting $\varepsilon = 1$, we get

$$\tilde{N}_m(r)r = (-m + \frac{1}{2})G_m(r) \quad ,$$

where $\tilde{N}_m$ is the derivative operator of G_m with regard to r and is symmetric which can be proved in a similar way as in Section 3. Let

$$\tilde{F}_m(r) = \frac{1}{-m + \frac{3}{2}} (G_m(r), r)$$

then

$$\left.\frac{d}{d\eta}\right|_{\eta=0} \tilde{F}_m(r + \eta v) = \frac{1}{-m + \frac{3}{2}} \{(\tilde{N}_m(r)v, r) + (G_m(r), v)\}$$

$$= \frac{1}{-m + \frac{3}{2}} \{(\tilde{N}_m(r)r, v) + (G_m(r), v)\}$$

$$= (G_m(r), v) \quad .$$

<u>Theorem 4.2</u> The evolution equation $r_t = JG_\ell$ has the Hamiltonian form

$$r_t = J \frac{\delta \tilde{F}_\ell}{\delta r} = A \frac{\delta \tilde{F}_{\ell-1}}{\delta r} \quad ,$$

and possesses two sets of conserved integrals: $\tilde{F}_m$, λ_n, $m, n = 1,2,...$ which satisfy the involutive relations:

$$\{\tilde{F}_m, \tilde{F}_n\} = \{\lambda_m, \tilde{F}_n\} = \{\lambda_m, \lambda_n\} = 0 \quad ,$$

where $\{\cdot, \cdot\}$ is the Poisson's bracket with respect to r:

$$\{F, K\} \stackrel{\Delta}{=} (\frac{\delta F}{\delta r}, J \frac{\delta K}{\delta r}) \quad .$$

<u>Remark</u>. The study of the Hamiltonian form for the isospectral

evolution equation of p is more complicated and is completed recently by Zhang Jinshun as $q(x) \equiv 0$. For $q(x) \not\equiv 0$, the problem is still open.

References

[1] P.D. Lax, SIAM Review, 18 (1976) 351-375.

[2] M.A. Naimark, Linear Differential Operators, Gostechizdat, 1954 (in Russian).

GLOBAL EXISTENCE OF THE SOLUTION OF THE NONLINEAR PARABOLIC EQUATION IN EXTERIOR DOMAINS

Chen Yunmei

Institute of Mathematics
Funda University
Shanghai, China

1. Introduction

In the recent years, a great deal of attention has been paid to the research on the global existence of the solution for nonlinear evolution equations. As far as we are aware the initial boundary value problem in interior domains has been studied in greater details, while that in exterior domains has not received the same amount attention. Recently, Y. Tsutsumi [4] discussed the following semilinear Schrödinger equation

$$(1.1)\qquad \begin{cases} u_t + i\Delta u = i\lambda |u|^p u & (t,x)\in Q \\ u(0,x) = \varphi(x) & x\in\Omega \\ u|_{\Sigma} = 0 & \end{cases}$$

where the domain Ω is the exterior domain of a compact set in R^n with smooth boundary Γ, $Q=(0,\infty)\times\Omega$, $\Sigma=(0,\infty)\times\partial\Omega$. λ is a real constant. He proved that when $n\geq 3$ and p is an even integer with $p \geq 2$, problem (1.1) has a unique global solution for small initial data under a certain assumption on the shape of the domain Ω.

In this paper we consider the following initial boundary value problem for the nonlinear parabolic equation in an exterior domain

$$
(1.2) \qquad \begin{cases} u_t - \Delta u = F(u, D_x u, D_x^2 u) & (t, x) \in Q \\ u(0, x) = \varphi(x) & x \in \Omega \\ u|_{\Sigma} = 0 & \end{cases}
$$

where Ω is the same as above. Our main result is the following

Theorem 1.1; Let $n \geq 3$ and N be an integer with $N \geq 3n + 3[\frac{n}{2}] + 23$. We assume that $F \in C^{2N}(R \times R^n \times R^{n^2})$ satisfying

$$
(1.3) \qquad |F(\lambda)| = O(|\lambda|^2) \qquad \text{near } \lambda = 0
$$

and $\varphi(x) \in H^{2N}(\Omega) \cap W^{2N,1}(\Omega)$. Then there exists a positive constant ε such that if $\varphi(x)$ satisfies

$$
(1.4) \qquad \| \varphi \|_{H^{2N}(\Omega)} + \| \varphi \|_{W^{2N,1}(\Omega)} < \varepsilon
$$

and the compatibility condition, problem (1.2) has a unique global solution $u(t, x)$, satisfying

$$
(1.5) \qquad \begin{cases} \dfrac{\partial^k u}{\partial t^k} \in C([0, +\infty); H^{2(N-k)}(\Omega) \cap H_0^1(\Omega)) \cap L^2(0, +\infty; H^{2(N-k)+1}(\Omega)) & 0 \leq k \leq N-1, \\ \dfrac{\partial^N u}{\partial t^N} \in C([0, +\infty); L^2(\Omega)) \cap L^2(0, +\infty; H_0^1(\Omega)) & \end{cases}
$$

Moreover, we have the following decay estimate

$$(1.6) \qquad \|u(t)\|_{L^{\infty}(\Omega)} = O(t^{-\frac{n}{2}}) \qquad \text{as } t \to +\infty .$$

Remark1.1: The compatibility condition means

$$(1.7) \qquad \left.\frac{\partial^j u}{\partial t^j}\right|_{t=0} \in H_0^1(\Omega) \qquad (0 \le j \le N-1).$$

It has been proved that problem (1.2) has a unique global solution when the space dimension $n \ge 1$ if Ω is a bounded open set in R^n and when $n \ge 3$ if $\Omega = R^n$ ([7], [8]).

Our plan in the present paper is as follows. In section 2 we prove the local existence and uniqueness theorem. In section3 we give the energy estimate. In section 4 we give the local energy decay for the homogeneous linear problem. In section 5 we give the decay estimates for the linear problem. Finally, in section 6 we establish a priori estimate by using the results of section 3 and 5. By the technique of Matsumura and Nishida[3], combining the priori estimate with the local existence theorem, we obtain Theorem1.1.

We give several notations. Let G be an open subset in R^n. For a nonnegative integer m we let $C^m(G)$ denote the space of the real valued functions on G that are m-time continuously differentiable. $C_0^m(G)$ denotes the space of $C^m(G)$ functions with compact support in G. For $1 \le p < \infty$, we let $L^p(G)$ denote the space of measurable functions on G which are p-integrable. Let $\overset{\infty}{L}(G)$ denote the space of bounded measurable functions on G. Let $\overset{m,p}{W}(G)$ denote the Sobolve space of functions all of whose derivatives of order m belong to $\overset{p}{L}(G)$. If p=2, we denote $\overset{m,2}{W}(G)$ by $\overset{m}{H}(G)$. Let $\overset{m}{H_0}(G)$ denote the completion of $\overset{m}{C_0}(G)$ in $\overset{m}{H}(G)$. Let X be a Banach space on G. We denote by $C([0,T];X)$ the space of coutinuous functions on $[0,T]$ valued in X. We denote $\frac{\partial u}{\partial t}$ and $\frac{\partial u}{\partial x_j}$ by u_t and u_{x_j} respectively. $D_x^m u$ represents the vector $\{(\frac{\partial}{\partial x})^\alpha u, |\alpha|=m\}$ where $\alpha=\{\alpha_1,\cdots\cdots,\alpha_n\}$ and $|\alpha|=\sum_{i=1}^n \alpha_i$ .Throughout this paper we denote by C various universal constants.

2. Local Existence and Uniqueness

In this section we shall prove the existence and uniqueness of the local solution for problem (1.2).

From (1.3) we know that there exists a positive constant γ_0, such that if $|\lambda|\leq\gamma_0$, assumption (1.3) hold. By Sobolev imbedding theorem we can choose a positive constant $E_0<\frac{1}{2}$, such that if $f\in \overset{[\frac{n}{2}]+1}{H}(\Omega)$ and satisfies $\|f\|_{H^{[\frac{n}{2}]+1}(\Omega)}\leq E_0$, then we have $\|f\|_{\overset{\infty}{L}(\Omega)}\leq\gamma_0$

Before we give the local existence theorem, we define the following space X_T by

(2.1) $$X_T=\Big\{(V_0(t,x),\cdots\cdots,V_N(t,x))\Big| \frac{\partial^K V_j}{\partial t^K}\in C([0,T];H^{2(N-K-j)}(\Omega)\cap H_0^1(\Omega))\cap L^2(0,T;H^{2(N-K-j)+1}(\Omega)),\ \ 0\leq K\leq N-j-1,\ 0\leq j\leq N-1,\ \ \frac{\partial^{N-j}V_j}{\partial t^{N-j}}\in C([0,T];L^2(\Omega))\cap L^2(0,T;H_0^1(\Omega)),\ \ 0\leq j\leq N\Big\}$$

where is equipped with the norm

(2.2) $$\|V\|_{X_T}^2 = \max_{0\le t\le T} \sum_{0\le j\le N} \sum_{0\le K\le N-j} \left\|\frac{\partial^K V_j}{\partial t^K}(t)\right\|^2_{H^{2(N-K-j)}(\Omega)} + \int_0^T \sum_{0\le j\le N} \sum_{0\le K\le N-j} \left\|\frac{\partial^K V_j}{\partial t^K}(t)\right\|^2_{H^{2(N-K-j)+1}(\Omega)} dt$$

We easily see that X_T is a Banach space.

For any $V=(v_0, v_1, \cdots\cdots, v_N)\in X_T$, set

(2.3) $$\begin{cases} f_0(t,x) = F(u, D_x u, D_x^2 u)\big|_{u=V_0} \\ f_1(t,x) = \left(\frac{\partial}{\partial t}F\right)(u, D_x u, D_x^2 u, u_t, D_x u_t, D_x^2 u_t)\big|_{u=V_0,\, u_t=V_1} \\ \cdots\cdots \\ f_j(t,x) = \left(\frac{\partial^j}{\partial t^j}F\right)(u, D_x u, D_x^2 u, \cdots\cdots; \partial_t^j u, D_x\partial_t^j u, D_x^2\partial_t^j u)\big|_{u=V_0,\cdots\cdots;\ \partial_t^j u = V_j} \\ \cdots\cdots \\ 0\le j\le N \end{cases}$$

and we denote $u_j(x)$ $(0\le j\le N)$ by

(2.4) $$\begin{cases} u_0 = \varphi \\ u_1 = \Delta u_0 + F(u, D_x u, D_x^2 u)\big|_{u=u_0} \\ \cdots\cdots \\ u_j = \Delta u_{j-1} + \left(\frac{\partial^{j-1}}{\partial t^{j-1}}F\right)(u, D_x u, D_x^2 u, \cdots\cdots \partial_t^{j-1}u, D_x\partial_t^{j-1}u, D_x^2\partial_t^{j-1}u)\big|_{u=u_0,\cdots\cdots;\ \partial_t^{j-1}u=u_{j-1}} \\ \cdots\cdots \\ 0\le j\le N \end{cases}$$

It is easily realized that if u is a smooth solution of problem (1.2) we have

(2.5) $$\frac{\partial^j u}{\partial t^j}(t,x)\Big|_{t=0} = u_j(x) \qquad 0\le j\le N.$$

For some positive constant $E\le E_0$, we define space $X_{T,E}$ by

(2.6) $$X_{T,E} = \left\{ (V_0(t,x), \cdots\cdots, V_N(t,x)) \,\middle|\, \begin{array}{l} V\in X_T,\ V_j(0,x)=u_j(x)\ (0\le j\le N), \\ \|V\|_{X_T} < E \end{array} \right\}$$

Theorem2.1: In problem (1.2) we assume that $F\in C^{2N}(R\times R^n\times R^{n^2})$ and satisfies (1.3), $\varphi\in H^{2N}(\Omega)$, where N is an integer with $N\geq[\frac{n}{2}]+2$. Then, there exist two positive constants T and δ (<1) such that for the suitably small constant E ($\leq E_0$), if $\varphi(x)$ satisfies

(2.7) $$\|\varphi\|_{H^{2N}(\Omega)}<\delta E$$

and the compatibility condition, problem (1.2) has a unique local solution u(t,x) satisfying

(2.8) $$U=(u,u_t,\cdots\cdots,\partial_t^N u)\in X_{T,E}.$$

In order to prove Theorem 2.1 we first prove the following lemmas.

Lemma 2.1: Let Ω be an exterior domain of a compact set in R^n and let u be the solution belonging to $H^1(\Omega)$ of the elliptic equation

(2.9) $$\begin{cases}\Delta u=f & x\in\Omega\\ u|_\Gamma=0\end{cases}$$

Let L be an arbitray nonnegative integer. If $f\in H^L(\Omega)$, then we have

(2.10) $$\|u\|_{H^{L+2}(\Omega)}\leq C(\|u\|_{L^2(\Omega)}+\|f\|_{H^L(\Omega)}),$$

where C is a positive constant depending only on n,L and Ω.

The above lemma is well known(see F.E.Browder [1]).

Lemma 2.2: In the problem

(2.11)
$$\begin{cases} u_t - \Delta u = f(t,x) & (t,x)\in Q \\ u(0,x) = \varphi(x) & x\in\Omega \\ u|_{\Sigma} = 0 \end{cases}$$

where Ω is the same as in problem (1.1), we assume that $f\in L^2(0,T;H^{-1}(\Omega))$ and $\varphi\in L^2(\Omega)$. Then, problem (2.11) has a unique solution $u(t,x)\in C([0,T];L^2(\Omega))\cap L^2(0,T;H_0^1(\Omega))$. Moreover, we have the following estimate

(2.12)
$$\max_{0\le t\le T}\|u(t)\|^2_{L^2(\Omega)} + \int_0^T \|u(t)\|^2_{H^1(\Omega)}dt \le Ce^T\left(\|\varphi\|^2_{L^2(\Omega)} + \int_0^T \|f(t)\|^2_{H^{-1}(\Omega)}dt\right).$$

The proof of this lemma can be obtained by Galerkin's method and energy integral.

Next, we prove Theorem 2.1.

Proof of Theorem 2.1: We fix two positive constant T and E ($<E_0$) to be determined later. Let J be the mapping which maps $V\in X_{T,E}$ into the solution $W=(w_1,\cdots\cdots,w_N)$ of the following problem

(2.13)
$$\begin{cases} \dfrac{\partial W_j}{\partial t} - \Delta W_j = f_j(t,x) & (t,x)\in Q_T=(0,T)\times\Omega \\ W_j(0,x) = u_j(x) & x\in\Omega \\ W_j|_{\Sigma_T} = 0 & 0\le j\le N \end{cases}$$

where $f_j(t,x)$ and $u_j(t,x)$ are determined by (2.3) and (2.4), respectively. We shall show that if δ , T and E are sufficiently small J is a contraction mapping of $X_{T,E}$ to itself.

Noticing $E\le E_0<\frac{1}{2}$ and $N\ge[\frac{n}{2}]+2$, using (1.3) and the estimate of the norms for the composite function and product of functions(see [8]), for any $V\in X_{T,E}$ we have

(2.14)
$$\begin{cases} \frac{\partial^k f_j}{\partial t^k} \in C([0,T]; H^{2(N-k-j-1)}(\Omega)) \cap L^2(0,T; H^{2(N-k-j)-1}(\Omega)), & 0\le k\le N-j-1,\ 0\le j\le N-1, \\ \frac{\partial^{N-j} f_j}{\partial t^{N-j}} \in L^2(0,T; H^{-1}(\Omega)) & 0\le j\le N \end{cases}$$

and

(2.15)
$$\max_{0\le t\le T} \sum_{j=0}^{N-1}\sum_{k=0}^{N-j-1} \| \frac{\partial^k f_j}{\partial t^k}(t)\|^2_{H^{2(N-k-j-1)}(\Omega)} + \int_0^T \sum_{j=0}^{N}\sum_{k=0}^{N-j} \| \frac{\partial^k f_j}{\partial t^k}(t)\|^2_{H^{2(N-k-j)-1}(\Omega)} \le C_1 E^4$$

where C_1 is a positive constant independent of T. Similarly, from (2.4) we get

(2.16)
$$\begin{cases} u_j(x)\in H^{2(N-j)}(\Omega)\cap H_0^1(\Omega), & 0\le j\le N-1 \\ u_N(x)\in L^2(\Omega) \end{cases}$$

From (2.3),(2.4) and (2.13) we get

(2.17)
$$\begin{cases} \frac{\partial}{\partial t}(\frac{\partial^k}{\partial t^k}w_j) - \Delta(\frac{\partial^k}{\partial t^k}w_j) = \frac{\partial^k}{\partial t^k} f_j & (t,x)\in Q_T \\ (\frac{\partial^k}{\partial t^k}w_j)(0,x) = \tilde{u}_{j,k} & x\in\Omega \\ \frac{\partial^k}{\partial t^k}w_j\big|_{\Sigma_T} = 0 & 0\le j\le N, \end{cases}$$

for each integer k with $0\le k\le N-j$, where $\tilde{u}_{j,k} = \Delta^k u_j + \sum_{\ell=0}^{k-1}\Delta^\ell \partial_t^{k-1-\ell} f_j\big|_{t=0}$.

Using Lemma 2.2, we conclude from (2.14) and (2.16) that

(2.18)
$$(\frac{\partial^{N-j}}{\partial t^{N-j}}w_j)(t,x) \in C([0,T]; L^2(\Omega)) \cap L^2(0,T; H_0^1(\Omega))$$

and

(2.19)
$$\max_{0\le t\le T} \| \frac{\partial^{N-j}}{\partial t^{N-j}} w_j(t)\|^2_{L^2(\Omega)} + \int_0^T \| \frac{\partial^{N-j}}{\partial t^{N-j}} w_j(t)\|^2_{H^1(\Omega)} dt$$
$$\le C e^T (\| \tilde{u}_{j,N-j}\|^2_{L^2(\Omega)} + \int_0^T \| \frac{\partial^{N-j}}{\partial t^{N-j}} f_j(t)\|^2_{H^{-1}(\Omega)} dt).$$

next, we shall prove that for each integer k with $0\le k\le N-j-1$, we have

(2.20)
$$\frac{\partial^k}{\partial t^k} w_j(t,x) \in C([0,T]; H^{2(N-k-j)}(\Omega)\cap H_0^1(\Omega)) \cap L^2(0,T; H^{2(N-k-j)+1}(\Omega)),$$
$$0\le j\le N-1$$

and

(2.21)
$$\max_{0\le t\le T}\|\frac{\partial^k}{\partial t^k}W_j(t)\|^2_{H^{2(N-k-j)}(\Omega)}+\int_0^T\|\frac{\partial^k}{\partial t^k}W_j(t)\|^2_{H^{2(N-k-j)+1}(\Omega)}dt$$
$$\le C(e^T+1)\Big(\sum_{i=k}^{N-j}\|\tilde{u}_{j,i}\|^2_{L^2(\Omega)}+\max_{0\le t\le T}\sum_{i=k}^{N-j-1}\|\frac{\partial^i}{\partial t^i}f_j(t)\|^2_{H^{2(N-i-j-1)}(\Omega)}$$
$$+\int_0^T\sum_{i=k}^{N-j}\|\frac{\partial^i}{\partial t^i}f_j(t)\|^2_{H^{2(N-i-j)-1}(\Omega)}dt\Big)$$

where C is a positive constant independent of T. Now we show that if (2.20) and (2.21) hold for an integer k with $1\le k\le N-j-1$ (2.20) and (2.21) also hold with k replaced by k-1. In fact, we easily see from (2.17) and Lemma 2.2 that

(2.22)
$$\frac{\partial^{k-1}}{\partial t^{k-1}}W_j\in C([0,T];L^2(\Omega))\cap L^2(0,T;H^1_0(\Omega))$$

and

(2.23)
$$\max_{0\le t\le T}\|\frac{\partial^{k-1}}{\partial t^{k-1}}W_j(t)\|^2_{L^2(\Omega)}+\int_0^T\|\frac{\partial^{k-1}}{\partial t^{k-1}}W_j(t)\|^2_{H^1(\Omega)}dt$$
$$\le Ce^T\Big(\|\tilde{u}_{j,k-1}\|^2_{L^2(\Omega)}+\int_0^T\|\frac{\partial^{k-1}}{\partial t^{k-1}}f_j(t)\|^2_{H^{-1}(\Omega)}dt\Big).$$

From (2.20) and (2.22)we get

(2.24)
$$\frac{\partial^{k-1}}{\partial t^{k-1}}W_j\in C([0,T];H^1(\Omega)).$$

From (2.17) we see that for all $t\in[0,T]$, $\frac{\partial^{k-1}}{\partial t^{k-1}}W_j$ satisfies

(2.25)
$$\begin{cases}\Delta\Big(\frac{\partial^{k-1}}{\partial t^{k-1}}W_j\Big)=\frac{\partial^k}{\partial t^k}W_j-\frac{\partial^{k-1}}{\partial t^{k-1}}f_j & x\in\Omega\\ \frac{\partial^{k-1}}{\partial t^{k-1}}W_j\Big|_\Gamma=0\end{cases}$$

Applying Lemma 2.1, we conclude from (2.14),(2.20) and (2.24) that

(2.26)
$$\frac{\partial^{k-1}}{\partial t^{k-1}}W_j(t,x)\in C([0,T];H^{2(N-k-j+1)}(\Omega)\cap H^1_0(\Omega))\cap L^2(0,T;H^{2(N-k-j+1)+1}(\Omega))$$

and we conclude from (2.21) and (2.23) that

(2.27)
$$\max_{0\le t\le T}\|\frac{\partial^{k-1}}{\partial t^{k-1}}W_j(t)\|^2_{H^{2(N-k-j+1)}(\Omega)}+\int_0^T\|\frac{\partial^{k-1}}{\partial t^{k-1}}W_j(t)\|^2_{H^{2(N-k-j+1)+1}(\Omega)}dt$$

$$\leq C\Big(\max_{0\leq t\leq T}\|\frac{\partial^{k-1}}{\partial t^{k-1}}w_j(t)\|^2_{L^2(\Omega)}+\int_0^T\|\frac{\partial^{k-1}}{\partial t^{k-1}}w_j(t)\|_{H^1(\Omega)}dt$$

$$+\max_{0\leq t\leq T}\|\frac{\partial^k}{\partial t^k}w_j(t)\|^2_{H^{2(N-k-j)}(\Omega)}+\int_0^T\|\frac{\partial^k}{\partial t^k}w_j(t)\|_{H^{2(N-k-j)+1}(\Omega)}dt$$

$$+\max_{0\leq t\leq T}\|\frac{\partial^{k-1}}{\partial t^{k-1}}f_j(t)\|^2_{H^{2(N-k-j)}(\Omega)}+\int_0^T\|\frac{\partial^{k-1}}{\partial t^{k-1}}f_j(t)\|^2_{H^{2(N-k-j)+1}(\Omega)}dt\Big)$$

$$\leq C(1+e^T)\Big(\sum_{i=k-1}^{N-j}\|\tilde{u}_{j,i}\|^2_{L^2(\Omega)}+\max_{0\leq t\leq T}\sum_{i=k-1}^{N-j-1}\|\frac{\partial^i}{\partial t^i}f_j(t)\|^2_{H^{2(N-i-j-1)}(\Omega)}$$

$$+\int_0^T\sum_{i=k-1}^{N-j}\|\frac{\partial^i}{\partial t^i}f_j(t)\|^2_{H^{2(N-i-j)-1}(\Omega)}dt\Big)$$

where C is a positive constant independent of T, which shows that (2.20) and (2.21) hold with k replaced by k-1. In the same way we can prove that (2.20) and (2.21) hold for k=N-j-1 from (2.18) and (2.19). Thus, an induction argument gives that (2.20) and (2.21) hold for each integer k with $0\leq k\leq N-j-1$. Combining these with (2.18) and (2.19) we obtain

(2.28) $$W\in X_T$$

(2.29) $$\|W\|^2_{X_T}\leq C_2(1+e^T)\Big(\sum_{j=0}^{N}\|u_j\|^2_{H^{2(N-j)}(\Omega)}+\max_{0\leq t\leq T}\sum_{j=0}^{N-1}\sum_{k=0}^{N-j-1}\|\frac{\partial^k}{\partial t^k}f_j(t)\|^2_{H^{2(N-k-j-1)}(\Omega)}+\int_0^T\sum_{j=0}^{N}\sum_{k=0}^{N-j}\|\frac{\partial^k}{\partial t^k}f_j(t)\|^2_{H^{2(N-k-j)-1}(\Omega)}dt\Big),$$

where C_2 is a positive constant independent of T.

Using the estimate of the norms for the composite function and product of functions and the fact $N\geq[\frac{n}{2}]+2$ and $E<\frac{1}{2}$ we obtain that there exists a positive constant $\delta<1$ such that if $\|\varphi\|_{H^{2N}(\Omega)}<\delta E$ the following inequality holds

(2.30) $$\|u_j\|_{H^{2(N-j)}(\Omega)}\leq E/\sqrt{6C_2N} \qquad 0\leq j\leq N.$$

Thus, if let T be sufficiently small such that $e^T<2$, and $E<\min(E_0,\frac{1}{\sqrt{12C_1C_2}})$ we have

(2.31) $$\|W\|^2_{X_T}\leq\frac{1}{2}E^2+3C_1C_2E^4<E^2.$$

Hence, J maps $X_{T,E}$ to itself. We easily see that $X_{T,E}$ is non-empty. When $\|\varphi\|_{H^{2N}(\Omega)}$ is sufficiently small, $(u_1(x),\cdots\cdots,u_N(x))e^{-t}$ $\in X_{T,E}$, where $u_j(x)(0\le j\le N)$ are defined by (2.4).

Next, we prove that if T and E are sufficiently small J is a contraction mapping of $X_{T,E}$ to itself. For any $V_1, V_2 \in$ $X_{T,E}$ let $W_1 = JV_1$, $W_2 = JV_2$, and $W = W_1 - W_2 = (w_0, \cdots; w_N)$. We have

(2.32) $$\begin{cases} \frac{\partial w_j}{\partial t} - \Delta w_j = f_j(V_1) - f_j(V_2) & (t,x)\in Q_T \\ w_j(0,x) = 0 & x\in\Omega \\ w_j\big|_{\Sigma_T} = 0 \qquad , & 0\le j\le N, \end{cases}$$

From (2.29) we have

(2.33) $$\|W\|^2_{X_T} \le C_2(1+e^T)\Big(\max_{0\le t\le T}\sum_{j=0}^{N-1}\sum_{K=0}^{N-j-1}\Big\|\frac{\partial^K}{\partial t^K}(f_j(V_1)-f_j(V_2))\Big\|^2_{H^{2(N-K-j-1)}(\Omega)} + \int_0^T\sum_{j=0}^{N}\sum_{K=0}^{N-j}\Big\|\frac{\partial^K}{\partial t^K}(f_j(V_1)-f_j(V_2))\Big\|^2_{H^{2(N-K-j)-1}(\Omega)}dt\Big).$$

Using the estimate of the norms for the composite function and product of functions and the fact $E\le E_0<\frac{1}{2}$, $N\ge[\frac{n}{2}]+2$ and (2.3) we get that if $e^T<2$ the following estimate holds

(2.34) $$\|W\|^2_{X_T}\le C_3(1+e^T)E^2\|V_1-V_2\|^2_{X_T}\le 3C_3E^2\|V_1-V_2\|^2_{X_T},$$

where C_3 is a positive constant independent of T. Thus, let E be sufficiently small such that $E\le\min(E_0, 1/\sqrt{12C_1C_2}, 1/\sqrt{6C_3})$ we obtain

(2.35) $$\|W\|^2_{X_T} < \frac{1}{2}\|V_1-V_2\|^2_{X_T}$$

This shows that there exist δ, T and E such that J is a contraction mapping of $X_{T,E}$ to itself. Then mapping J has a unique fixed point $U=(u_0,u_1,\cdots\cdots,u_N)\in X_{T,E}$. $u_0(t,x)$ is a local solution of problem(1.2). The uniqueness follows from the similar inequality as (2.35). The Proof of Theorem 2.1 is complete.

3. The Energy Estimate

In this section we shall establish the energy estimate for the local solution of problem (1.2).

Lemma3.1: Let N be an integer with $N \geq [\frac{n}{2}]+2$. We assume that $F \in C^{2N}(R \times R^n \times R^{n^2})$ and $\varphi(x) \in H^{2N}(\Omega)$ satisfy (1.3) and the compatibility, respectively. Then, there exists a positive constant $E(<E_0)$ such that if u(t,x) is the solution belonging to $\bigcap_{K=0}^{N} C^K([0,T]; H^{2(N-K)}(\Omega))$ of problem (1.2) and satisfies

(3.1) $$\sum_{|\alpha|+2j \leq 2N} \| D_x^\alpha \partial_t^j u(t) \|^2_{L^2(\Omega)} \leq E^2, \qquad 0 \leq t \leq T,$$

the following energy inequality holds

(3.2) $$\sum_{|\alpha|+2j \leq 2N} \| D_x^\alpha \partial_t^j u(t) \|^2_{L^2(\Omega)} \leq C_N \| \varphi \|_{H^{2N}(\Omega)} e^{C_N \int_0^t \sum_{|\alpha|+2j \leq 2[\frac{N}{2}]+3} \| D_x^\alpha \partial_t^j u(\tau) \|_{L^\infty(\Omega)} d\tau}, \qquad 0 \leq t \leq T,$$

where C_N is a positive constant independent of T.

Proof: For each integer L with $1 \leq L \leq N$, $\partial_t^L u$ satisfies

(3.3) $$\begin{cases} \partial_t(\partial_t^L u) = \Delta(\partial_t^L u) + \partial_t^L F & (t,x) \in Q_T \\ (\partial_t^L u)(0,x) = \Delta^L \varphi + \sum_{j=0}^{L-1} \Delta^{L-1-j} \partial_t^j F(0,x) & x \in \Omega \\ \partial_t^L u |_{\Sigma_T} = 0 & \end{cases} .$$

Taking the inner product with $\partial_t^L u$ on the both side of (3.3) we get

(3.4) $$\frac{1}{2} \frac{d}{dt} \| \partial_t^L u(t) \|^2_{L^2(\Omega)} + \| \nabla \partial_t^L u(t) \|^2_{L^2(\Omega)} \leq \int_\Omega \partial_t^L u \cdot \partial_t^L F \, dx .$$

Next, we shall estimate the right side of (3.4). If we let $F = u_{ij}u_{k\ell}$, $u_{ij}u_k$, $u_{ij}u$, u_iu_j, uu_j and u^2 where i,j,k and l are integers with $1 \le i,j,k,l \le n$, using Hölder inequality we obtain the following estimates (3.5)-(3.10), respectively.

(3.5)
$$\int_\Omega \partial_t^L u \, \partial_t^L u^2 \, dx \le C \|\partial_t^L u(t)\|_{L^2(\Omega)} \sum_{j \le L} \|\partial_t^j u(t)\|_{L^2(\Omega)} \sum_{j \le [\frac{L}{2}]} \|\partial_t^j u(t)\|_{L^\infty(\Omega)},$$

(3.6)
$$\int_\Omega \partial_t^L u \cdot \partial_t^L (u u_j) dx \le C \|\partial_t^L u(t)\|_{L^2(\Omega)} \{ \|u(t)\|_{L^\infty(\Omega)} \|\partial_t^L u_j(t)\|_{L^2(\Omega)} + \sum_{K \le L} \|\partial_t^K u(t)\|_{L^2(\Omega)} \sum_{K \le [\frac{L}{2}]} \|\partial_t^K u_j(t)\|_{L^\infty(\Omega)} + \sum_{K \le L-1} \|\partial_t^K u_j(t)\|_{L^2(\Omega)} \cdot \sum_{K \le [\frac{L}{2}]} \|\partial_t^K u(t)\|_{L^\infty(\Omega)} \},$$

(3.7)
$$\int_\Omega \partial_t^L u \cdot \partial_t^L (u_i u_j) dx \le C \|\partial_t^L u(t)\|_{L^2(\Omega)} \{ \|\partial_t^L u_j(t)\|_{L^2(\Omega)} \|u_i(t)\|_{L^\infty(\Omega)} + \|\partial_t^L u_i(t)\|_{L^2(\Omega)} \|u_j(t)\|_{L^\infty(\Omega)} + \sum_{K \le L-1} \|\partial_t^K u_i(t)\|_{L^2(\Omega)} \sum_{K \le [\frac{L}{2}]} \|\partial_t^K u_j(t)\|_{L^\infty(\Omega)} + \sum_{K \le L-1} \|\partial_t^K u_j(t)\|_{L^2(\Omega)} \sum_{K \le [\frac{L}{2}]} \|\partial_t^K u_i(t)\|_{L^\infty(\Omega)} \}$$

(3.8)
$$\int_\Omega \partial_t^L u \cdot \partial_t^L (u u_{ij}) dx = \int_\Omega \partial_t^L u \cdot u \cdot \partial_t^L u_{ij} dx + \int_\Omega \partial_t^L u [\partial_t^L (u u_{ij}) - u \partial_t^L u_{ij}] dx$$
$$\le C \{ \|\partial_t^L u_i(t)\|_{L^2(\Omega)} \|\partial_t^L u_j(t)\|_{L^2(\Omega)} \|u(t)\|_{L^\infty(\Omega)} + \|\partial_t^L u(t)\|_{L^2(\Omega)} \|\partial_t^L u_j(t)\|_{L^2(\Omega)} \|u_i(t)\|_{L^\infty(\Omega)} + \|\partial_t^L u(t)\|_{L^2(\Omega)} \sum_{K \le L} \|\partial_t^K u(t)\|_{L^2(\Omega)} \sum_{K \le [\frac{L}{2}]} \|\partial_t^K u_{ij}(t)\|_{L^\infty(\Omega)} + \|\partial_t^L u(t)\|_{L^2(\Omega)} \sum_{K \le L-1} \|\partial_t^K u_{ij}(t)\|_{L^2(\Omega)} \sum_{K \le [\frac{L}{2}]} \|\partial_t^K u(t)\|_{L^\infty(\Omega)} \},$$

(3.9)
$$\int_\Omega \partial_t^L u \cdot \partial_t^L (u_K u_{ij}) dx \le \int_\Omega \partial_t^L u (u_K \partial_t^L u_{ij} + u_{ij} \partial_t^L u_K) dx + \int_\Omega \partial_t^L u [\partial_t^L (u_K u_{ij}) - u_K \partial_t^L u_{ij} - u_{ij} \partial_t^L u_K] dx$$
$$\le C \{ \|\partial_t^L u_j(t)\|_{L^2(\Omega)} \|\partial_t^L u_i(t)\|_{L^2(\Omega)} \|u_K(t)\|_{L^\infty(\Omega)} + \|\partial_t^L u(t)\|_{L^2(\Omega)} \|\partial_t^L u_j(t)\|_{L^2(\Omega)} \cdot \|u_{Ki}(t)\|_{L^\infty(\Omega)} + \|\partial_t^L u(t)\|_{L^2(\Omega)} \|\partial_t^L u_K(t)\|_{L^2(\Omega)} \|u_{ij}(t)\|_{L^\infty(\Omega)} +$$

$$\|\partial_t^L u(t)\|_{L^2(\Omega)}\Big(\sum_{\ell\le L-1}\|\partial_t^\ell u_k(t)\|_{L^2(\Omega)}\sum_{\ell\le[\frac{L}{2}]}\|\partial_t^\ell u_{ij}(t)\|_{L^\infty(\Omega)}$$

$$+\sum_{\ell\le L-1}\|\partial_t^\ell u_{ij}(t)\|_{L^2(\Omega)}+\sum_{\ell\le[\frac{L}{2}]}\|\partial_t^\ell u_k(t)\|_{L^\infty(\Omega)}\Big)\Big\},$$

(3.10) $$\int_\Omega \partial_t^L u\cdot\partial_t^L(u_{ij}u_{k\ell})dx \le C\{\|\partial_t^L u(t)\|_{L^2(\Omega)}\|\partial_t^L u_\ell(t)\|_{L^2(\Omega)}\|u_{ijk}(t)\|_{L^\infty(\Omega)}$$

$$+\|\partial_t^L u_k(t)\|_{L^2(\Omega)}\|\partial_t^L u_\ell(t)\|_{L^2(\Omega)}\|u_{ij}(t)\|_{L^\infty(\Omega)}+\|\partial_t^L u(t)\|_{L^2(\Omega)}\|\partial_t^L u_j(t)\|_{L^2(\Omega)}$$

$$\|u_{k\ell i}(t)\|_{L^\infty(\Omega)}+\|\partial_t^L u_i(t)\|_{L^2(\Omega)}\|\partial_t^L u_j(t)\|_{L^2(\Omega)}\|u_{k\ell}(t)\|_{L^\infty(\Omega)}+\|\partial_t^L u(t)\|_{L^2(\Omega)}$$

$$\sum_{j\le L-1}\|\partial_t^j u_{k\ell}(t)\|_{L^2(\Omega)}\sum_{k\le[\frac{L}{2}]}\|\partial_t^k u_{ij}(t)\|_{L^\infty(\Omega)}+\|\partial_t^L u(t)\|_{L^2(\Omega)}\sum_{k\le L-1}\|\partial_t^k u_{ij}(t)\|_{L^2(\Omega)}\cdot$$

$$\sum_{j\le[\frac{L}{2}]}\|\partial_t^j u_{k\ell}(t)\|_{L^\infty(\Omega)}\},$$

where C is a positive constant dependent on L but not on T. Using the estimate of the norms for the composite function and the product of functions and the fact $E<\frac{1}{2}$, $N\ge[\frac{n}{2}]+2$ and (3.1) we conclude that if F satisfies (1.3) we have

(3.11) $$\int_\Omega \partial_t^L u\cdot\partial_t^L F\,dx \le C_4\{\sum_{|\alpha|+2j\le 2L}\|D_x^\alpha\partial_t^j u(t)\|^2_{L^2(\Omega)}\sum_{|\alpha|+2j\le 2[\frac{L}{2}]+3}\|D_x^\alpha\partial_t^j u(t)\|_{L^\infty(\Omega)}$$

$$+\|D_x\partial_t^L u(t)\|^2_{L^2(\Omega)}\|u(t)\|_{W^{3,\infty}(\Omega)}\},$$

where C_4 is a positive constant independent of T. By the imbedding theorem we obtain from (3.1) and $N\ge[\frac{n}{2}]+2$

(3.12) $$\|u(t)\|_{W^{3,\infty}(\Omega)}\le C_5\|u(t)\|_{H^{[\frac{n}{2}]+4}(\Omega)}\le C_5E$$

where C_5 is a positive constant independent of T. Thus, if let E be sufficiently small such that $C_4C_5E<\frac{1}{2}$, from (3.4) and (3.11) we obtain

(3.13)
$$\frac{d}{dt}\|\partial_t^L u(t)\|^2_{L^2(\Omega)}+\|D_x\partial_t^L u(t)\|^2_{L^2(\Omega)}$$
$$\leq C\sum_{|\alpha|+2j\leq 2L}\|D_x^\alpha\partial_t^j u(t)\|^2_{L^2(\Omega)}\sum_{|\alpha|+2j\leq 2[\frac{L}{2}]+3}\|D_x^\alpha\partial_t^j u(t)\|_{L^\infty(\Omega)}.$$

On the other hand, noticing (1.2),(1.3) and the fact $\varphi(x)\in \overset{2N}{H}(\Omega)$ and $N\geq[\frac{n}{2}]+2$ and using the estimate of the norms for the composite function and the product of functions we can obtain in proper order that

(3.14)
$$\|\partial_t^L u(0)\|_{L^2(\Omega)}\leq C\|\varphi\|_{H^{2L}(\Omega)},\qquad 1\leq L\leq N.$$

Thus, from (3.13) and (3.14) we get

(3.15)
$$\|\partial_t^L u(t)\|^2_{L^2(\Omega)}$$
$$\leq C\{\|\varphi\|^2_{H^{2L}(\Omega)}+\int_0^t\sum_{|\alpha|+2j\leq 2L}\|D_x^\alpha\partial_\tau^j u(\tau)\|^2_{L^2(\Omega)}\sum_{|\alpha|+2j\leq 2[\frac{L}{2}]+3}\|D_x^\alpha\partial_\tau^j u(\tau)\|_{L^\infty(\Omega)}d\tau\},\quad 0\leq t\leq T$$

The above estimate was obtained for $1\leq L\leq N$. We easily obtain (3.15) for L=0 by the same way. Therefore (3.15) holds for each integer L with $0\leq L\leq N$. Next, we shall show that if for an integer J with $0\leq J\leq N-1$ we have

(3.16)
$$\sum_{|\alpha|\leq 2J}\|D_x^\alpha\partial_t^{N-J}u(t)\|^2_{L^2(\Omega)}\leq C\{\|\varphi\|^2_{H^{2N}(\Omega)}+\sum_{|\alpha|+2j\leq 2N}\|D_x^\alpha\partial_t^j u(t)\|_{L^2(\Omega)}\cdot$$
$$\sum_{|\alpha|+2j\leq N+1}\|D_x^\alpha\partial_t^j u(t)\|^2_{L^\infty(\Omega)}+\int_0^t\sum_{|\alpha|+2j\leq 2N}\|D_x^\alpha\partial_\tau^j u(\tau)\|^2_{L^2(\Omega)}\sum_{|\alpha|+2j\leq 2[\frac{N}{2}]+3}\|D_x^\alpha\partial_\tau^j u(\tau)\|_{L^\infty(\Omega)}d\tau\},\quad 0\leq t\leq T,$$

where C is a positive constant independent of T, the inequality (3.16) also holds with J replaced by J+1. In fact, we easily see that for $0\leq t\leq T$ $\partial_t^{N-J-1}u$ satisfies

(3.17)
$$\begin{cases}\Delta(\partial_t^{N-J-1}u)=\partial_t^{N-J}u-\partial_t^{N-J-1}F & x\in\Omega\\ \partial_t^{N-J-1}u|_{\partial\Omega}=0\end{cases}$$

By Lemma2.1 we have

(3.18) $$\sum_{|\alpha|\le 2(J+1)} \| D_x^\alpha \partial_t^{N-J-1} u(t) \|^2_{L^2(\Omega)} \le C \{ \| \partial_t^{N-J-1} u(t) \|^2_{L^2(\Omega)} + \sum_{|\alpha|\le 2J} \| D_x^\alpha \partial_t^{N-J} u(t) \|^2_{L^2(\Omega)} + \sum_{|\alpha|\le 2J} \| D_x^\alpha \partial_t^{N-J-1} F(t) \|^2_{L^2(\Omega)} \}, \quad 0\le t\le T.$$

From (1.3) we have

(3.19) $$\sum_{|\alpha|\le 2J} \| D_x^\alpha \partial_t^{N-J-1} F(t) \|^2_{L^2(\Omega)} \le C \sum_{|\alpha|+2j\le 2N} \| D_x^\alpha \partial_t^j u(t) \|^2_{L^2(\Omega)} \sum_{|\alpha|+2j\le N+1} \| D_x^\alpha \partial_t^j u(t) \|^2_{L^\infty(\Omega)}.$$

Thereby, from (3.15),(3.16) and (3.18) we obtain that (3.16) holds with J replaced by J+1. From(3.15) we already know that (3.16) holds for J=0. An induction argument gives

(3.20) $$\sum_{|\alpha|+2j\le 2N} \| D_x^\alpha \partial_t^j u(t) \|^2_{L^2(\Omega)} \le C_6 \{ \| \varphi \|^2_{H^{2N}(\Omega)} + \sum_{|\alpha|+2j\le 2N} \| D_x^\alpha \partial_t^j u(t) \|^2_{L^2(\Omega)} \sum_{|\alpha|+2j\le N+1} \| D_x^\alpha \partial_t^j u(t) \|^2_{L^\infty(\Omega)} + \int_0^t \sum_{|\alpha|+2j\le 2N} \| D_x^\alpha \partial_\tau^j u(\tau) \|^2_{L^2(\Omega)} \sum_{|\alpha|+2j\le 2[\frac{N}{2}]+3} \| D_x^\alpha \partial_\tau^j u(\tau) \|_{L^\infty(\Omega)} d\tau \} \quad 0\le t\le T.$$

By (3.1) and the imbedding theorem we get

(3.21) $$\sum_{|\alpha|+2j\le N+1} \| D_x^\alpha \partial_t^j u(t) \|^2_{L^\infty(\Omega)} \le C \sum_{|\alpha|+2j\le 2N} \| D_x^\alpha \partial_t^j u(t) \|^2_{L^2(\Omega)} \le C_7 E^2.$$

Thus, taking

(3.22) $$E = \min (E_0, \ 1/2C_4C_5, \ 1/2\sqrt{C_6C_7})$$

we have

(3.23) $$\sum_{|\alpha|+2j\le 2N} \| D_x^\alpha \partial_t^j u(t) \|^2_{L^2(\Omega)} \le C_N \{ \| \varphi \|^2_{H^{2N}(\Omega)} + \int_0^t \sum_{|\alpha|+2j\le 2N} \| D_x^\alpha \partial_\tau^j u(\tau) \|^2_{L^2(\Omega)} \sum_{|\alpha|+2j\le 2[\frac{N}{2}]+3} \| D_x^\alpha \partial_\tau^j u(\tau) \|_{L^\infty(\Omega)} d\tau \},$$

where C_N is a positive constant dependent on n,N and Ω, but not on T. By Gronwall's inequality we conclude (3.2). The Proof of Lemma 3.1 is complete.

4. The Local Energy Decay Estimate

In this section we shall investigate the local energy decay for solutions of the following parabolic equation

$$(4.1)\qquad \begin{cases} u_t - \Delta u = 0 & (t,x)\in Q \\ u(0,x) = \varphi(x) & x\in\Omega \\ u|_\Sigma = 0 \end{cases}$$

where the domain Ω is the same as in problem(1.1). We denote the solution of problem(4.1) by $U(t)\varphi$, where $U(t)$ is the evolution operator associated with problem (4.1), namely,

$$(4.2)\qquad U(t) = \frac{1}{2\pi i}\int_{-d+i\infty}^{-d-i\infty} e^{-\tau t}(\tau+\Delta)^{-1}d\tau\,, \qquad d>0.$$

In this section we denote by $L_a^2(\Omega)$ and $H_a^2(\Omega)$ the closed subspaces of $L^2(\Omega)$ and $H^2(\Omega)$, respectively, consisting of functions that vanish for $|x|>a$. Let $H_e^2(\Omega)$ be the Banach space $\{u;\ e^{-|x|^2}u(x)\in H^2(\Omega)\}$ with the norm $\|u\|_{H_e^2(\Omega)} = \|e^{-|x|^2}u\|_{H^2(\Omega)}$. Let $L(X,Y)$ be the Banach space consisting of all bounded linear operators from Banach space X to Banach space Y. We denote its norm by $\|\cdot\|_{L(X,Y)}$. Let R be a positive constant such that $\partial\Omega\subset\{x\in R^n;\ |x|<R\}$. For $r>R$ we write $\Omega_r = \{x\in\Omega\ ;|x|<r\}$.

Lemma 4.1: Let $n\geq 3$ and Ω be the same as in problem(1.1) Then, for two positive constants a and b with $a,b>R$ there exists a positive constant C such that

$$(4.3)\qquad \|U(t)\|_{L(L_a^2(\Omega),\,L^2(\Omega_b))} \leq Ct^{-\frac{n}{2}}, \qquad t\geq 1,$$

where C is a positive constant dependent only on n,a,b and Ω.

We may assume that $b \geq a > R+1$.

Proof: In order to estimate U(t) we first investigate the resolvent $(\tau+\Delta)^{-1}$. Let D be the entire complex plane if n is odd and the Riemann surface on which the function lnk is single-valued if n is even. Let D^+ be the region $\{k;\ k\in D, 0<\arg k<\pi, k\neq 0\}$ The resolvent $(k^2+\Delta)^{-1}$ is a $L(L^2(\Omega); H^2(\Omega))$ -valued analytic function with respect to $k\in D^+$. About $(k^2+\Delta)^{-1}$ we have the following two lemmas.

Lemma 4.2: Let $n\geq 3$. Then the resolvent $(k^2+\Delta)^{-1}$ admits a meromorphic extension to D as a $L(L^2_a(\Omega);\ H^2_e(\Omega))$-valued function and the set of all poles of the meromorphic extension has no limit point in D.

We also denote the extension by $(k^2+\Delta)^{-1}$.

The Lemma has proved by B.R.Vainberg [6].

Lemma 4.3: Let $n\geq 3$. Then there exists a positive constant $\delta_1<1$ such that

(1) If n is odd,

$$(4.4) \qquad (k^2+\Delta)^{-1} = \sum_{j=0}^{\infty} B_{2j} k^{2j} + \sum_{j=\frac{n-1}{2}}^{\infty} B_{2j+1} k^{2j+1}$$

in the region $W=\{k; k\in D, |k|<\delta_1\}$, where the operators B_j (j=0,1,……) $\in L(L^2_a(\Omega), H^2_e(\Omega))$ and the expansion (4.4) converges uniformly and absolutely in the operator norm;

(2) If n is even,

$$(4.5)\qquad (K^2+\Delta)^{-1}=\sum_{m=0}^{\infty}\sum_{j=0}^{\infty}B_{mj}\,(K^{n-2}\ln K)^m K^{2j}$$

in the region $W=\{k;\ k\in D,|k|<\delta_1,-\frac{\pi}{2}<\arg k<\frac{3\pi}{2}\}$,where the operator $B_{mj}(m,j=0,1,\cdots\cdots)\in L(L^2_a(\Omega);H^2_e(\Omega))$ and the expansion (4.5) converges uniformly and absolutely in operator norm.

This Lemma has proved by Y.Tsutsumi [5].

Remark4.1: As a consequence of Lemma 4.3, the meromorphic extension $(K^2+\Delta)^{-1}$ has no pole and is bounded in a neighbourhood of k=0.

Next, we shall translate tne results on $(K^2+\Delta)^{-1}$ into those on $(\tau+\Delta)^{-1}$,and complete the proof of Lemma 4.1.

Proof of Lemma4.1: Taking $\delta_0<\frac{\delta_1}{2}$,we consider the region D_K on the k-plane, which is hatched in Figure 1. From Remark4.1 we can choose δ_0 so small that $(K^2+\Delta)^{-1}$ has no pole in the region D_K. Under the mapping $\tau=k^2$, the region D_K is taken one to one onto the region D_τ on the τ-plane, which is hatched in Figure 2.

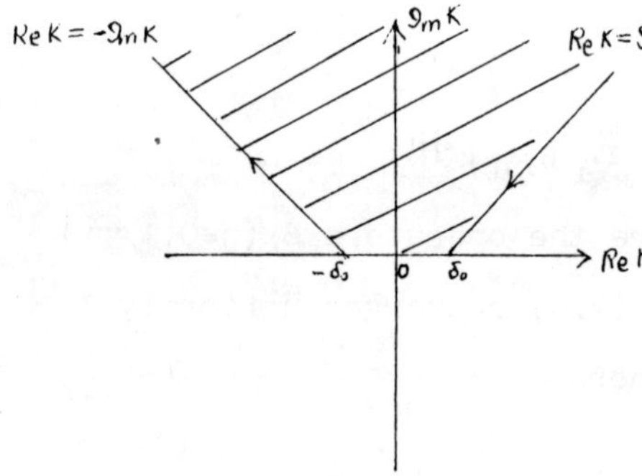

Fig.1 . k-plane

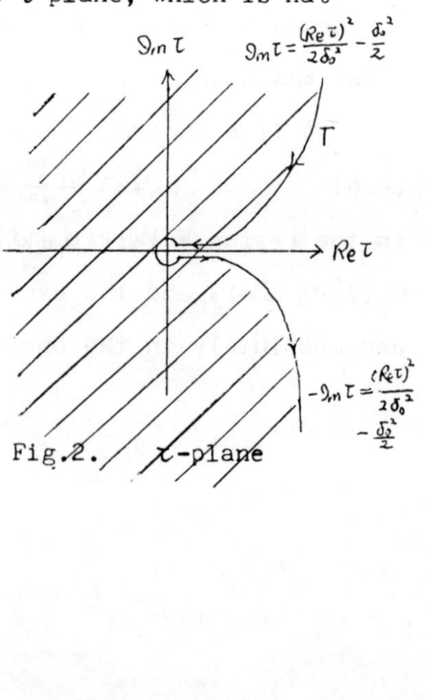

Fig.2. τ-plane

It is well known that there exist three positive constants ξ, η and M such that

(4.6) $$\|(\lambda-\Delta)^{-1}\|_{L(L^2(\Omega);\,L^2(\Omega))} \leq \frac{M}{|\lambda|}$$

for all $\lambda\in\{\lambda;\ |\arg(\lambda-\xi)|<\frac{\pi}{2}+\eta\}$. Thus, there exist two positive constants A and M such that

(4.7) $$\|(\tau+\Delta)^{-1}\|_{L(L^2_a(\Omega),\,L^2(\Omega_b))} \leq \frac{M}{|\tau|}$$

for all $\tau\in\{D_\tau,\ |\tau|>A\}$, where constants A and M do not depend on τ.

Now, we consider the following integral

(4.8) $$U(t)-I = \frac{1}{2\pi i}\int_{-d+i\infty}^{-d-i\infty} e^{-\tau t}\{(\tau+\Delta)^{-1}-\frac{I}{\tau}\}d\tau,\quad d>0.$$

Since we have

(4.9) $$(\tau+\Delta)^{-1}-\frac{I}{\tau} = -\frac{1}{\tau}(\tau+\Delta)^{-1}\Delta\ ,$$

we obtain from (4.7)

(4.10) $$\|(\tau+\Delta)^{-1}-\frac{I}{\tau}\|_{L(H^2_a(\Omega)\cap H^1_0(\Omega);\,L^2(\Omega_b))}$$
$$\leq \frac{1}{|\tau|}\|(\tau+\Delta)^{-1}\|_{L(L^2_a(\Omega),\,L^2(\Omega_b))}\|\Delta\|_{L(H^2_a(\Omega)\cap H^1_0(\Omega);\,L^2_a(\Omega))} \leq C|\tau|^{-2}$$

for all $\tau\in\{D_\tau,\ |\tau|>A\}$. Therefore, the integral in (4.8) converges absolutely in $L(H^2_a(\Omega)\cap H^1_0(\Omega);\,L^2(\Omega_b))$. By (4.10) and the Cauchy theorem we can shift the contour of the integral in (4.8) into the contour Γ as in Figure 2.

Since

(4.11) $$\frac{1}{2\pi i}\int_\Gamma e^{-\tau t}\frac{I}{\tau}d\tau = I\ ,$$

we obtain from (4.8)

(4.12) $$U(t) = \frac{1}{2\pi i}\int_\Gamma e^{-\tau t}(\tau+\Delta)^{-1}d\tau\ .$$

Finally, we estimate the integral in (4.12). We denote by Γ_A^+ and Γ_A^- the parabolic parts of Γ which are situated on the upper half plane and the lower half plane, respectively. We denote by Γ_B^+ and Γ_B^- the straight line parts of Γ which are situated on the upper half plane and the lower half plane, respectively. We denote by Γ_c the circular part of Γ. By Lemma 4.3 we easily see

(4.13) $$\lim_{\varepsilon\to 0}\left\|\frac{1}{2\pi i}\int_{\Gamma_c} e^{-\tau t}(\tau+\Delta)^{-1}d\tau\right\|_{L(L_a^2(\Omega),L^2(\Omega_b))}=0 .$$

By Lemma4.3 and (4.7) we know that $\|(\tau+\Delta)^{-1}\|_{L(L_a^2(\Omega),L^2(\Omega_b))}$ is bounded on Γ_A^+ and Γ_A^-. Thus, noticing the analytic expression of Γ_A^+ and Γ_A^- we obtain

(4.14) $$\left\|\frac{1}{2\pi i}\int_{\Gamma_A^+\cup\Gamma_A^-} e^{-\tau t}(\tau+\Delta)^{-1}d\tau\right\|_{L(L_a^2(\Omega),L^2(\Omega_b))}\leq Ce^{-ct}, \quad t\geq 1.$$

By Lemma4.3 we can obtain that if n is odd,

(4.15) $$(\tau+\Delta)^{-1}=B_1(\tau)+\tau^{\frac{n-2}{2}}B_2(\tau)$$

for all $\tau\in\{D, |\tau|<\delta_0^2\}$, where $B_1(\tau)$ and $B_2(\tau)$ are $L(L_a^2(\Omega), H^2(\Omega_b))$ -valued holomorphic functions. If n is even,

(4.16) $$(\tau+\Delta)^{-1}=B_3(\tau)+B_4\tau^{\frac{n-2}{2}}\ln\sqrt{\tau}+B_5(\tau)\tau^{\frac{n-2}{2}}$$

for all $\tau\in\{D, |\tau|<\delta_0^2, -\pi<\arg\tau<3\pi\}$, where $B_3(\tau)$ is a $L(L_a^2(\Omega), H^2(\Omega_b))$ -valued holomorphic function. $B_5(\tau)$ is a $L(L_a^2(\Omega), H^2(\Omega_b))$-valued bounded continuous function. B_4 is a bounded operator from $L_a^2(\Omega)$ to $H^2(\Omega_b)$.

Therefore, if n is odd we have

(4.17) $$\left\|\frac{1}{2\pi i}\int_{\Gamma_B^+\cup\Gamma_B^-} e^{-\tau t}(\tau+\Delta)^{-1}d\tau\right\|_{L(L_a^2(\Omega),L^2(\Omega_b))}$$

$$\leq c\int_0^{\delta_0^2} e^{-\zeta t}\zeta^{\frac{n-2}{2}}d\zeta \leq ce^{-ct}+ct^{-\frac{n-1}{2}}\int_0^{\delta_0^2}e^{-\zeta t}\frac{d\zeta}{\sqrt{\zeta}}$$

$$\leq Ct^{-\frac{n}{2}}, \qquad t\geq 1.$$

If n is even we have

(4.18)
$$\left\|\frac{1}{2\pi i}\int_{\Gamma_B^+\cup\Gamma_B^-} e^{-\tau t}(\tau+\Delta)^{-1}d\tau\right\|_{L(L_a^2(\Omega),\,L^2(\Omega_b))}$$

$$\leq c\int_0^{\delta_0^2}(e^{-\zeta t}\zeta^{\frac{n-2}{2}}\ln\sqrt{\zeta e^{i2\pi}}-e^{-\zeta t}\zeta^{\frac{n-2}{2}}\sqrt{\zeta})d\zeta$$

$$\leq c\int_0^{\delta_0^2}e^{-\zeta t}\zeta^{\frac{n-2}{2}}d\zeta \leq Ct^{-\frac{n}{2}}, \qquad t\geq 1.$$

Noticing (4.12)-(4.14) again we conclude (4.2). The proof of Lemma4.1 is complete.

5. The Decay Estimate

In this section we shall discuss the time decay estimate of the solution for the following initial boundary value problem for the parabolic equation

(5.1)
$$\begin{cases} u_t-\Delta u=f(t,x) & (t,x)\in Q \\ u(0,x)=\varphi(x) & x\in\Omega \\ u|_{\Sigma}=0 \end{cases},$$

where Ω is the same as in problem (1.1), $\varphi(x)$ and f(t,x) are so smooth that all norms of φ and f which will appear in lemmas stated below are bounded. In addtion we assume that φ and f satisfy the compatibility condition.

For the sake of convenience, we introduce the following notations: Let g(t,x) be a function defined on $[0,+\infty)\times\Omega$

and $h(x)$ be a function defined on Ω . For all integers $L \geq n+1$, all real numbers $k \geq 0$ and all real numbers p with $1 \leq p \leq \infty$, we define $[g,p,k,L,\Omega](t)$ and $Q(h,g,k,L)(t)$ as follows:

$$(5.2) \qquad [g,p,k,L,\Omega](t) = \sup_{0\leq\tau\leq t}(1+\tau)^k \sum_{|\alpha|+2j\leq L} \|D_x^\alpha \partial_t^j g(\tau)\|_{L^p(\Omega)},$$

$$(5.3) \qquad Q(h,g,k,L)(t) = \|h\|_{W^{L,1}(\Omega)} + \|h\|_{H^{L-n-1}(\Omega)} + [g,1,k,L,\Omega](t) + [g,2,k,L-n-1,\Omega](t)$$

Lemma 5.1: Let $n \geq 3$. Then, for each nonnegative integer L, the solution u of problem (5.1) satisfies the following estimates

$$(5.4) \qquad [u,\infty,\tfrac{n}{2},2L,\Omega](t) \leq CQ(\varphi,f,\tfrac{n}{2},2L+2n+4)(t), \quad t \geq 0,$$

$$(5.5) \qquad [u,2,\tfrac{n}{4},2L,\Omega](t) \leq CQ(\varphi,f,\tfrac{n}{2},2L+3[\tfrac{n}{2}]+5)(t), \quad t \geq 0,$$

where C is a positive constant depending only on n, L and Ω. In addition for any multi-index α and any integer $j \geq 0$ with $\frac{|\alpha|}{2} + j \geq \frac{n}{2}$, we have

$$(5.6) \qquad \|D_x^\alpha \partial_t^j u(t)\|_{L^2(\Omega)} \leq C(1+t)^{-\frac{n}{4}} Q(\varphi,f,\tfrac{n}{4},|\alpha|+2j+n+3)(t), \quad t \geq 0,$$

where C is a positive constant depending only on n, α, j and Ω.

Our method of the proof of Lemma 5.1 follows [4], where he regard the problem in an exterior domain as perturbation of problem in entire space. In order to prove Lemma5.1 we first give several lemmas.

Lemma 5.2: Let $n \geq 3$ and $u(t,x)$ be the smooth solution of the following Cauchy problem

$$(5.7) \qquad \begin{cases} u_t - \Delta u = f(t,x) & (t,x) \in R^+ \times R^n \\ u(0,x) = \varphi(x) & x \in R^n \end{cases} .$$

Then, for each integer $L \geq 0$, the following inequalities hold

$$(5.8) \qquad [u, \infty, \tfrac{n}{2}, 2L, R^n](t) \leq C \{ \|\varphi\|_{W^{2L+n+1,1}(R^n)} + [f, 1, \tfrac{n}{2}, 2L+n+1, R^n](t) \}, \qquad t \geq 0,$$

$$(5.9) \qquad [u, 2, \tfrac{n}{4}, 2L, R^n](t) \leq C \{ \|\varphi\|_{W^{2L+[\frac{n}{2}]+1,1}(R^n)} + [f, 1, \tfrac{n}{2}, 2L+[\tfrac{n}{2}]+1, R^n](t) \}, \qquad t \geq 0,$$

where C is a positive constant depending only on n and L. In addition for any multi-index α and any integer $j \geq 0$ with $\frac{|\alpha|}{2} + j \geq \frac{n}{2}$, we have

$$(5.10) \qquad \| D_x^\alpha \partial_t^j u(t) \|_{L^2(R^n)} \leq C(1+t)^{-\frac{n}{4}} \{ \|\varphi\|_{H^{|\alpha|+2j}(R^n)} + [f, 2, \tfrac{n}{4}, |\alpha|+2j, R^n](t) \}, \qquad t \geq 0,$$

where C is a positive constant depending only on n, α, j.

Proof: Let W(t) denote the evolution operator associated with the following heat equation

$$(5.11) \qquad \begin{cases} u_t = \Delta u & (t,x) \in R^+ \times R^n \\ u(0,x) = v(x) & x \in R^n \end{cases} .$$

Namely. the solution of problem (5.11) is expressed as

$$(5.12) \qquad u(t,x) = W(t)v \triangleq \left(\frac{1}{2\sqrt{\pi t}}\right)^n \int_{R_\xi^n} e^{-\frac{|x-\xi|^2}{4t}} v(\xi) d\xi$$

We easily obtain that for any multi-index α the following inequalities hold

(5.13) $$\| D_x^\alpha (W(t)v)\|_{L^\infty(R^n)} \le C\, t^{-(\frac{n}{2}+\frac{|\alpha|}{2})} \|v\|_{L^1(R^n)}, \quad t \ge 1,$$

(5.14) $$\| D_x^\alpha (W(t)v)\|_{L^1(R^n)} \le C\, t^{-\frac{|\alpha|}{2}} \|v\|_{L^1(R^n)}, \quad t \ge 1,$$

(5.15) $$\| D_x^\alpha (W(t)v)\|_{L^2(R^n)} \le C\, t^{-\frac{|\alpha|}{2}} \|v\|_{L^2(R^n)}, \quad t \ge 1.$$

From (5.13) and (5.14) we obtain

(5.16) $$\| D_x^\alpha (W(t)v)\|_{L^2(R^n)} \le C\, t^{-(\frac{n}{4}+\frac{|\alpha|}{2})} \|v\|_{L^1(R^n)}, \quad t \ge 1.$$

Moreover, by the imbedding theorem we have

(5.13') $$\| D_x^\alpha (W(t)v)\|_{L^\infty(R^n)} \le C \| D_x^\alpha v\|_{L^\infty(R^n)} \le C \|v\|_{W^{|\alpha|+n+1,1}(R^n)}, \quad t > 0,$$

(5.15') $$\| D_x^\alpha (W(t)v)\|_{L^2(R^n)} \le \| D_x^\alpha v\|_{L^2(R^n)}, \quad t > 0.$$

(5.16') $$\| D_x^\alpha (W(t)v)\|_{L^2(R^n)} \le C \|v\|_{W^{|\alpha|+[\frac{n}{2}]+1,1}(R^n)}, \quad t > 0.$$

Noticing equality (5.11) we conclude

(5.17) $$\| D_x^\alpha \partial_t^j (W(t)v)\|_{L^\infty(R^n)} \le C(1+t)^{-(\frac{n}{2}+\frac{|\alpha|}{2}+j)} \|v\|_{W^{|\alpha|+2j+n+1,1}(R^n)}, \quad t > 0,$$

(5.18) $$\| D_x^\alpha \partial_t^j (W(t)v)\|_{L^2(R^n)} \le C(1+t)^{-(\frac{n}{4}+\frac{|\alpha|}{2}+j)} \|v\|_{W^{|\alpha|+2j+[\frac{n}{2}]+1,1}(R^n)}, \quad t > 0,$$

(5.19) $$\| D_x^\alpha \partial_t^j (W(t)v)\|_{L^2(R^n)} \le C(1+t)^{-(\frac{|\alpha|}{2}+j)} \|v\|_{H^{|\alpha|+2j}(R^n)}, \quad t > 0.$$

On the other hand, by Duhamel's principle for the solution we have

(5.20) $$D_x^\alpha \partial_t^j u(t,x) = D_x^\alpha \partial_t^j (W(t)\varphi) + \int_0^t D_x^\alpha \partial_t^j (W(t-s) f(s))\, ds.$$

Therefore, for all $t \ge 0$ the following inequalities hold

(5.21) $$\| D_x^\alpha \partial_t^j u(t)\|_{L^\infty(R^n)} \le C(1+t)^{-(\frac{n}{2}+\frac{|\alpha|}{2}+j)} \|\varphi\|_{W^{|\alpha|+2j+n+1,1}(R^n)}$$
$$+ C\int_0^t (1+t-s)^{-(\frac{n}{2}+\frac{|\alpha|}{2}+j)} (1+s)^{-\frac{n}{2}} ds \cdot [f, 1, \tfrac{n}{2}, |\alpha|+2j+n+1, R^n](t),$$

(5.22) $$\| D_x^\alpha \partial_t^j u(t)\|_{L^2(R^n)} \le C(1+t)^{-(\frac{n}{4}+\frac{|\alpha|}{2}+j)} \|\varphi\|_{W^{|\alpha|+2j+[\frac{n}{2}]+1,1}(R^n)}$$
$$+ C\int_0^t (1+t-s)^{-(\frac{n}{4}+\frac{|\alpha|}{2}+j)} (1+s)^{-\frac{n}{2}} ds \cdot [f, 1, \tfrac{n}{2}, |\alpha|+2j+[\tfrac{n}{2}]+1, R^n](t),$$

$$\|D_x^\alpha \partial_t^j u(t)\|_{L^2(R^n)} \le C(1+t)^{-(\frac{|\alpha|}{2}+j)} \|\varphi\|_{H^{|\alpha|+2j}(R^n)} \tag{5.23}$$
$$+ C\int_0^t (1+t-s)^{-(\frac{|\alpha|}{2}+j)} (1+s)^{-\frac{n}{4}} ds \cdot [f, 2, \tfrac{n}{4}, |\alpha|+2j, R^n](t).$$

Noticing $n \geq 3$, we immediately conclude (5.8)-(5.10). The proof of Lemma 5.2 is complete.

Lemma 5.3: Let Ω be the same as in problem (1.1) and $u(x)$ be a solution in $H^1(\Omega)$ of

$$\begin{cases} \Delta u = f & x \in \Omega \\ u|_{\partial\Omega} = 0 \end{cases} \tag{5.24}$$

Let L be an arbitrary nonnegative integer and r_1 and r_2 be two positive constants such that $r_2 < r_1$ and $\partial\Omega \subset \{x \in R^n, |x| < r_2\}$. Then if $f \in H^L(\Omega)$ $u(x)$ satisfies

$$\sum_{|\alpha| \le L+2} \|D_x^\alpha u\|_{L^2(\Omega_{r_2})} \le C(\|u\|_{L^2(\Omega_{r_1})} + \sum_{|\alpha| \le L} \|D_x^\alpha f\|_{L^2(\Omega_{r_1})}), \tag{5.25}$$

where C is a positive constant depending only on L, r_1, r_2, n and Ω.

The proof of this lemma refer to D.Gilberg & N.S.Tradinger [2].

Now, we prove Lemma 5.1.

Proof of Lemma 5.1: We extend $\varphi(x)$ and $f(t,x)$ from Ω to R^n, in the same regularity class. Note that relevant Sobolev norms of extensions of $\varphi(x)$ and $f(t,\cdot)$ can be bounded by corresponding Sobolev norms of $\varphi(x)$ and $f(t,\cdot)$

multiplied by a constant independent of φ, $f(t,\cdot)$ and t. We also denote extended functions by $\varphi(x)$ and $f(t,x)$, respectively. Now let $u_1(t,x)$ be the solution of the Cauchy problem

(5.26)
$$\begin{cases} \partial_t u_1 = \Delta u_1 + f(t,x) & (t,x)\in R^+\times R^n \\ u_1(0,x) = \varphi(x) \end{cases}$$

By Lemma5.2 and the property of the extension we obtain

(5.27)
$$[u_1, \infty, \tfrac{n}{2}, 2L, R^n](t) \leq C Q(\varphi, f, \tfrac{n}{2}, 2L+n+1)(t), \quad t\geq 0,$$

(5.28)
$$[u_1, 2, \tfrac{n}{4}, 2L, R^n](t) \leq C Q(\varphi, f, \tfrac{n}{2}, 2L+[\tfrac{n}{2}]+1)(t), \quad t\geq 0,$$

where C is a positive constant depending only on n,L and Ω. In addition for $\frac{|\alpha|}{2}+j\geq \frac{n}{2}$ we have

(5.29)
$$\|D_x^\alpha \partial_t^j u_1(t)\|_{L^2(R^n)} \leq C(1+t)^{-\frac{n}{4}}\{\|\varphi\|_{H^{|\alpha|+2j}(\Omega)} + [f, 2, \tfrac{n}{4}, |\alpha|+2j, \Omega](t)\}, \quad t\geq 0,$$

where C is a positive constant depending only on n,α,j,Ω.

Now we fix a positive constant γ, such that $\partial\Omega\subset\{x\in R^n; |x|<\gamma\}$. We choose a function $\tilde{\varphi}(x)\in C^\infty(R^n)$, such that $\tilde{\varphi}(x)=1$ for $|x|>\gamma+2$ and $\tilde{\varphi}(x)=0$ for $|x|<\gamma+1$. We define $u_2(t,x)$ and $u_3(t,x)$ by

(5.30)
$$u_2(t,x) = \tilde{\varphi}(x)\, u_1(t,x),$$

(5.31)
$$u_3(t,x) = u(t,x) - u_2(t,x).$$

From (5.27) and (5.28) we easily know

(5.32)
$$[u_2, \infty, \tfrac{n}{2}, 2L, \Omega](t) \leq C Q(\varphi, f, \tfrac{n}{2}, 2L+n+1)(t), \quad t\geq 0,$$

(5.33) $$[u_2, 2, \tfrac{n}{4}, 2L, \Omega](t) \leq C Q(\varphi, f, \tfrac{n}{2}, 2L+[\tfrac{n}{2}]+1)(t), \quad t \geq 0,$$

where C is a positive constant depending only on n,L and Ω. Moreover, we have for any multi-index α and any integer $j \geq 0$

(5.34) $$\begin{aligned} \| D_x^\alpha \partial_t^j u_2(t) \|_{L^2(\Omega)} &= \| D_x^\alpha \partial_t^j u_2(t) \|_{L^2(R^n)} \\ &\leq \| \tilde{\varphi} (D_x^\alpha \partial_t^j u_1)(t) \|_{L^2(R^n)} + C \sum_{\substack{|\alpha_1|+|\alpha_2| \leq |\alpha| \\ |\alpha_1| > 0}} \| (D_x^{\alpha_1} \tilde{\varphi})(D_x^{\alpha_2} \partial_t^j u_1)(t) \|_{L^2(\Omega_{y+2})} \\ &\leq C (\| D_x^\alpha \partial_t^j u_1(t) \|_{L^2(R^n)} + \sum_{|\alpha_2| \leq |\alpha|-1} \| D_x^{\alpha_2} \partial_t^j u_1(t) \|_{L^2(\Omega_{y+2})}) \\ &\leq C (\| D_x^\alpha \partial_t^j u_1(t) \|_{L^2(R^n)} + \| D_x^\alpha \partial_t^j u_1(t) \|_{L^2(\Omega_{y+2})}) \leq C \| D_x^\alpha \partial_t^j u_1(t) \|_{L^2(R^n)}, \end{aligned}$$

where C is a positive constant depending only on n,α,j. Therefore, from (5.29) and (5.34) we obtain that when $\frac{|\alpha|}{2} + j \geq \frac{n}{2}$ for all $t \geq 0$ we have

(5.35) $$\begin{aligned} &\| D_x^\alpha \partial_t^j u_2(t) \|_{L^2(\Omega)} \\ &\leq C(1+t)^{-\frac{n}{4}} \{ \| \varphi \|_{H^{|\alpha|+2j}(\Omega)} + [f, 2, \tfrac{n}{4}, |\alpha|+2j, \Omega](t) \}, \quad t \geq 0, \end{aligned}$$

where C is a positive constant depending only on n,α,j,Ω.

Next, we shall estimate $u_3(t,x)$. It is easily known that $u_3(t,x)$ satisfies

(5.36) $$\begin{cases} \partial_t u_3 = \Delta u_3 + h(t,x) & (t,x) \in Q \\ u_3(0,x) = (1-\tilde{\varphi}(x)) \varphi(x) & x \in \Omega \\ u_3 |_\Sigma = 0 & \end{cases},$$

where

(5.37) $$h(t,x) = g(t,x) + (1-\tilde{\varphi}(x)) f(t,x),$$

(5.38) $$g(t,x) = 2 \nabla \tilde{\varphi} \cdot \nabla u_1 + u_1 \Delta \tilde{\varphi}.$$

Noticing (5.27) we have

(5.39) $$\operatorname{Supp} g \subset [0,+\infty)\times\{x\in R^n,\ \nu+1\le |x|\le \nu+2\}$$

(5.40) $$[g,\infty,\tfrac{n}{2},2L,\Omega](t)\le CQ(\varphi.f,\tfrac{n}{2},2L+n+2)(t),\quad t\ge 0.$$

Moreover, using the method by which we get (5.27), we obtain that for the solution $u_1(t,x)$ the following inequality holds

(5.41) $$[u_1,\infty,\tfrac{n}{4},2L,R^n](t)\le CQ(\varphi.f,\tfrac{n}{4},2L+n+1)(t),\quad t\ge 0.$$

Thus, from (5.38) we obtain

(5.42) $$[g,\infty,\tfrac{n}{4},2L,\Omega](t)\le CQ(\varphi,f,\tfrac{n}{4},2L+n+2)(t),\quad t\ge 0.$$

On the other hand, from (5.36) we know that for all integers $L\geq 1$ $\partial_t^L u_3(t,x)$ satisfies

(5.43) $$\begin{cases}\partial_t(\partial_t^L u_3)=\Delta(\partial_t^L u_3)+\partial_t^L h & (t,x)\in Q\\ (\partial_t^L u_3)(0,x)=\Delta^L((1-\widetilde{\varphi})\varphi)+\sum_{j=0}^{L-1}\Delta^{L-1-j}\partial_t^j h(0,x), & x\in\Omega,\\ \partial_t^L u_3|_\Sigma=0\end{cases}$$

So, we have

(5.44) $$\partial_t^L u_3(t,x)=U(t)\{\Delta^L((1-\widehat{\varphi})\varphi)+\sum_{j=0}^{L-1}\Delta^{L-1-j}\partial_t^j h(0,x)\}$$
$$+\int_0^t U(t-s)(\partial_s^L h)(s)\,ds,$$

where U(t) is a evolution operator defined by (4.2). By Lemma4.1 (set $a=b=\nu+2$), (5.40) and (5.39) we obtain

(5.45) $$\|\partial_t^L u_3(t)\|_{L^2(\Omega_{\nu+2})}\le C(1+t)^{-\frac{n}{2}}\{\sum_{|\alpha|\le 2L}\|D_x^\alpha\varphi\|_{L^2(\Omega)}+$$
$$\sum_{j=0}^{L-1}\sum_{|\alpha|\le 2(L-1-j)}\|D_x^\alpha\partial_t^j h(0,\cdot)\|_{L^2(\Omega)}\}+\int_0^t(1+t-s)^{-\frac{n}{2}}(1+s)^{-\frac{n}{2}}ds\cdot$$
$$[h,2,\tfrac{n}{2},2L,\Omega](t)\le C(1+t)^{-\frac{n}{2}}\{\|\varphi\|_{H^{2L}(\Omega)}+$$

$$[g,\infty,\tfrac{n}{2},2L,\Omega](t)+[f,2,\tfrac{n}{2},2L,\Omega](t)\}$$

$$\leq C(1+t)^{-\frac{n}{2}}Q(\varphi,f,\tfrac{n}{2},2L+n+2)(t), \qquad t\geq 0,$$

where C is a positive constant depending only on n,L and Ω. By the same way we can obtain that (5.45) holds for L=0. Thus, (5.45) holds for all integers L≥0.Next, we shall show that if for an integer J with 0≤J≤L-1 we have

(5.46)
$$\sum_{|\alpha|\leq 2J}\|D_x^\alpha \partial_t^{L-J}u_3(t)\|_{L^2(\Omega_{y+2-\frac{J}{L}})} \leq C(1+t)^{-\frac{n}{2}}Q(\varphi,f,\tfrac{n}{2},2L+n+2)(t), \qquad t\geq 0,$$

the inequality (5.46) also holds with J replaced by J+1. In fact, we easily see that for all t≥0 $\partial_t^{L-J-1}u_3(t,x)$ satisfies

(5.47)
$$\begin{cases}\Delta(\partial_t^{L-J-1}u_3)=\partial_t(\partial_t^{L-J-1}u_3)-\partial_t^{L-J-1}h & x\in\Omega,\\ \partial_t^{L-J-1}u_3|_{\partial\Omega}=0 & .\end{cases}$$

By (5.39),(5.40),(5.45),(5.46) and Lemma5.3 we obtain

(5.48)
$$\sum_{|\alpha|\leq 2(J+1)}\|D_x^\alpha\partial_t^{L-J-1}u_3(t)\|_{L^2(\Omega_{y+2-\frac{J+1}{L}})}\leq C\{\|\partial_t^{L-J-1}u_3(t)\|_{L^2(\Omega_{y+2-\frac{J}{L}})}+\sum_{|\alpha|\leq 2J}\|D_x^\alpha\partial_t^{L-J}u_3(t)\|_{L^2(\Omega_{y+2-\frac{J}{L}})}+\sum_{|\alpha|\leq 2J}\|D_x^\alpha\partial_t^{L-J-1}h(t)\|_{L^2(\Omega_{y+2-\frac{J}{L}})}\}\leq C(1+t)^{-\frac{n}{2}}Q(\varphi,f,\tfrac{n}{2},2L+n+2)(t).$$

From (5.45) we already know that (5.46) holds for J=0 . Thereby, an induction argument gives

(5.49)
$$[u_3,2,\tfrac{n}{2},2L,\Omega_{y+1}](t)\leq CQ(\varphi,f,\tfrac{n}{2},2L+n+2)(t), \qquad t\geq 0.$$

It implies

(5.50)
$$[u_3,1,\tfrac{n}{2},2L,\Omega_{y+1}](t)\leq CQ(\varphi,f,\tfrac{n}{2},2L+n+2)(t), \qquad t\geq 0.$$

Moreover, by the imbedding theorem and (5.49) we conclude

(5.51)
$$[u_3,\infty,\tfrac{n}{2},2L,\Omega_{y+1}](t)\leq CQ(\varphi,f,\tfrac{n}{2},2L+3[\tfrac{n}{2}]+4)(t), \quad t\geq 0.$$

The positive C in (5.49)-(5.51) depends only on n,L and Ω.

Similarly, by Lemma4.1 and (5.39),(5.42) we obtain that for all $t\geq 0$, the following inequality holds

(5.52) $$[u_3, 2, \tfrac{n}{4}, 2L, \Omega_{\nu+1}](t) \leq CQ(\varphi, f, \tfrac{n}{4}, 2L+n+2)(t), \quad t\geq 0,$$

where C is a positive constant depending only on n.L and Ω.

Finally, we shall evaluate $u_3(t,x)$ for $|x|>\nu+1$. We choose a function $\tilde{\psi}(x)\in C^\infty(R^n)$ such that $\tilde{\psi}(x)=1$ for $|x|>\nu+\frac{3}{4}$ and $\tilde{\psi}(x)=0$ for $|x|<\nu+\frac{1}{4}$. Set

(5.53) $$u_4(t,x)=\tilde{\psi}(x)u_3(t,x).$$

It is easily known that $u_4(t,x)$ satisfies

(5.54) $$\begin{cases} \partial_t u_4 = \Delta u_4 + K(t,x) & (t,x)\in R^+\times R^n \\ u_4(0,x)=\tilde{\psi}(1-\tilde{\varphi})\varphi \end{cases}$$

where

(5.55) $$K(t,x) = -2\nabla\tilde{\psi}\cdot\nabla u_3 - u_3\Delta\tilde{\psi} + \tilde{\psi}R.$$

Applying Lemma5.2, from (5.39),(5.40),(5.49)-(5.50) and the property of the extension we have

(5.56) $$\begin{aligned}[u_4, \infty, \tfrac{n}{2}, 2L, \Omega](t) &= [u_4, \infty, \tfrac{n}{2}, 2L, R^n](t) \\ &\leq C\{\|\varphi\|_{W^{2L+n+1,1}(\Omega)} + [K, 1, \tfrac{n}{2}, 2L+n+1, R^n](t)\} \\ &\leq C\{\|\varphi\|_{W^{2L+n+1,1}(\Omega)} + [u_3, 1, \tfrac{n}{2}, 2L+n+2, \Omega_{\nu+1}](t) \\ &\quad + [g, \infty, \tfrac{n}{2}, 2L+n+1, \Omega](t) + [f, 1, \tfrac{n}{2}, 2L+n+1, \Omega](t)\} \\ &\leq CQ(\varphi, f, \tfrac{n}{2}, 2L+2n+4)(t), \qquad t\geq 0,\end{aligned}$$

Similarly, applying Lemma5.2 from (5.39),(5.40),(5.52)

and the property of the extension we obtain that

$$[u_4, 2, \tfrac{n}{4}, 2L, \Omega](t) \leq C\,Q(\varphi, f, \tfrac{n}{2}, 2L+3[\tfrac{n}{2}]+5)(t), \quad t \geq 0. \tag{5.57}$$

The positive constant C in (5.56) and (5.57) depends only on n, L, Ω. Moreover, from (5.10),(5.42),(5.52) and the property of the extension we obtain that when $\frac{|\alpha|}{2} + j \geq \frac{n}{2}$ the following inequality holds

$$\begin{aligned} \|D_x^\alpha \partial_t^j u_4(t)\|_{L^2(\Omega)} &= \|D_x^\alpha \partial_t^j u_4(t)\|_{L^2(R^n)} \\ &\leq C(1+t)^{-\frac{n}{4}} \{ \|\varphi\|_{H^{|\alpha|+2j}(\Omega)} + [K, 2, \tfrac{n}{4}, |\alpha|+2j, R^n](t) \} \\ &\leq C(1+t)^{-\frac{n}{4}} \{ \|\varphi\|_{H^{|\alpha|+2j}(\Omega)} + [u_3, 2, \tfrac{n}{4}, |\alpha|+2j+1, \Omega_{y+1}](t) \\ &\quad + [g, \infty, \tfrac{n}{4}, |\alpha|+2j, \Omega](t) + [f, 2, \tfrac{n}{4}, |\alpha|+2j, \Omega](t) \} \\ &\leq C(1+t)^{-\frac{n}{4}} Q(\varphi, f, \tfrac{n}{4}, |\alpha|+2j+n+3)(t), \end{aligned} \tag{5.58}$$

where C is a positive constant depending only on n, α, j, Ω.

Therefore, from (5.49),(5.51),(5.56) and (5.57) we conclude

$$[u_3, \infty, \tfrac{n}{2}, 2L, \Omega](t) \leq C\,Q(\varphi, f, \tfrac{n}{2}, 2L+2n+4), \quad t \geq 0, \tag{5.59}$$

$$[u_3, 2, \tfrac{n}{4}, 2L, \Omega](t) \leq C\,Q(\varphi, f, \tfrac{n}{2}, 2L+3[\tfrac{n}{2}]+5)(t), \quad t \geq 0, \tag{5.60}$$

The positive constant C in (5.59) and (5.60) depends only on n, L and Ω. Moreover, from (5.52) and (5.58) we conclude that when $\frac{|\alpha|}{2} + j \geq \frac{n}{2}$, we have

$$\|D_x^\alpha \partial_t^j u_3(t)\|_{L^2(\Omega)} \leq C(1+t)^{-\frac{n}{4}} Q(\varphi, f, \tfrac{n}{4}, |\alpha|+2j+n+3)(t), \quad t \geq 0, \tag{5.61}$$

where C is a positive constant depending only on n, α, j, Ω.

From (5.32),(5.33),(5.35) and (5.59)-(5.61) we obtain the conclusion of Lemma5.1.

6. The Priori Estimate and The Proof of Theorem1.1

In order to prove Theorem1.1 we first establish the following priori estimate.

Lemma6.1: Let the assumptions of Theorem1.1 hold. Moreover, let u be the solution of problem (1.2) satisfying (2.8) and energy estimate (3.2). Then there exist two positive constants ε_0 and K_N such that if $\varphi(x)$ satisfies

$$\|\varphi\|_{W^{2N,1}(\Omega)} + \|\varphi\|_{H^{2N}(\Omega)} < \varepsilon_0 , \tag{6.1}$$

we have

$$\|u(t)\|_{H^{2N}(\Omega)} \leq K_N \|\varphi\|_{H^{2N}(\Omega)} , \tag{6.2}$$

for $0 \leq t \leq T$. Here constants ε_0 and K_N do not depend on T.

Proof: For each integer $N \geq 3n+3[\frac{n}{2}]+23$, set

$$M_{\bar{s}}(t) = [u, \infty, \tfrac{n}{2}, 2[\tfrac{N}{2}]+3, \Omega](t) + [u, 2, \tfrac{n}{4}, \bar{s}, \Omega](t), \tag{6.3}$$

where $\bar{s}$ is a positive integer to be determined later. From (5.4) and (1.3) we obtain that for the local solution u of problem (1.2) the following estimate holds on $[0,T]$

$$\begin{aligned} [u, \infty, \tfrac{n}{2}, 2[\tfrac{N}{2}]+3, \Omega](t) &\leq CQ(\varphi, F, \tfrac{n}{2}, 2[\tfrac{N}{2}]+2n+7)(t) \\ &\leq C\Big\{\|\varphi\|_{W^{2[\frac{N}{2}]+2n+7,1}(\Omega)} + \|\varphi\|_{H^{2[\frac{N}{2}]+n+6}(\Omega)} + [u, 2, \tfrac{n}{4}, 2[\tfrac{N}{2}]+2n+9, \Omega](t)^2 \\ &\quad + [u, 2, \tfrac{n}{4}, 2[\tfrac{N}{2}]+n+8](t)[u, \infty, \tfrac{n}{2}, [\tfrac{2[\frac{N}{2}]+n+6}{2}]+2, \Omega](t)\Big\}. \end{aligned} \tag{6.4}$$

Since $N \geq 3n+3[\frac{n}{2}]+23$, we obtain $[\frac{2[\frac{N}{2}]+n+6}{2}]+2 \leq 2[\frac{N}{2}]+3$ and $2[\frac{N}{2}]+2n+7 \leq 2N$.

Thereby, from (6.1) and (6.3) we obtain

(6.5) $$[u,\infty,\tfrac{n}{2},2[\tfrac{N}{2}]+3,\Omega](t)\le C\{\varepsilon_0+M_{\bar s}(t)^2\},$$

where we take

(6.6) $$\bar s = 2[\tfrac{N}{2}]+2n+9 .$$

Moreover, from (5.5) we know that for $0\le t\le T$ $u(t,x)$ satisfies

(6.7) $$\begin{aligned}[u,2,\tfrac{n}{4},\bar s,\Omega](t) &\le CQ(\varphi,F,\tfrac{n}{2},\bar s+3[\tfrac{n}{2}]+5)(t)\\ &\le C\{\|\varphi\|_{W^{\bar s+3[\frac{n}{2}]+5,1}(\Omega)}+\|\varphi\|_{H^{\bar s+[\frac{n}{2}]+4}(\Omega)}\\ &\quad+[F,1,\tfrac{n}{2},\bar s+3[\tfrac{n}{2}]+5,\Omega](t)+[F,2,\tfrac{n}{2},\bar s+[\tfrac{n}{2}]+4,\Omega](t)\}.\end{aligned}$$

Noticing (1.3) we have

(6.8) $$\begin{aligned}[F,1,\tfrac{n}{2},\bar s+3[\tfrac{n}{2}]+5,\Omega](t) &\le C\{[u,2,\tfrac{n}{4},\bar s,\Omega](t)^2\\ &\quad+[u,2,\tfrac{n}{4},[\tfrac{\bar s+3[\frac{n}{2}]+5}{2}]+2,\Omega](t)\cdot\\ &\quad\sup_{0\le\tau\le t}(1+\tau)^{\frac{n}{4}}\sum_{\bar s-1\le|\alpha|+2j\le\bar s+3[\frac{n}{2}]+7}\|D_x^\alpha\partial_\tau^j u(\tau)\|_{L^2(\Omega)}\}.\end{aligned}$$

From (6.6),(6.3) and the fact that $N\ge 3n+3[\frac{n}{2}]+23$, we obtain that the right side of (6.8) do not exceed

(6.9) $$C\{M_{\bar s}(t)^2+M_{\bar s}(t)\sup_{0\le\tau\le t}(1+\tau)^{\frac{n}{4}}\sum_{\bar s-1\le|\alpha|+2j\le\bar s+3[\frac{n}{2}]+7}\|D_x^\alpha\partial_\tau^j u(\tau)\|_{L^2(\Omega)}\}.$$

Noticing the fact that $\bar s-1>\frac{n}{2}$, from (5.6) and (3.2) we obtain

(6.10) $$\begin{aligned}&\sup_{0\le\tau\le t}(1+\tau)^{\frac{n}{4}}\sum_{\bar s-1\le|\alpha|+2j\le\bar s+3[\frac{n}{2}]+7}\|D_x^\alpha\partial_\tau^j u(\tau)\|_{L^2(\Omega)}\\ &\le CQ(\varphi,F,\tfrac{n}{4},\bar s+3[\tfrac{n}{2}]+n+10)(t)\\ &\le C\{\|\varphi\|_{W^{\bar s+3[\frac{n}{2}]+n+10,1}(\Omega)}+\|\varphi\|_{H^{\bar s+3[\frac{n}{2}]+9}(\Omega)}\\ &\quad+[u,2,0,\bar s+3[\tfrac{n}{2}]+n+12,\Omega](t)[u,2,\tfrac{n}{4},[\tfrac{\bar s+3[\frac{n}{2}]+n+10}{2}]+2,\Omega](t)\\ &\quad+[u,\infty,\tfrac{n}{2},[\tfrac{\bar s+3[\frac{n}{2}]+9}{2}]+2,\Omega](t)[u,2,0,\bar s+3[\tfrac{n}{2}]+11,\Omega](t)\}\end{aligned}$$

$$\leq C\{\|\varphi\|_{W^{\bar{s}+3[\frac{n}{2}]+n+10,1}(\Omega)} + \|\varphi\|_{H^{\bar{s}+3[\frac{n}{2}]+9}(\Omega)}$$

$$+ [u, 2, \tfrac{n}{4}, [\tfrac{\bar{s}+3[\frac{n}{2}]+n+10}{2}]+2, \Omega](t)\cdot\|\varphi\|_{H^{\bar{s}+3[\frac{n}{2}]+n+12}(\Omega)}\cdot$$

$$\exp(c\int_0^t(1+\tau)^{-\frac{n}{2}}d\tau\cdot[u,\infty,\tfrac{n}{2}, 2[\tfrac{\bar{s}+3[\frac{n}{2}]+n+12}{4}]+3, \Omega](t))$$

$$+ [u,\infty,\tfrac{n}{2}, [\tfrac{\bar{s}+3[\frac{n}{2}]+9}{2}]+2, \Omega](t)\|\varphi\|_{H^{\bar{s}+3[\frac{n}{2}]+11}(\Omega)}$$

$$\exp(c\int_0^t(1+\tau)^{-\frac{n}{2}}d\tau\cdot[u,\infty,\tfrac{n}{2}, 2[\tfrac{\bar{s}+3[\frac{n}{2}]+11}{4}]+3, \Omega](t)\}.$$

Since $s=2[\frac{N}{2}]+2n+9$ and $N\geq 3n+3[\frac{n}{2}]+23$, we get

(6.11)
$$\bar{s}+3[\tfrac{n}{2}]+n+12 \leq 2N;$$
$$[\tfrac{\bar{s}+3[\frac{n}{2}]+n+10}{2}]+2 \leq \bar{s};$$
$$2[\tfrac{\bar{s}+3[\frac{n}{2}]+n+12}{4}]+3 \leq 2[\tfrac{N}{2}]+3.$$

Therefore, from (6.8)-(6.10) we conclude

(6.12)
$$[F, 1, \tfrac{n}{2}, \bar{s}+3[\tfrac{n}{2}]+5, \Omega](t)$$
$$\leq C\{M_{\bar{s}}(t)^2+\mathcal{E}_0 M_{\bar{s}}(t)(1+M_{\bar{s}}(t)e^{cM_{\bar{s}}(t)})\}, \quad 0\leq t\leq T.$$

Moreover, from (3.2) we obtain

(6.13)
$$[F, 2, \tfrac{n}{2}, \bar{s}+[\tfrac{n}{2}]+4, \Omega](t)$$
$$\leq C[u,\infty,\tfrac{n}{2}, [\tfrac{\bar{s}+[\frac{n}{2}]+4}{2}]+2, \Omega](t)[u,2,0,\bar{s}+[\tfrac{n}{2}]+6,\Omega](t)$$
$$\leq C[u,\infty,\tfrac{n}{2}, [\tfrac{\bar{s}+[\frac{n}{2}]+4}{2}]+2, \Omega](t)\|\varphi\|_{H^{\bar{s}+[\frac{n}{2}]+6}(\Omega)}$$
$$\exp(c\int_0^t(1+\tau)^{-\frac{n}{2}}d\tau\cdot[u,\infty,\tfrac{n}{2}, 2[\tfrac{\bar{s}+[\frac{n}{2}]+6}{4}]+3, \Omega](t),$$
$$0\leq t\leq T.$$

Since $s=2[\frac{N}{2}]+2n+9$ and $N\geq 3n+3[\frac{n}{2}]+23$, we know that $\bar{s}+[\frac{n}{2}]+6\leq 2N$, $2[\frac{\bar{s}+[\frac{n}{2}]+6}{4}]+3\leq 2[\frac{N}{2}]+3$ and $[\frac{\bar{s}+[\frac{n}{2}]+4}{2}]+2\leq 2[\frac{N}{2}]+3$. Thus, we have

(6.13')
$$[F, 2, \tfrac{n}{2}, \bar{s}+[\tfrac{n}{2}]+4, \Omega](t)$$
$$\leq C\mathcal{E}_0 M_{\bar{s}}(t)e^{cM_{\bar{s}}(t)} \leq C(\mathcal{E}_0+\mathcal{E}_0 M_{\bar{s}}(t)^2 e^{cM_{\bar{s}}(t)}).$$

From (6.7),(6.12) and (6.13') we conclude

(6.14)
$$[u, z, \tfrac{n}{4}, \bar{s}, \Omega](t) \leq C\{\varepsilon_0 + M_{\bar{s}}(t)^2 + \varepsilon_0 M_{\bar{s}}(t)(1 + M_{\bar{s}}(t)e^{CM_{\bar{s}}(t)})\}, \qquad 0 \leq t \leq T.$$

Finally, from (6.3),(6.5) and(6.14) we obtain

(6.15)
$$M_{\bar{s}}(t) \leq C\{\varepsilon_0 + M_{\bar{s}}(t)^2 + \varepsilon_0 M_{\bar{s}}(t)(1 + M_{\bar{s}}(t)e^{CM_{\bar{s}}(t)})\}, \quad 0 \leq t \leq T.$$

where C is a positive constant independent of T.

Set $X = M_{\bar{s}}(t)$, $f(X) = C\{\varepsilon_0 + X^2 + \varepsilon_0 X(1 + Xe^{CX})$ and $g(X) = f(X) - X$. (6.15) expresses $g(X) \geq 0$. Because there exists a positive constant b(b depend only on the constant C in (6.15) but not on T) such that when $0 \leq X \leq b$, $g'_X(X) = 2CX + C\varepsilon_0 + C^2\varepsilon_0 X^2 e^{CX} + 2C\varepsilon_0 Xe^{CX} - 1 \leq -\frac{1}{2} + C\varepsilon_0$ and $g(0) = C\varepsilon_0 > 0$, we can obtain when ε_0 is sufficiently small g(X) has a positive zero point M . Here M do not depend on T. Again because when ε_0 is sufficiently small, $M_{\bar{s}}(0) < M$ and $M_{\bar{s}}(t)$ is a continuous function, we obtain

(6.16)
$$M_{\bar{s}}(t) \leq M.$$

From (3.2) we conclude

(6.17)
$$\sum_{|\alpha|+2j \leq 2N} \| D_x^\alpha \partial_t^j u(t) \|^2_{L^2(\Omega)} \leq C_N \|\varphi\|^2_{H^{2N}(\Omega)} e^{C_N \int_0^t (1+\tau)^{-\frac{n}{2}} d\tau \cdot M_{\bar{s}}(t)} \leq K_N^2 \|\varphi\|^2_{H^{2N}(\Omega)}.$$

where $K_N = (C_N e^{4MC_N})^{1/2}$ which is independent of T. It follows that (6.2) holds. The proof of Lemma 6.1 is complete.

By the technique of Matsumura and Nishida, taking

$$\varepsilon = min(\delta E, \delta E/K_N, \varepsilon_0) \tag{6.18}$$

and applying Theorem2.1 and Lemma6.1 repeatly, we can obtain if $\varphi(x)$ satisfies (1.4), problem (1.2) has a unique global solution such that $U=(u,u_t,\cdots\cdots,\partial_t^N u)\in X_{\infty,E}$. Namely, the solution u satisfies (1.5). The decay estimate (1.6) follows from (6.3) and (6.16). The proof of Theorem1.1 is complete.

REFERENCE

(1) F.E.Browder, On the Spectral Theory of Elliptic Differential Operators, I.Math.Annalen, 142(1961) 22-130.

(2) D.Gilbarg & N.S.Trudinger, Elliptic Partial Differential Equations of Second Order, Springer-Verlag, Berlin, Heidelberg, New York, (1977).

(3) A.Matsumura & T.Nishida, The Initial Value Problem for the Equations of Motion of Viscous and Heat-Conductive Gases, J.Math.Kyoto Univ. 20 (1980) 67-104.

(4) Y.Tsutsumi, Global Solution of the Nonlinear Schrödinger Equation in Exterior Domains. Comm. in P.D.E. 8(12) (1983) 1337-1374.

(5) Y.Tsutsumi, Local Energy Decay of the Solution to the Free Schrödinger Equation in Exterior Domains, (Preprint).

(6) B.R.Vainberg, On the Analytical Properties of the Resolvent for a Certain Class of Operator-Pencils, Math. USSR.Sbornic 6 (1968) 241-273.

(7) Zheng Songmu & Chen Yunmei, Global Existence for Nonlinear Parabolic Equation, (Preprint).

(8) Zheng Songmu, Global Existence of the Initial Boundary Value Problem for Nonlinear Parabolic Equation,(Preprint).

HIGHER DIMENSIONAL SUBSONIC FLOW

Biao Ou and Guangchang Dong
Zhejiang University

The existence and uniqueness of steady irrotational subsonic flow of a perfect gas past a given profile has been studied extensively in the two dimensional case. In Ref. 1, Bers proved the existence and uniqueness of plane subsonic flow around a given profile. For higher space dimensions, Finn and Gilbarg[2] proved the existence and uniqueness in three dimensionals provided the velocity was not too large (the maximum Mach number less than 0.7). Dong[3] proved the existence and uniqueness in arbitrary dimensionals. The proof is given by means of the method of *a priori* estimate of solutions and the Leray-Schauder fixed point theorem. Ou[4] extends the results of Ref. 3 to the case of subsonic flow with source by variational method. Hence it is easier to use the direct method for the approximation solution.

The results and proofs of Ref. 4 can be sketched as follows.

The velocity potential $\varphi(x)$ in $\mathbb{R}^n$ satisfies the equation

$$\sum_{i=1}^{n} \frac{\partial}{\partial x_i}\left(\rho \frac{\partial \varphi}{\partial x_i}\right) = f(x)$$

where $\rho = \rho(q)$ is the density of gas and $q = \left(\sum_{i=1}^{n} u_i^2\right)^{\frac{1}{2}}$, $u_i = \dfrac{\partial \varphi}{\partial x_i}$ $(1 \le i \le n)$. $f(x)$ is the source or sink which represents the effect of chemical reaction or electric-magnetic force or gas mixture. Hence we have

$$\sum_{i,j=1}^{n} a_{ij} \frac{\partial^2 \varphi}{\partial x_i \partial x_j} = f(x) \quad , \quad a_{ij} = \rho\delta_{ij} + \frac{\rho'}{q} \frac{\partial \varphi}{\partial x_i} \frac{\partial \varphi}{\partial x_j} \quad ,$$

$$\delta_{ij} = \begin{cases} 1 \ (i = j) \\ 0 \ (i = j) \end{cases} .$$

The eigenvalues of (a_{ij}) are $\lambda_1 = \rho + \rho' q$, $\lambda_2 = \ldots = \lambda_n = \rho$. Let $\rho(0) = 1$, $\rho'(0) = 0$ and $\rho + \rho' q$ be monotone decreasing and q_c be the root of $\rho + \rho' q$ (called critical speed). The profile Γ is a bounded closed surface. Assume that the curvature of Γ is bounded. The region outside Γ is denoted by Ω.

The subsonic flow $\varphi(x)$ around Γ satisfies

$$(*) \begin{cases} \sum_{i=1}^{n} \frac{\partial}{\partial x_i} (\rho \frac{\partial \varphi}{\partial x_i}) = f(x) \quad , \quad x \in \Omega \quad , \\ \left.\frac{\partial \varphi}{\partial N}\right|_{\Gamma} = 0 \quad , \quad N \text{ is the outer normal of } \Gamma, \\ \nabla\varphi|_{\infty} = u^{\infty} \quad , \quad u^{\infty} \text{ is a given constant vector (called incoming flow)} \quad , \\ \sup_{\bar{\Omega}} |\nabla\varphi| < q_c \quad . \end{cases}$$

It is easy to see that the necessary condition for the existence of $u(x)$ is

$$\int_{\Omega} f(x)dx = - \int_{\Gamma} \rho \frac{\partial \varphi}{\partial N} dS = 0 \quad .$$

The main result of Ref. 4 is as follows.

Assume that

$$\int_\Omega f(x)dx = 0 \quad , \quad |f(x)| \le \frac{C}{|x|^{n+z}} \quad , \quad |Df| \le \frac{C}{|x|^{\frac{n}{2}+z}}$$

$$(z > 0, \quad \{x = 0\} \notin \bar{\Omega}) \quad .$$

Assume that there exists a subsonic flow $\varphi^*(x)$ for a special incoming flow $u^\infty = u_0^*$. Then a family of subsonic flows U exist, for the incoming flow varying continuously from u_0^*, and there is a sequence $u_m \in U$ such that

$$\lim_{m\to\infty} \sup_{\bar{\Omega}} |\nabla\varphi_m| = q_c \quad .$$

The subsonic flow solution is unique and depends continuously on u^∞.

The proof for existence is first to reduce (*) to the variational problem

$$\delta \int_\Omega [G(q^2) + f\varphi]dx = 0 \quad , \quad G(q^2) = \int \rho q dq$$

and secondly to find the minimum of the functional

$$I(\varphi) \equiv \int_\Omega [\tilde{G}(q^2) - \tilde{G}(q_0^2) - 2G'(q_0^2)u_0(\nabla\varphi - u_0) + f(\varphi - u_0 x)]dx$$
$$+ 2G'(q_0^2)u_0 \int_\Gamma NdS = \min_{\varphi\in\Phi(u_0)}$$

where $\tilde{G} \equiv G$ is the subsonic part. We substitute G by $\tilde{G}$ such that $\sum_{i,j=1}^{n} a_{ij} \frac{\partial^2\varphi}{\partial x_i \partial x_j} = f$ is uniformly elliptic in $\bar{\Omega}$, where $N = (\cos(N, x_1), \ldots, \cos(N, x_n))$ and $\Phi(u_0)$ is a Banach space such that $\varphi - u_0 x \in W^{1,2}_{loc}(\bar{\Omega})$ and the norm of $\varphi \in \Phi(u_0)$ is defined by

$$\|\varphi\| = \left[\int_\Omega |\nabla(\varphi - u_0 x)|^2 dx + \int_\Omega \frac{1}{|x|^{n+2}} (\varphi - u_0 x)^2 dx\right]^{\frac{1}{2}} \qquad (z > 0) \quad .$$

In Ref. 4 it is proved that $I(\varphi)$ is bounded below in $\Phi(u_0)$ and there exists only one $\varphi_0 \in \Phi(u_0)$ such that $I(\varphi_0) = \min$. Then the interior regularity, boundary regularity and regularity at ∞ of φ_0 is studied, and it is shown that φ_0 depends continuously on u^∞.

References

1. L. Bers, "Existence and uniqueness of subsonic flow pass a given profile", Commun. Pure Appl. Math. 7 (1954) 441-504.

2. R. Finn and D. Gilbarg, "Three-dimensional subsonic flow, and asymptotic estimates for elliptic partial differential equations", Acta Math. 98 (1957) 265-296.

3. G.C. Dong, "Three-dimensional subsonic flows and their boundary value problems extended to higher dimensions", Journal of Zhejiang University 1 (1979) 36-61. Also see M.R.C. Tech. Sum. Report #2193 (1981), University of Wisconsin.

4. B. Ou, "Higher dimensional subsonic flow around a given profile with source", preprint.

Application of Generalized Distance in the Study of Stability of Motion

Guan Ke-ying

(Department of Mathematics, Beijing Institute of Aeronautics and Astronautics)

In the study of the nonlinear stability of the single soliton solution of the K-dV equation

$$u_t + 6uu_x + u_{xxx} = 0$$

Benjamin introduced a special measure--"generalized distance" to describe the difference between any two solutions $u_1(x,t)$ and $u_2(x,t)$

$$d(u_1, u_2) = \inf_{\xi\in\mathbb{R}}\left\{\int_{-\infty}^{\infty}\left[(\tau_\xi u_1 - u_2)^2 + \left(\frac{\partial \tau_\xi u_1}{\partial x} - \frac{\partial u_2}{\partial x}\right)^2\right] dx\right\}^{\frac{1}{2}} \qquad (1)$$

where τ_ξ is an operator of the shift transformation

$$\tau_\xi u(x,t) = u(x-\xi,t) \quad , \quad \xi\in\mathbb{R}$$

The reason why this measure is used comes from the following facts. The soliton moves with a constant velocity while its shape is invariable. But, if there is a small diturbance at the initial time, its shape will change, and the disturbed solution will break at last into two parts: the first part, the main part, is still a soliton approximate to the original one, but its velocity is usually different to the original one's; the other part is formed by several very small solitons and a decayed wave which can be neglected. Therefore, the stability of the original soliton is meant by that the shape and the size of the main soliton of the disturbed solution is approximate to the original one's. In order to measure this approximation, we must neglect the difference of the positions between these two solitons. Hence, the shift transformation is introduced to the distance to make these two solitons coincide almost at each time.

The generalized distance has following properties:

I. It is non-negative and symmetric, i.e.,

$$d(u_1, u_2) = d(u_2, u_1) \geq 0$$

II. It satisfies the triangle inequality:

$$d(u_1, u_2) \leq d(u_1, u_3) + d(u_2, u_3)$$

III. It is a homogeneous form of degree 1, i.e.,

$$d(\alpha u_1, \alpha u_2) = |\alpha| d(u_1, u_2), \quad \alpha \in \mathbb{R} \text{ (or } \mathbb{C})$$

IV. If $u_1 = u_2$, then $d(u_1, u_2) = 0$.

But, if $d(u_1, u_2) = 0$, it does not need to imply $u_1 = u_2$.

The generalized distance is similar to the seminorm on a functional space, but it is more general, since it is possible that

$$d(u_1 - u_2, 0) \neq d(u_1, u_2)$$

according to the generalized distance, while according to the seminorm

$$\|u_1 - u_2 - 0\| = \|u_1 - u_2\| .$$

The generalized distance can be widely used in the study of stability of motion system, especially when the motion system is complicated. The reason is that, usually in a complicated motion system, only one part of motion quantities is stable, while the other part is instable. In this situation, the generalized distance can be used to measure only the difference between such parts of quantities that they are important to the stability. The other part of motion quantities which is not important to the stability can be neglected by using the generalized distance. The generalized distance (1) given by Benjamin measures only the difference of the shapes and the sizes between ,any two solutions, since it is important to the stability of soliton. What is neglected is the difference of the positions, which is not important to the stability.

Consider two further examples:

Example 1.

The stability of the single soliton solution

$$\Psi_s(x, t) = \varphi_s(x, t) \cdot \exp[i \theta_s(x, t)]$$

$$\begin{cases} \varphi_s(x,t) = A \cdot \text{sech}[A(xBBt)] \\ \theta_s(x,t) = B/2 \cdot (x-Ct) \end{cases}$$

of the nonlinear Schrödinger equation

$$i\frac{\partial \Psi}{\partial t} + \frac{\partial^2 \Psi}{\partial x^2} + 2|\Psi|^2\Psi = 0 \tag{2}$$

where $A,B,C \in \mathbb{R}$, $A=0$, $2BC=B^2-4A^2$.

This problem is a little more difficult than the K-dV equation, since its solution is complex valued. Through a physical consideration, I introduced the following generalized distance between any two complex valued solutions $\Psi_1(x,t) = \varphi_1(x,t) \cdot \exp[i\theta_1(x,t)]$ and $\Psi_2(x,t) = \varphi_2(x,t) \cdot \exp[i\theta_2(x,t)]$ where φ and θ are real valued functions,

$$d(\Psi_1, \Psi_2) = \inf_{\xi \in \mathbb{R}} \left(\int_{-\infty}^{\infty} \left\{ (\tau_\xi \varphi_1 - \varphi_2)^2 + \left(\frac{\partial \tau_\xi \varphi_1}{\partial x} - \frac{\partial \varphi_2}{\partial x} \right)^2 + \left[\tau_\xi \varphi_1 \left(\frac{\partial \tau_\xi \theta_1}{\partial x} - \frac{V_1}{2} \right) - \varphi_2 \left(\frac{\partial \theta_2}{\partial x} - \frac{V_2}{2} \right) \right]^2 \right\} dx \right)^{\frac{1}{2}} \tag{3}$$

where

$$\tau_\xi f(x,t) = f(x-\xi, t)$$

and

$$V_i = \frac{d}{dt}\langle x \rangle_i \equiv \frac{d}{dt} \frac{\int_{-\infty}^{\infty} \Psi_i^* x \Psi_i \, dx}{\int_{-\infty}^{\infty} \Psi_i^* \Psi_i \, dx} \qquad i=1,2.$$

This distance reflects the difference of the shapes and the sizes by the first two terms in the integration of (3) and the difference of the "internal energies" between the two solutions by the third term. What are neglected are the differences of their positions, mean velocities and phases. Just by this generalized distance, we have introduced an appropriate Lyapunov function and proved the nonlinear stability of the single soliton solution of the NLS equation.[2]

Example 2. The stability of the equilibrium rotation of a a rigid body with a cavity filled with liquid.

The motion is described with the following quantities

$$u(t) = \{\vec{\gamma}(t), \vec{\omega}(t), \vec{v}(\vec{x},t)\}$$

which are measured in the coordinate system

$$\{0, x_1, x_2, x_3\}$$

which rotates with a constant angular velocity

$$\vec{\Omega}_0 = (0, 0, \Omega_0).$$

The 3X3 tensor

$$\overleftrightarrow{\gamma}(t) = \begin{pmatrix} \gamma_{11}(t), & \gamma_{12}(t), & \gamma_{13}(t) \\ \gamma_{21}(t), & \gamma_{22}(t), & \gamma_{23}(t) \\ \gamma_{31}(t), & \gamma_{32}(t), & \gamma_{33}(t) \end{pmatrix}$$

is used to describe the relation between the coordinate system $\{0, x_1, x_2, x_3\}$ and the coordinate system $\{0, x_1', x_2', x_3'\}$ which is fixed in the rotating body, i.e.,

$$x_l' = \gamma_{lm}(t) x_m.$$

$\vec{\omega}(t)$ is the angular velocity of the body, $\vec{v}(\vec{x}, t)$ is the velocity of the liquid at the position $\vec{x}$ and time t in the cavity. At the initial time t_0, these quantities are

$$u^{(0)}(t_0) = \{\overleftrightarrow{I}, 0, 0\}.$$

The system of motion equations is

$$\begin{cases} \dfrac{d}{dt}\overleftrightarrow{\gamma}(t) = \overleftrightarrow{\gamma}(t)\cdot\overleftrightarrow{\epsilon} \\ \dfrac{d}{dt}[\vec{\omega}\cdot\overleftrightarrow{J}_1] + \{\vec{\Omega}_0 \times [\vec{\omega}\cdot\overleftrightarrow{J}_1] + \dfrac{d}{dt}[\vec{\Omega}_0\cdot\overleftrightarrow{J}_1] \\ \qquad = -\iint_S \vec{r}\times(-p_1\overleftrightarrow{I} + \overleftrightarrow{\tau})\cdot\vec{n}\,ds - Mgh(\gamma_{32}, -\gamma_{31}, 0) \\ \qquad\quad - \{\vec{\Omega}_0 \times [\vec{\Omega}_0 \cdot (\overleftrightarrow{J}_1 + \overleftrightarrow{J}_2)]\} \\ \dfrac{d}{dt}\vec{v} + 2\vec{\Omega}_0\times\vec{v} = \dfrac{1}{\rho}\nabla(-p_1\overleftrightarrow{I} + \overleftrightarrow{\tau}) \\ \nabla\cdot\vec{v} = 0 \\ \vec{v}\big|_s = \vec{\omega}\times\vec{r}\big|_s \end{cases} \tag{4}$$

where

$$\overleftrightarrow{\epsilon} = \begin{pmatrix} 0, & \omega_3, & -\omega_1 \\ -\omega_3, & 0, & \omega_2 \\ \omega_1, & -\omega_2, & 0 \end{pmatrix}$$

$\overleftrightarrow{J}_1$ and $\overleftrightarrow{J}_2$ are respectively the inertia tensors of the body and the liquid

$$\overleftrightarrow{J}_i = \overleftrightarrow{\gamma}^T \begin{pmatrix} A_i, & 0, & 0 \\ 0, & B_i, & 0 \\ 0, & 0, & C_i \end{pmatrix} \overleftrightarrow{\gamma}, \qquad i = 1, 2$$

where A_i, B_i, C_i are positive constants. The components of the tensor $\overleftrightarrow{\tau}$ are

$$\tau_{ij} = \mu\left(\frac{\partial V_i}{\partial x_j} + \frac{\partial V_j}{\partial x_i}\right)$$

M is the total mass of the body and the liquid, h the height of the center of gravity of this system, and g is the gravitational acceleration.

The system of equations of motion (4) is very complicated. It includes the operations of ordinary derivatives, the partial derivatives and the integrations, and it is strong nonlinear.

By a physical analysis, we noticed that among these motion quantities, only one part can be stable under certain conditions, i.e., the direction of the rotation--described by $\gamma_{33}(t)$, the angular velocity of the body and the velocity of the liquid. But the angles of the rotation of the body described by $\gamma_{11}(t)$ and $\gamma_{12}(t)$ are instable in any cases. Based on this analysis, we introduced the following generalized distance between any two groups of motion quantities $u^{(1)}(t) = \{\overleftrightarrow{\gamma}^{(1)}(t), \vec{\omega}^{(1)}(t), \vec{v}^{(1)}(\vec{x},t)\}$ and $u^{(2)}(t) = \{\overleftrightarrow{\gamma}^{(2)}(t), \vec{\omega}^{(2)}(t), \vec{v}^{(2)}(x,t)\}$

$$d(u^{(1)}(t), u^{(2)}(t)) = \Big[(\gamma_{33}^{(1)}(t) - \gamma_{33}^{(2)}(t))^2 + (\vec{\omega}^{(1)}(t) - \vec{\omega}^{(2)}(t))^2 + \iiint (\vec{v}_i^{(1)}(\vec{x},t) - \vec{v}^{(2)}(\vec{x},t))^2\, dv\Big]^{\frac{1}{2}} \qquad (5)$$

Just by this distance, we have obtained a suitable Lyapunov function and proved the nonlinear stability of the equilibrium rotation of this system.[3]

The ordinary distance, the norm or the seminorm of the functional space do not work in treating the above-mentioned examples.

The generalized distance which is introduced and used first by Benjamin will play important role more and more in the further study of stability of motion.

REFERENCES

[1] T.B.Benjamin,F.R.S.,Proc. Roy. Soc. Lond., Series A, Vol.328, (1972)153-183.

[2] Guan Ke-ying, Stability of Single Soliton Solution of Nonlinear Schrödinger Equation, to appear.

[3] Qin Yuan-xun, Guan Ke-ying, Li Li, Stability of the Equilibrium Rotation of a Rigid Body with a Cavity Filled with Liquid, KEXUE TONGBAO, No.4(1984), 198-201

Penalty-Nonconforming Finite Element Method for Stokes Equations

Han Houde
Department of Mathematics
Peking University
Beijing, China

§ 1. Introduction

Consider the following boundary value problem of Stokes equations

$$(1.1)\qquad \begin{cases} -\mu \Delta \vec{u} + \nabla p = \vec{f} & \text{in } \Omega \\ \operatorname{div} \vec{u} = 0 & \text{in } \Omega \\ \vec{u} = 0 & \text{on } \partial\Omega \end{cases}$$

where $\mu > 0$ is the viscosity, $\vec{u} = (u_1, u_2 \cdots u_n)$ is the velocity field, p is the pressure, $\vec{f}$ is the body force density, and Ω is an open bounded domain in $\mathbb{R}^n$, $\partial\Omega$ is the boundary of Ω satisfying Lipschitz condition.

Let $V = (H_0^1(\Omega))^n$, $M = \{ q \in H^0(\Omega), \int_\Omega q \, dx = 0 \}$, where $H_0^1(\Omega)$, $H^0(\Omega)$ denote the usual Sobolev spaces. Then boundary value problem (1.1) is equivalent to the following variational problem:

Find $(\vec{u}, p) \in V \times M$, such that

$$(1.2)\qquad \begin{cases} a(\vec{u}, \vec{v}) + b(\vec{v}, p) = \langle \vec{f}, \vec{v} \rangle & \forall \vec{v} \in V \\ b(\vec{u}, q) = 0 & \forall q \in M \end{cases}$$

where $$a(\vec{u},\vec{v}) = \mu \sum_{i,j=1}^{2} \int_\Omega \frac{\partial u_i}{\partial x_j}\frac{\partial v_i}{\partial x_j}\,dx,$$

$$b(\vec{v},q) = -\int_\Omega (\mathrm{div}\,\vec{v})\,q\,dx = -(\mathrm{div}\,\vec{v},q),$$

$$\langle \vec{f},\vec{v}\rangle = \int_\Omega \vec{f}\cdot\vec{v}\,dx.$$

A direct finite-element approximation of problem (1.2) leads to the so-called mixed finite element method using conforming and nonconforming finite elements which has been studied extensively in the literature[1-5]. An alternative formulation of (1.2) is provided by the exterior penalties where (1.2) is replaced by a family of perturbations consisting of unconstrained problems depending on a penalty parameter $\varepsilon > 0$. Then a penalty approximation of the variational problem (1.2) consists of finding $(\vec{u}_\varepsilon, p_\varepsilon) \in V \times M$, such that

$$(1.3)\quad \begin{cases} a(\vec{u}_\varepsilon,\vec{v}) + b(\vec{v},p_\varepsilon) = \langle \vec{f},\vec{v}\rangle & \forall \vec{v}\in V \\ b(\vec{u}_\varepsilon,q) - \varepsilon(p_\varepsilon,q) = 0 & \forall q \in M \end{cases}$$

We can eliminate the pressure p_ε from the last equation and obtain the penalty approximation of the variational problem (1.2) containing only unknown functions $\vec{u}_\varepsilon$:

$$(1.3)^*\quad \begin{cases} a(\vec{u}_\varepsilon,\vec{v}) + \varepsilon^{-1}(\mathrm{div}\,\vec{u}_\varepsilon, \mathrm{div}\,\vec{v}) = \langle \vec{f},\vec{v}\rangle \quad \forall \vec{v}\in V \\ p_\varepsilon = -\frac{1}{\varepsilon}\,\mathrm{div}\,\vec{u}_\varepsilon \end{cases}$$

Finite element methods based on variational problem (1.3)* have been proposed by several authors[5-9]. On the basis of numerical experiments, they have determined that it is necessary to use "reduced integration" of the penalty terms in formulation (1.3)* in order to obtain results which are physically reasonable. These reduced-integration-penalty schemes also have been studied in mathematics, and the error estimate is given. We refer to the work of Oden, Kikuchi and Song[10].

In this paper, nonconforming finite elements are applied to penalty finite element approximation of Stokes equations. An abstract error estimate is given. For nonconforming triangular elements, in particular, the analysis shows that the "reduced integration" technique is not necessary in the integration of the penalty term on each element. It means that a loss of precision is avoided in this case. Therefore the optimal error estimate is given.

§2 An abstract error estimate

Let V_h, M_h be two finite dimensional spaces, and $M_h \subset M$; in general, V_h is not a subspace of V. Suppose $V_h \subset (H^0(\Omega))^n$, We extend the definitions of $a(\vec{u}, \vec{v})$ and $b(\vec{v}, q)$ to $(V_h \cup V) \times (V_h \cup V)$ and $(V_h \cup V) \times M$ respectively. Let $a_h(\vec{u}, \vec{v})$ and $b_h(\vec{v}, q)$ denote those extensions, and

$$a_h(\vec{u}, \vec{v}) = a(\vec{u}, \vec{v}) \qquad \forall\ \vec{u}, \vec{v} \in V,$$

$$b_h(\vec{v}, q) = b(\vec{v}, q) \qquad \forall\ \vec{v} \in V,\ q \in M.$$

Furthermore, assume that

(1) there are three constants $\alpha > 0$, $A > 0$, $B > 0$ such that

$$(2.1) \quad \begin{cases} a_h(\vec{v}_h, \vec{v}_h) \geq \alpha \|\vec{v}_h\|_h^2 & \forall\ \vec{v}_h \in V_h \\ |a_h(\vec{u}_h, \vec{v}_h)| \leq A \|\vec{u}_h\|_h \|\vec{v}_h\|_h & \forall\ \vec{u}_h, \vec{v}_h \in V_h \\ |b_h(\vec{v}_h, q_h)| \leq B \|\vec{v}_h\|_h \|q_h\|_M & \forall\ \vec{v}_h \in V_h,\ q_h \in M_h \end{cases}$$

(2) there is an operator $\rho_h : V_h \to M_h$ satisfying

(2.2) $$-b_h(\vec{v}_h, q_h) = (\rho_h \vec{v}_h, q_h) \qquad \forall q_h \in M_h,$$

(3) there is a constant C indepent of h, such that

(2.3) $$\|\vec{v}\|_{0,\Omega} \leq C \|\vec{v}_h\|_h \qquad \forall \vec{v}_h \in V_h$$

(4) there exists a constant $\beta_h > 0$, such that

(2.4) $$\sup_{\vec{v} \in V_h} \frac{b_h(\vec{v}_h, q_h)}{\|\vec{v}_h\|_h} \geq \beta_h \|q_h\|_M \qquad \forall q_h \in M_h.$$

We consider the nonconforming finite element approximation of (1.3).

Find $(\vec{u}_h^\varepsilon, p_h^\varepsilon) \in V_h \times M_h$, such that

(2.5) $$\begin{cases} a_h(\vec{u}_h^\varepsilon, \vec{v}_h) + b_h(\vec{v}_h, p_h^\varepsilon) = \langle \vec{f}, \vec{v}_h \rangle & \forall \vec{v}_h \in V_h, \\ b_h(\vec{u}_h^\varepsilon, q_h) - \varepsilon(p_h^\varepsilon, q_h) = 0 & \forall q_h \in M_h. \end{cases}$$

By the condition (2.2), the problem (2.5) is equivalent to the following problem

(2.5)* $$\begin{cases} a_h(\vec{u}_h^\varepsilon, \vec{v}_h) + \frac{1}{\varepsilon}(\rho_h \vec{v}_h, \rho_h \vec{u}_h^\varepsilon) = \langle \vec{f}, \vec{v}_h \rangle & \forall \vec{v}_h \in V_h \\ p_h^\varepsilon = -\frac{1}{\varepsilon} \rho_h \vec{u}_h^\varepsilon \end{cases}$$

By Lax-Milgram theorem[11], the problem (2.5)* has a unique solution $(\vec{u}_h^\varepsilon, p_h^\varepsilon)$. We have

Theorem 2.1 Suppose that hypothesises (1)-(4) hold and $(\vec{u}_h^\varepsilon, p_h^\varepsilon)$ is the solution to problem (2.5)*, then the following error estimate holds

$$\|\vec{u}-\vec{u}_h^\varepsilon\|_h + \beta_h \|p-p_h^\varepsilon\|_M \leqslant \left(1+\frac{1}{\beta_h}\right)\Big\{\inf_{\vec{v}_h\in V_h}\|\vec{u}-\vec{v}_h\|_h +$$

(2.6)

$$+\inf_{q_h\in M_h}\|p-q_h\|_M + \|E_h\|_{V_h'} + \frac{\varepsilon}{\beta_h}\|\vec{f}\|_{0,\Omega}\Big\},$$

where $(\vec{u}, p)$ is the solution of problem (1.2) and

$$\|E_h\|_{V_h'} = \sup_{\vec{v}_h\in V_h}\frac{|a_h(\vec{u},\vec{v}_h)+b_h(\vec{v}_h,p)-\langle\vec{f},\vec{v}_h\rangle|}{\|\vec{v}_h\|_h}.$$

The proof can be found in [12].

§3. Nonconforming triangular elements

In this section we shall confine ourself to the case n=2. Moreover, suppose Ω is an open convex polygen. $\bar{\Omega}$ is divided into some triangles $\{K\}$. Let $\mathcal{T}_h$ denote this triangulation satisfying

(1) $\bar{\Omega} = \sum_{K\in\mathcal{T}_h} K$

(2) for each distinct K_1 and $K_2 \in \mathcal{T}_h$, either $K_1 \cap K_2$ is empty or K_1 and K_2 have a common vertex or K_1 and K_2 have a common side.

(3) let $h_K = diam(K)$,

$$\rho_K = \sup\{ \text{diam}(S);\ S \text{ is a circle contained in } K \},$$

$$h = \max_{K \in \mathcal{T}_h} \{ h_K \}, \quad \text{and}$$

$$\rho = \min_{K \in \mathcal{T}_h} \{ \rho_K \};$$

suppose $h/\rho \leq \sigma$, where $\sigma > 0$ is a constant.

We choose the nonconforming linear triangular element to form space V_h. Let N_0 denote the set of the midpoints of the sides of $\mathcal{T}_h$ on the boundary of Ω and N_1 denote the set of the midpoints of the sides of $\mathcal{T}_h$ in the interior of Ω. Suppose

$$\mathring{S}_h = \{ v \mid v|_K \in P_1(K),\ \forall K \in \mathcal{T}_h,\ v \text{ is continuous on } N_1 \text{ and } v(b) = 0,\ \forall b \in N_0 \},$$

$$M_h = \{ q \mid q|_K \in P_0(K),\ \int_\Omega q\, dx = 0 \},$$

$$V_h = \mathring{S}_h \times \mathring{S}_h ,$$

where $P_k(K)$ denote the space of all polynomials of degree $\leq k$ on element K. Obviously, M_h is a subspace of M, but V_h is not a subspace of V. Therefore we need to define the bilinear forms:

$$a_h(\vec{u}, \vec{v}) = \mu \sum_{K \in \mathcal{T}_h} \sum_{i,j=1}^{2} \int_K \frac{\partial u_i}{\partial x_j} \frac{\partial v_i}{\partial x_j}\, dx, \quad \forall \vec{u}, \vec{v} \in V_h \cup V,$$

$$b_h(\vec{v}, p) = -\sum_{K \in \mathcal{T}_h} \int_K (\text{div}\, \vec{v})\, p\, dx, \quad \forall v \in V_h \cup V,\ q \in M.$$

Let $\|\vec{v}_h\|_h = (a_h(\vec{v}_h, \vec{v}_h))^{1/2}$, $\forall \vec{v}_h \in V_h$. It is straightforward to see that hypothesis (2.1) holds. In this case, operater ρ_h is given by

(3.1) $$\rho_h \vec{v}_h = \operatorname{div} \vec{v}_h \ , \ \text{on } K \ , \ \forall K \in \mathcal{T}_h \ , \ \forall \vec{v}_h \in V_h \, .$$

Obviously, $\rho_h \vec{v}_h \in M_h$ and

(3.2) $$(\rho_h \vec{v}, q_h) = -b_h(\vec{v}_h, q_h), \quad \forall \vec{v}_h \in V_h, q_h \in M_h \, .$$

Hence, penalty-nonconforming finite element approximate problem (2.5)* is redued to

(3.3) $$\begin{cases} a_h(\vec{u}_h^\varepsilon, \vec{v}_h) + \frac{1}{\varepsilon} \sum_{K \in \mathcal{T}_h} (\operatorname{div} \vec{u}_h^\varepsilon, \operatorname{div} \vec{v}_h)_K = \langle \vec{f}, \vec{v}_h \rangle, \ \forall \vec{v}_h \in V_h \\ p_h^\varepsilon = -\frac{1}{\varepsilon} \operatorname{div} \vec{v}_h^\varepsilon \quad \text{on } K, \quad \forall K \in \mathcal{T}_h \end{cases}$$

It is clear that problem (3.3) is the nonconforming finite element approximation of problem (1.3)* and the "reduced integration" technique is not necessary for this case. We have

Theorem 3.1 Suppose that $(\vec{u}, p)$ is the solution of problem (1.2) satisfying $\vec{u} \in (H^2(\Omega))^2$, $p \in H^1(\Omega)$ then the following inequality holds

$$\| \vec{u} - \vec{u}_h^\varepsilon \|_h + \| p - p_h^\varepsilon \|_M \leq C \{ h(|\vec{u}|_{2,\Omega} + |p|_{1,\Omega}) + \varepsilon \| \vec{f} \|_{0,\Omega} \},$$

where C is a constant independent of h, $(\vec{u}_h^\varepsilon, p_h^\varepsilon)$ is the solution to (3.3).

Similarly, we can use nonconforming quadratic triangular element to form the approximate space V_h for solving the penalty variational problem (1.3)*. In this case, the "reduced integration" technique is not necessary in the integration of the penalty term on each element and the optimal error estimate is given.

References

[1] M. Crouzeix and P.A.Raviart, Conforming and nonconforming finite element methods for solving the stationary Stokes equations. R.A.I.R.O. 3(1974) 33-76.

[2] M.Fortin, An analysis of the convergence of mixed finite element methods R.R.I.R.O. 11(4) (1977) 341--354.

[3] H.Han, Nonconforming elements in the mixed finite element method, J. Comput. Math. Vol. 2, No. 3(1984) 223-233.

[4] R.Temam , Navier Stokes Equations, North-Holland, Amsterdam, 1977.

[5] V.Girault and P.A.Raviart, Finite Element Approximation of the Navier-Stokes Equations, Lecture Notes in Mathematics 749, Springer, New York, 1979.

[6] I.Fried, Finite element analysis of incompressible material by residual energy balancing. Internal. J. Solids and Structures 10(1974) 993-1002.

[7] T.J.R.Hughes, R.L.Taylor and J.F.Levy, A finite element method for incompressible viscous flows. Second Internal. Symp. On Finite Element Methods in Flow Problems. S. Margherita Lique. Italy 1976.

[8] J.N.Reddy, On the accuracy and existence of solutions to primitive variable models of viscous incompressible fluids, Internal. J. Engng. Sci. 16(1978) 921-923.

[9] O.C.Zienkiewicz, R.L.Taylor and J.M.Too, Reduced integration technique in general analysis of plates and shells. Internal. J. Numer. Meths. Engrg. 3(1971) 275-290.

[10] J. T. Oden, N. Kikuchi and Y. J. Song, Penalty-finite element methods for the analysis of stokesion flows, Computer Methods in Applied Mechanics and Engineering 31(1982) 297-329.

[11] P. G. Ciarlet, The finite element method for elliptic

problems, North-Holland, 1978.
[12] H. Han, An analysis of penalty-nonconforming finite element method for Stokes equations, to appear.

Nonlocal boundary value problems and homogenization of boundary conditions

Li Ta-tsien (Li Da-qian)

Fudan University

Shanghai, China

1. Introduction

Let Ω be a bounded domain in R^n with piecewise smooth boundary Γ, we consider a linear self-ajoint elliptic equation of second order in Ω, without loss of generality, here we only consider the Laplace equation

$$(1.1)\qquad \Delta u=0 \qquad \left(\Delta = \sum_{i=1}^{n} \frac{\partial^2 u}{\partial x_i^2} \right).$$

If we prescribe on Γ_0, a part of Γ, the following boundary condition

$$\begin{cases} (1.2) & u = c \text{ (unknown constant) on } \Gamma_0, \\ (1.3) & \displaystyle\int_{\Gamma_0} \frac{\partial u}{\partial n}\, dS = A_0 \text{ (given constant)}, \end{cases}$$

where $\vec{n}$ is the unit outward normal vector on Γ_0, and on the remainder of Γ, i.e. on $\Gamma\backslash\Gamma_0$, we prescribe the boundary condition of Dirichlet type or of Neumann type etc., then we get a nonlocal boundary value problem which will be discussed in this paper. In the evolution case, for the heat equation, the wave equation, the Schrödinger equation and the pseudoparabolic equation etc., we can similarly consider this kind of nonlocal boundary value problems, in this paper, however, we only concentrate our attention on the stationary case.

There are a lot of important problems in practice, which lead us to study this kind of nonlocal boundary value problems. For instance, the boundary condition for the temperature on the surface of a cable, for the electrostatic potential on the surface of a charged electrical conductor, for the potential on an electrode in the electrical log, for the stress function in the torsion problem on the boundary of an inner hole of an elastic rod with a multiply connected section etc., are all of this kind of nonlocal boundary conditions.

In these examples, the first condition (1.2) means that on Γ_0 the temperature, the potential or the stress function etc. is a constant to be determined, while the second condition (1.3) means that the total (heat, electric,$\cdots$) flux is given. From this point of view, the boundary condition (1.2)-(1.3) is called the equivalued surface boundary condition or the total flux boundary condition, and then the corresponding nonlocal boundary value problem is called the equivalued surface boundary value problem or the total flux boundary value problem.

By means of a variational principle, there is no difficulty to get the existence and uniqueness theorem for this kind of nonlocal boundary value problems, and then the finite element method can be used to give the corresponding numerical solutions.

We now return to the problem of electrical log in the oil exploitation. The potential u essentially satisfies the Laplace equation (1.1) in the domain, the nonlocal boundary condition (1.2)-(1.3) on the surface Γ_0 of an electrode, where A_0 is the total electric current emitting from the electrode, and the boundary condition of Dirichlet type and (or) of Neumann type on the remainder of the boundary Γ .

Recently, in order to fit better the surface of the well, a piecewise electrode has been used in the oil field. In this case, the surface Γ_0 of the original electrode is divided into two regular subsets Γ_0^ε and $\tilde{\Gamma}_0^\varepsilon$, where Γ_0^ε is the surface of insulating material, while $\tilde{\Gamma}_0^\varepsilon$, the surface of the piecewise elactrode, is composed of a number of connected surfaces $\tilde{\Gamma}_{0,i}^\varepsilon$ $(i=1,\cdots,m(\varepsilon))$:

(1.4) $$\tilde{\Gamma}_0^\varepsilon = \bigcup_{i=1}^{m(\varepsilon)} \tilde{\Gamma}_{0,i}^\varepsilon,$$

in which every $\tilde{\Gamma}^{\varepsilon}_{0,i}$ denotes the surface of a connected piece of electrode. Since there is a short-circuit among these connected pieces of electrode, the potential u must be a constant on the whole $\tilde{\Gamma}^{\varepsilon}_{0} = \cup\, \tilde{\Gamma}^{\varepsilon}_{0,i}$; on the other hand, the total electric current emitting from $\tilde{\Gamma}^{\varepsilon}_{0}$ is still the given constant A_0 . Thus, we get the boundary condition on $\Gamma_0 = \Gamma^{\varepsilon}_{0} \cup \tilde{\Gamma}^{\varepsilon}_{0}$ as follows:

(1.5) $$\frac{\partial u}{\partial n} = 0 \qquad \text{on} \quad \Gamma^{\varepsilon}_{0} \ ,$$

(1.6) $$u=c \qquad \text{(unknown constant) on} \quad \tilde{\Gamma}^{\varepsilon}_{0} \ ,$$

(1.7) $$\int_{\tilde{\Gamma}^{\varepsilon}_{0}} \frac{\partial u}{\partial n}\, dS = A_0 \quad \text{(given constant).}$$

The problem under consideration is still well-posed, noticing (1.4), however, it is very complicated to solve this problem directly by, say, the finite element method, since we need a great number of finer elements in the neighbourhood of Γ_0 . In order to reduce this complexity, it is natural to ask if the complicated boundary condition (1.5)-(1.7) on Γ_0 can be replaced approximately by a simpler and unified boundary condition and, if it is possible, what is the reduced boundary condition. A similar question for partial differential equations, called the problem of homogenization of partial differential equations, has been exhaustively studied by many authors (for instance, see [1]). Therefore, we call the preceding problem the problem of homogenization of boundary conditions which is first presented and studied in [2].

In what follows, we shall discuss three different kinds of homogenization problems of boundary conditions.

Let

(1.8) $$\Gamma = \Gamma_0 \cup \Gamma_1 \cup \Gamma_2 \ .$$

For any given $\varepsilon>0$, we consider a partition of Γ_0 :

(1.9) $$\Gamma_0 = \Gamma^{\varepsilon}_{0} \cup \tilde{\Gamma}^{\varepsilon}_{0} \ ,$$

where $\tilde{\Gamma}_0^\varepsilon$ is composed of a number of connected surfaces $\tilde{\Gamma}_{0,i}^\varepsilon$ $(i=1,\cdots,m(\varepsilon))$:

$$(1.10)\qquad \tilde{\Gamma}_0^\varepsilon = \bigcup_{i=1}^{m(\varepsilon)} \tilde{\Gamma}_{0,i}^\varepsilon \quad .$$

For example, when Γ_0 is a regular domain on a hyperplane, we can take this partition according to a ε-periodic srtucture (see the figure 1).

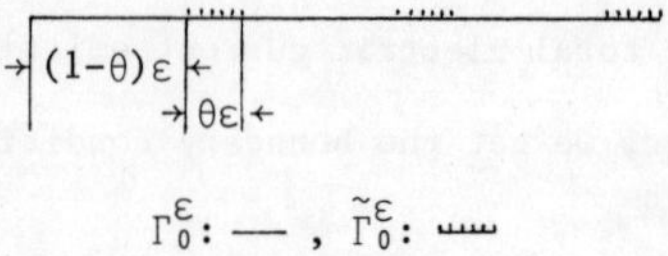

Figure 1

2. <u>Case I</u>

We now consider the following nonlocal boundary value problem on Ω :

$$(I_\varepsilon)\qquad \begin{cases} \Delta u_\varepsilon = 0 & \text{in} \quad \Omega \;, \\ \dfrac{\partial u_\varepsilon}{\partial n} = g_\varepsilon & \text{on} \quad \Gamma_0^\varepsilon , \\ u_\varepsilon = c_\varepsilon \text{(unknown constant)} & \text{on} \quad \tilde{\Gamma}_0^\varepsilon \;, \\ \displaystyle\int_{\tilde{\Gamma}_0^\varepsilon} \frac{\partial u_\varepsilon}{\partial n} dS = \int_{\tilde{\Gamma}_0^\varepsilon} g_\varepsilon dS + F, & \\ u_\varepsilon = 0 & \text{on} \quad \Gamma_1 \;, \\ \dfrac{\partial u_\varepsilon}{\partial n} = 0 & \text{on} \quad \Gamma_2 \;, \end{cases}$$

where g_ε is a given function on Γ_0 and F is a given constant.

In the problem of electrical log with piecewise electrode, $g_\varepsilon \equiv 0$ and F is equal to the total electric current A_0 emitting from the piecewise electrode.

We give the following hypotheses:

(H1). Let

$$(2.1)\qquad \chi_\varepsilon = \begin{cases} 1, & \text{on } \tilde{\Gamma}_0^\varepsilon\ , \\ 0, & \text{on } \Gamma_0^\varepsilon \end{cases}$$

be the characteristic function of $\tilde{\Gamma}_0^\varepsilon$ on Γ_0 . Suppose that for any sequence $\{\chi_{\varepsilon'}\}$ there always exists a subsequence $\{\chi_{\varepsilon''}\}$ such that

$$(2.2)\qquad \chi_{\varepsilon''} \to \chi \qquad \text{weakly * in } L^\infty(\Gamma_0) \text{ as } \varepsilon\to 0$$

and

$$(2.3)\qquad \chi \neq 0 \qquad \text{a.e. on } \Gamma_0\ .$$

(H2). g_ε is bounded in the dual space of $H^{\frac{1}{2}}(\Gamma_0)$.

(H3). There exists a constant G such that

$$(2.4)\qquad \int_{\Gamma_0} g_\varepsilon dS \to G \qquad \text{as} \qquad \varepsilon\to 0.$$

Here, hypothesis (H1) actually gives a restriction on the geometrical structure of $\tilde{\Gamma}_0^\varepsilon$ as $\varepsilon\to 0$. In fact, it follows from (H1) that there exists a constant $\alpha > 0$ such that for any $\varepsilon>0$,

$$(2.5)\qquad \text{mes } \tilde{\Gamma}_0^\varepsilon \text{ on } \Gamma_0 \geq \alpha > 0.$$

particularly, in the case of periodic structure as shown in the figure 1, it can be proved that the whole sequence

$$(2.6)\qquad \chi_\varepsilon \to \theta \neq 0 \quad \text{weakly * in } L^\infty(\Gamma_0) \text{ as } \varepsilon\to 0,$$

then (H1) is satisfied.

We have the following

Theorem 1: Under assumptions (H1)-(H3), as $\varepsilon\to 0$ the solution u_ε of problem (I_ε) converges weakly in $H^1(\Omega)$ to the solution u of the following problem:

$$\text{(I)}\quad\begin{aligned}&\Delta u = 0, && \text{in}\quad \Omega\ ,\\ &u = c \text{ (unknown constant) on } \Gamma_0\ ,\\ &\int_{\Gamma_0} \frac{\partial u}{\partial n}\, dS = F+G\ ,\\ &u = 0 && \text{on}\quad \Gamma_1\ ,\\ &\frac{\partial u}{\partial n} = 0 && \text{on}\quad \Gamma_2\ .\end{aligned}$$

This theorem tells us that in this case the homogenized boundary condition on Γ_0 is a nonlocal boundary condition of the kind (1.2)-(1.3). As an application to the problem of electrical log, instead of the piecewise electrode $\tilde{\Gamma}_0^\varepsilon$, we can approximately regard the whole surface Γ_0 as an electrode and then there is no difficulty to use the finite element method to get a numerical solution.

The sketch of the proof of Theorem 1 is the following.

By means of a variational principle and noticing (H2) and (2.5), we get the uniform boundedness of u_ε in $H^1(\Omega)$ and of c_ε , then by weak compactness, there exist subsequences $\{u_{\varepsilon'}\}$ and $\{c_{\varepsilon'}\}$ such that, as $\varepsilon'\to 0$,

$$u_{\varepsilon'} \to u \quad \text{weakly in } H^1(\Omega)$$

and

$$c_{\varepsilon'} \to c.$$

The next step is to show that u is actually a solution of problem (I). Here, the essential point is to prove

$$\text{(2.7)}\qquad u=c \qquad \text{on}\quad \Gamma_0\ .$$

By the trace theorem, we have

$$u_{\varepsilon'}|_{\Gamma_0} \to u|_{\Gamma_0} \qquad \text{weakly in } L^2(\Gamma_0),$$

then, using (H1), we get

$$\chi_{\varepsilon''} u_{\varepsilon''}|_{\Gamma_0} \to \chi\, u|_{\Gamma_0} \qquad \text{weakly in } L^2(\Gamma_0).$$

On the other hand, we have

$$\chi_{\varepsilon''} u_{\varepsilon''}|_{\Gamma_0} = c_{\varepsilon''}\chi_{\varepsilon''} \to c\chi \quad \text{weakly* in } L^{\infty}(\Gamma_0).$$

Hence,

$$\chi u|_{\Gamma_0} = c\chi \quad ,$$

then (2.7) follows immediately from (2.3). Q.E.D.

3. Case II

We now consider the following boundary value problem on Ω :

$$(\text{II}_\varepsilon) \quad \begin{cases} \Delta u_\varepsilon = 0 & \text{in} \quad \Omega \ , \\ \dfrac{\partial u_\varepsilon}{\partial n} = g_\varepsilon & \text{on} \quad \Gamma_0^\varepsilon \ , \\ u_\varepsilon = 0 & \text{on} \quad \tilde{\Gamma}_0^\varepsilon \\ u_\varepsilon = 0 & \text{on} \quad \Gamma_1 \ , \\ \dfrac{\partial u_\varepsilon}{\partial n} = h & \text{on} \quad \Gamma_2 \ , \end{cases}$$

where g_ε is a given function on Γ_0 , h is a given function on Γ_2 , for example, we may suppose $h \varepsilon L^2(\Gamma_2)$.

We have the following

Theorem 2: Under assumptions (H1)-(H2), as $\varepsilon \to 0$, the solution u_ε of problem (II_ε) converges weakly in $H^1(\Omega)$ to the solution u of the following problem:

$$(\text{II}) \quad \begin{cases} \Delta u = 0 & \text{in} \quad \Omega \ , \\ u = 0 & \text{on} \quad \Gamma_0 , \\ u = 0 & \text{on} \quad \Gamma_1 , \\ \dfrac{\partial u}{\partial n} = h & \text{on} \quad \Gamma_2 . \end{cases}$$

In this case, the homogenized boundary condition on Γ_0 is of Dirichlet type.

4. Case III

Now, recalling (1.10), we consider the following nonlocal boundary value problem on Ω :

$$(\mathrm{III}_\varepsilon)\quad \begin{cases} \Delta u_\varepsilon = 0 & \text{in } \Omega , \\ \dfrac{\partial u_\varepsilon}{\partial n} = g_\varepsilon & \text{on } \Gamma_0^\varepsilon , \\ u_\varepsilon = c_{\varepsilon,i} \quad \text{(unknown constant) on} & \tilde{\Gamma}_{0,i}^\varepsilon , \\ \displaystyle\int_{\tilde{\Gamma}_{0,i}^\varepsilon} \frac{\partial u_\varepsilon}{\partial n} dS = \int_{\tilde{\Gamma}_{0,i}^\varepsilon} g_\varepsilon dS & (i=1,\cdots,m(\varepsilon)), \\ u_\varepsilon = 0 & \text{on } \Gamma_1 , \\ \dfrac{\partial u_\varepsilon}{\partial n} = 0 & \text{on } \Gamma_2 , \end{cases}$$

where g_ε is a given function on Γ_0 .

We give the following hypotheses:

(H4). $g_\varepsilon \to g$ weakly in the dual space of $H^{\frac{1}{2}}(\Gamma_0)$ as $\varepsilon\to 0$.

(H5). Introducing the following sets of functions:

$$V_\varepsilon=\{\psi\,|\,\psi\varepsilon H^1(\Omega),\ \ \psi|_{\tilde{\Gamma}_{0,i}^\varepsilon}=\text{constant } (i=1,\cdots,m(\varepsilon)),\ \ \psi|_{\Gamma_1}=0\} ,$$

$$V=\{\psi\,|\,\psi\varepsilon H^1(\Omega),\ \ \psi|_{\Gamma_1}=0 \} ,$$

suppose that for any $\psi\varepsilon V$, there exists a sequence $\{\varepsilon_n\}$ and $\psi_n\varepsilon V_{\varepsilon_n}$ such that, as $n\to\infty$, $\varepsilon\to 0$ and

$$\psi_n\to\psi \qquad \text{strongly in } H^1(\Omega).$$

Hypothesis (H5) also gives a restriction on the geometrical structure of $\tilde{\Gamma}_0^\varepsilon$ as $\varepsilon\to 0$. For instance, it follows from (H5) that on Γ_0 the measure of each connected component $\tilde{\Gamma}_{0,i}^\varepsilon$ of $\tilde{\Gamma}_0^\varepsilon$ must converge to zero as $\varepsilon\to 0$. In the case of periodic structure as shown in the figure 1, particularly, it can be proved that hypothesis (H5) is certainly satisfied.

We have the following

Theorem 3: Under assumptions (H4)-(H5), as $\varepsilon \to 0$, the solution u_ε of problem (III_ε) converges weakly in $H^1(\Omega)$ to the solution u of the following problem:

$$
\text{(III)} \quad
\begin{cases}
\Delta u = 0 & \text{in} \quad \Omega \ , \\
\dfrac{\partial u}{\partial n} = g & \text{on} \quad \Gamma_0 \ , \\
u = 0 & \text{on} \quad \Gamma_1 \ , \\
\dfrac{\partial u}{\partial n} = 0 & \text{on} \quad \Gamma_2 \ .
\end{cases}
$$

In this case, the homogenized boundary condition on Γ_0 is then of Neumann type.

The proof of Theorems 2 and 3 is similar to that of Theorem 1.

In conclusion, the homogenized boundary condition for different kinds of problems may be quite different, it may be a nonlocal boundary condition, or a boundary conditon of Dirichlet type or of Neumann type etc.

References

1 J.L.Lions, Some Methods in the Mathematical Analysis of Systems and their Control, Science Press, Beijing 1981.

2 A. Damlamian and Li Ta-tsien,Homogénéisation sur le bord pour des problèmes elliptiques, C.R.Acad. Sc. Paris, t. 299, Série I, n^0 17 (1984), 859-862.

THE APPROXIMATION OF BRANCH SOLUTION OF THE NAVIER-STOKES EQUATIONS*

Li Kaitai, Mei Zhen, Zhang Chengdian

Department of Mathematics
Xi'an Jiaotong University, China

Some essential results of branch solution and its approximation of the Navier-Stokes equations are presented. The error estimation is in good agreement with the approximation of regular solutions. Furthermore, the existence and distribution of turning point solution of stationary Navier-Stokes equations is investigated.

1. The Auxiliary Propositions

Throughout the discussion, we shall frequently use the following notations:

V, W, H: Banach spaces;

$L_k(V, W)$: Banach space of bounded k-linear operators from V to W with $L_1(V, W)$;

$\|\cdot\|$, etc.: Various norms in spaces V, W etc.;

h: Positive parameter accumulating to 0;

$C, C_1\ldots,$ $k, k_1\ldots$: Various constants independent of h;

v, x, and w, y: Elements in V or W, H;

$\Lambda \subset R$: Subset of parameter;

T, F, C and T_h, F_h: Mappings and their corresponding approximations;

---

* Project supported by the Science Fund of the Chinese Academy of Sciences.

$D_\lambda^i D_u^j F$ ... etc: i-th and j-th partial derivatives w.r.t. λ and u respectively;

$D^k F$: k-th total derivative of F.

Assume $T(\cdot)$, $T_h(\cdot): \Lambda \to L(W, V)$, $G(\cdot, \cdot): \Lambda \times V \to W$ are C^m $(m \geq 2)$ bounded mappings. We consider the following non-linear equations in a neighborhood $S(\lambda_0, u_0)$ of a point $(\lambda_0, u_0) \in \Lambda \times V$:

$$F(\lambda, u) = 0 \tag{1.1}$$

$$F_h(\lambda, u) = 0 \tag{1.2}$$

where $F(\lambda, u) = u + T(\lambda)G(\lambda, u)$, $F_h(\lambda, u) = u + T_h(\lambda)G(\lambda, u)$.

We suppose that:

h1) $F(\lambda_0, u_0) = 0$;

h2) $T(\lambda) \in L(W, V)$ is a compact operator for all $\lambda \in \Lambda$;

h3) $\lim_{h \to 0,\ \lambda \in \Lambda} \sup \| D^k(T_h(\lambda) - T(\lambda)) \| = 0$, $0 \leq k \leq m-1$

$\sup_{\lambda \in \Lambda} \| D^m T_h(\lambda) \| \leq C_0$;

h4) Range $(DF(\lambda_0, u_0)) = V$, dim Ker $(DF(\lambda_0, u_0)) = 1$

Remark 1.1 The hypothesis (h4) can be rewritten as follows:

1) $D_u F(\lambda_0, u_0)$ is an isomorphism on V;

or 2) dim Ker $(D_u F(\lambda_0, u_0)) = 1$, and $D_\lambda F(\lambda_0, u_0) \notin$ Range $(D_u F(\lambda_0, u_0))$.

Theorem 1.1 Under the hypotheses (h1)-(h4), there exist constants h_0, $t_0 \geq 0$ and neighborhood $S(\lambda_0, u_0)$ of (λ_0, u_0) such that for $h \leq h_0$:

i) Each of the equations (1.1) and (1.2) possesses a unique C^m class branch solution $(u(t), \lambda(t))$ and $(u_h(t), \lambda_h(t))$, $|t| \leq t_0$ respectively, $(u(0), \lambda(0)) = (u_0, \lambda_0)$.

ii) $\lim_{h \to 0} \sup_{|t| \leq t_0} \{ \| u^{(k)}(t) - u_h^{(k)}(t) \|_V + | \lambda^{(k)} - \lambda_h^{(k)}(t) | \} = 0$,

$0 \leq k \leq m-1$

$$\| u^{(k)}(t) - u_h^{(k)}(t) \|_V + |\lambda^{(k)}(t) - \lambda_h^{(k)}(t)|$$

$$\le c \sum_{i=0}^{k} \| \frac{d^i}{dt^i} (T(\lambda) - T_h(t)) G(\lambda, u) \| \qquad \text{for all} \qquad |t| \le t_0 \quad ,$$

$k = 0, 1, \ldots, m-1$

iii) $\sup_{|t| \le t_0} \| D^m u_h(t) \| \le C.$

2. The Bifurcation of Navier-Stokes Equation

Now, we consider the steady viscous incompressible flow problems.

$$\begin{aligned} -\nu\Delta u + (u\nabla)u - \nabla p = f \quad & \text{in} \quad \Omega \\ \nabla \cdot u = 0 \quad & \\ u = 0 \quad & \text{on} \quad \Gamma = \partial\Omega \\ \int_\Omega p dx = 0 \quad & \end{aligned} \tag{2.1}$$

where $\nu > 0$ is the viscosity, u is the velocity field and p is the hydrostatic pressure, $f \in (L^{4/3}(\Omega))^n$, $n \le 4$ is the unit body force, $\Omega \subset R^n$ is a bounded conic domain.

Set $V = H_0^1(\Omega)^n \times L_0^2(\Omega)$, $W = L^{4/3}(\Omega)^n \times L^2(\Omega)$, $W_0 = L^{4/3}(\Omega)^n \times \{0\}$. Obviously $V \subset W \subset V' = H^{-1}(\Omega)^n \times L^2(\Omega)$, $W_0 \subset W$. Define mapping $G: R \times V \to W$ as $G(\lambda, (u, p)) = ((u \cdot \nabla)u - f, 0)$. It is well known that the weak form of problem (2.1) is

$$a(\lambda; (u, p), (v, q)) + \langle (v, q), G(\lambda, (u, p)) \rangle = 0 \qquad \forall (V, q) \in V \tag{2.2}$$

where $\lambda = 1/\nu$, and

$$a(\lambda, (u, p), (v, q)) = \lambda^{-1} \int_\Omega \nabla u \cdot \nabla v \, dx + \int_\Omega (p \nabla \cdot v + q \nabla \cdot u) dx \quad .$$

The corresponding approximate problem is

$$a(\lambda;\ (u_h,\ p_h),\ (v,\ q)) + \langle (v,\ q),\ G(\lambda,\ (u_h,\ p_h))\rangle = 0 \quad , \quad \forall (v,\ q) \in V_h \tag{2.3}$$

where

$$V_h = X_h \times L_h \quad , \quad X_h \subset H_0^1(\Omega)^n \cap Y_h \quad , \quad Y_h \subset H^2(\Omega)^n \quad ,$$

$$L_h \subset L_0^2(\Omega)$$

are finite element spaces and satisfy the conditions (H1)-(H3) in [9]. Now we introduce the operator $T:(g,\ 0) \in W_0 \to (u,\ p) \in V$ as the solution of the Stokes problem below:

$$a(\lambda;\ (u,\ p),\ (v,\ q)) = \langle (v,\ q),\ (g,\ 0)\rangle \quad , \quad \forall (v,\ q) \in V \quad . \tag{2.4}$$

It follows from Sobolev imbedding theorem that $T(\lambda)$ is compact. The approximate operator $T_h(\lambda):(g,\ 0) \in W_0 \to (u_h,\ p_h) \in V_h$ is defined implicitly by:

$$a(\lambda;\ (u_h,\ p_h),\ (v,\ q)) = \langle (v,\ q),\ (q,\ 0)\rangle \quad , \quad \forall (v,\ q) \in V_h \quad . \tag{2.5}$$

From [10] we know that problem (2.5) has a unique solution for all $\lambda \in R_+$, $(g,\ 0) \in W_0$ under the hypotheses (H1)-(H3).

<u>Lemma 2.1</u> Under the hypotheses (H1)-(H3), we have

i) T, T_h, G are all of class C^m; (2.6)

ii) $\lim\limits_{h \to 0} \sup\limits_{\lambda \in \Lambda} \| D_\lambda^\ell (T_h(\lambda) - T(\lambda))\| = 0 \quad , \quad \ell = 0,1\ldots m$ (2.7)

for all bounded subsets $\Lambda \subset R_+$.

iii) Furthermore, if $T(1)(g,\ 0) \in H^{m+1}(\Omega)^n \times H^m(\Omega)$ we get

$$\| T_h(\lambda)(g,\ 0) - T(\lambda)(g,\ 0)\| \le ch^m \| g \|_{m-1}$$

Lemma 2.2 The problems (2.2) and (2.3) are equivalent to the following respectively:

Find $(u, p) \in V$ such that

$$F(\lambda, (u, p)) = (u, p) + T(\lambda) \cdot G(\lambda, (u, p)) = 0 \qquad (2.8)$$

and find $(u_h, p_h) \in V_h$ satisfying

$$F_n(\lambda, (u_h, p_h)) = (u_h, p_h) + T_h(\lambda)G(\lambda, (u_h, p_h)) = 0 \quad .$$

Theorem 2.1 Under the hypotheses (h1)-(h3) in [4], if $(\lambda_0, (u_0, p_0))$ in $R \times V$ is a nonsingular point or a simple limit point of (2.2), (cf. [2]) then there exist constants $t_0, h_0 > 0$ and a neighborhood $S(\lambda_0, (u_0, p_0))$ of $(\lambda_0, (u_0, p_0))$ such that for $h \leq h_0$, there are two mappings $(\lambda(t), (u(t), p(t)), (\lambda_h(t)), (u_h(t), p_h(t))$: $[-t_0, t_0] S(\lambda_0, (u_0, p_0))$ satisfying

i) They are solutions of (2.2) and (2.3) respectively;

ii) $$\lim_{h\to 0} \sup_{|t| \leq t_0} (|\lambda_h^{(\ell)}(t) - \lambda^{(\ell)}(t)| + \|(u_h^{(\ell)}(t), p_h^{(\ell)}(t)) - (u^{(\ell)}, p^{(\ell)})\|_V) = 0 \quad , \quad 0 \leq \ell \leq m$$

iii) If $u(t) \in H^m(\Omega)^n$, $f \in H^{m-1}(\Omega)^n$, then

$$\sup_{|t| \leq t_0} \{|\lambda_h^{(\ell)}(t) - \lambda^{(\ell)}(t)| + \|(u_h^{(\ell)}(t), p_h^{(\ell)}(t)) - (u^{(\ell)}, p^{(\ell)})\|_V\} \leq kh^m \quad , \quad 0 \leq \ell \leq m \quad .$$

Proof: From the definition of (λ_0, u_0) and the statement (2.4), (2.6), (2.7), we know the condition (h1)-(h4) are all satisfied. So using Theorem 1.1 we can easily derive the conclusion.

Theorem 2.2 Suppose the conditions (H1)-(H3). If $(\lambda_0, (u_0, p_0))$ is a nondegenerate turning point of (2.2), then for all $h \leq h_0$, we have:

1) The approximate problem (2.3) possesses a nondegenerate turning point $(\lambda_h^0, u_h^0, p_h^0)$;

2) $$\lim_{h\to 0} (|\lambda_h^0 - \lambda_0| + \|(u_h, p_h^0) - (u_0, p_0)\|_V) = 0$$

3) If $u(t) \in H^m(\Omega)^n$, $f \in H^{m-1}(\Omega)^n$, then

$$|\lambda_h^0 - \lambda_0| + \| (u_h^0, p_h^0) - (u_0, p)\|_V \le K_m h^m .$$

The proof can be accomplished by simple deduction.

3. The Penalty Approximation of Branch Solutions

Using the same notations and operators as Section 2, we still consider the stationary incompressible Navier-Stokes equations. Now we shall choose $H = H_0^1(\Omega)^n$, $V = \{v \in H,\ \mathrm{div}\ v = 0\}$, $W = L^{4/3}(\Omega)^n$. Then the problem (2.1) is equivalent to:

Find $u \in V$, such that

$$a_0(u, v) + \lambda < (u \cdot \nabla)u - f, v > = 0 \quad , \quad \forall v \in V$$

where

$$a_0(u, v) = \int_\Omega \nabla u \cdot \nabla v dx \quad , \quad \lambda = 1/\mu .$$

Define $G: \Lambda \times H \to W$ as $G(\lambda, u) = (u \cdot \nabla)u - f$ and $S: W \to V$ by:

$$a_0(S_g, v) = < g, v > \quad , \quad \forall v \in V \quad , \quad g \in W .$$

It is easy to see that S is compact. Thus, if we define the operator $T(\lambda)$: $W \to V$ for all $\lambda \in \Lambda$ as $T(\lambda)g = \lambda S_g$, $g \in W$, we know that problem (2.1) becomes:

Find $u \in H$, such that

$$F(\lambda, u) = u + T(\lambda)G(\lambda, u) = 0 . \tag{3.1}$$

Introduce the finite element spaces $X_h \subset H_0^1(\Omega)^n \cap Y_n$, $Y_h \subset H_0^2(\Omega)^n$ and $L_n \subset L_0^2(\Omega)$ satisfying the hypotheses (H1)-(H3) in [4]. Let ρ_h be the orthogonal projection from $L_0^2(\Omega)$ to L_h. Then we define the operator $T_h(\lambda)$: $g \in W \to T_h(\lambda)g = u_h \in X_h$ as the solution of (3.2) below:

$$a_0(u_h, v_h) + \varepsilon^{-1}(\rho_h \nabla \cdot u_h, \rho_h \nabla \cdot v_h) = \lambda < g, v_h > \quad , \quad \forall v_h \in X_h . \tag{3.2}$$

Hence, similarly to Lemma 2.1, we have

Lemma 3.1 Under the above hypotheses, there exist constants h_0 and $C > 0$ such that for $h \le h_0$,

$$|T(\lambda)g - T_h(\lambda)g|_{1,\Omega} \le C(h + \varepsilon)\|g\|_W \quad , \quad \forall\lambda \in \Lambda \quad . \tag{3.3}$$

Furthermore, if $T(\lambda)g \in H^{m-1}(\Omega)^n$, then

$$|T(\lambda)g - T_h(\lambda)g|_{1,\Omega} \le C(h^m + \varepsilon)\|g\|_{m-1} \quad . \tag{3.4}$$

Now given a function $\varepsilon(h)$: $R_+ \to R_+$ as $\lim_{h\to 0} \varepsilon(h) = 0$, let $\tilde{T}_h(\lambda) = T_h^{\varepsilon(h)}(\lambda)$ and define the approximate problem (3.2):

Find $u \in H$ such that

$$F_h(\lambda, u) = u + \tilde{T}_h(\lambda)G(\lambda, u) = 0 \quad . \tag{3.5}$$

Theorem 3.1 Let $(\lambda_0, u_0) \in \Lambda \times H$ be a nonsingular point or simple limit point of (3.1). Then under the hypotheses (H1), (H2), there exist constants h_0, t_0, $K_m > 0$ such that for $h \le h_0$ small enough, there is a unique pair of C^∞ mappings $\{(\lambda(t), u(t));\ |t| \le t_0,\ \lambda(0) = \lambda_0,\ u(0) = u_0\}$ and $(\lambda_h(t), u_h(t));\ |t| \le t_0\}$ satisfying:

i) They are the solutions of problems (3.1) and (3.2) respectively.

ii) We have the error estimate:

$$|\lambda_h^m(t) - \lambda^{(m)}(t)| + \|u_h^{(m)}(t) - u^{(m)}(t)\|_H \le Km(h + \varepsilon(h))$$

iii) Furthermore, if $f \in H^{m-1}(\Omega)^n$, $u(t) \in H^m(\Omega)^n$, then

$$|\lambda_h^{(m)}(t) - \lambda^{(m)}(t)| + \|u_h^{(m)}(t) - u^{(m)}(t)\|_H \le K_m(h^m + \varepsilon(h)) \quad .$$

Proof: By the definition of $T(\lambda)$, $T_h(\lambda)$, $G(\lambda, u)$ and Lemma 3, we see the conditions in Theorem 1.1 are satisfied. Thus the first part of the Theorem hold. Referring to error estimation, we have

$$|\lambda_h^{(m)}(t) - \lambda^{(m)}| + \| u_h^{(m)}(t) - u^{(m)}(t)\|_H$$

$$\leq K_m \sum_{i=0}^{m} \| \frac{d^i}{dt^i} (T(\lambda) - T_h(\lambda))G(\lambda, u)\|_H \quad .$$

So, using Lemma 3.1 and the C^m boundedness of mapping G, we complete the proof.

4. The Improvement of Error Estimation

Suppose the hypotheses (h1)-(h4) hold, and in addition, we have (h5) H is a Banach space and $V \subsetneq H$ with continuous imbedded; (h6) there exists a constant C such that along the branch solution $(\lambda(t), u(t))$ of exact problem, we have

$$\| D_\lambda^{k-\ell} D_u^\ell G(\lambda(t), u(t))(v, v_i)\|_{L_\ell(V)} \leq C\| v_1 \|_H \sum_{i=1}^{\ell-1} \| v_i \|_V$$

for $|t| \leq t_0$, $k = 1,\ldots,m-1$, $1 \leq \ell \leq k$,

$v_1, v_2,\ldots,v_\ell \in V$.

Theorem 4.1 Suppose the hypotheses (h1)-(h6) are satisfied. Then for the exact and approximate branches of solutions defined by Theorem 1.1, there exist constants t_0, h_0, C such that for $0 \leq k \leq m-1$, $|t| \leq t_0$, $h \leq h_0$ we have

$$\| u^{(k)}(t) - u_h^{(k)}(t)\|_H + |\lambda^{(k)}(t) - \lambda_h^{(k)}(t)| \leq C\Big\{ \sum_{i=0}^{K} \| \frac{d^i}{dt^i} (T(\lambda) - T_h(\lambda))G(\lambda, u)\|_H + \| (T(\lambda) - T_h(\lambda))G(\lambda(t), u(t))\|_V^2\Big\} \quad .$$

Remark 4.1 The estimate of $|\lambda^{(k)}(t) - \lambda_h^{(k)}(t)|$ is improved.

Remark 4.2 For the Navier-Stokes equation, hypothesis (h6) can be verified by the regularity properties of its exact solutions.

Remark 4.3 From Sections 2 and 3, we see that if the Stokes

problem is well approximated, we can obtain a better approximation of the Navier-Stokes equation. Furthermore, if we have the L^2 or L^∞ error estimation for the Stokes problem, using Theorem 4.1 we can get the same error bound for the nonlinear ones.

5. Numerical Analysis of Turning Point Solution of Stationary Navier-Stokes Equations

We still consider the problem (2.1) with $\lambda = 1/Re$. We know that there is at least one solution of (2.1) for each given $R_e > 0$. But the distribution of the solution with Re is very complicated. Now our aim is to investigate the turning point solution of (2.1) using FEM.

Spaces V, W and W_0 are defined as before. The variational form of (2.1) is given as

$$\begin{aligned} \lambda a_0(u, v) + a(u; u, v) + b(u, v) &= g(v) \\ b(v, u) &= 0 \qquad \forall v \in V \end{aligned} \tag{5.1}$$

where $u = (u, p)$, $v = (v, q)V$, $\lambda = 1/Re$. Functional a_0, b, g, a_1 are defined as:

$$a_0(u, v) = \int_\Omega \nabla u \cdot \nabla v dx$$

$$a(u; v, w) = \int_\Omega (u \cdot \nabla)v \cdot w dx \quad , \quad \forall u, v, w \in V$$

$$b(u, v) = \int_\Omega p\nabla \cdot v dx \quad , \quad g(v) = \int_\Omega g \cdot v \, dx \quad . \tag{5.2}$$

From $a_1(u, u)$ we define an element in V' for each fixed $u \in V$:

$$< G(u), v > = a_1(u; u, v) \quad , \quad \forall v \in V \quad . \tag{5.3}$$

Obviously, $G: V \to V'$ is a quadratic mapping.

Define mapping $T: g = (g, 0) \in W_0 \to Tg = (u, p) \in V$ as a solution of normal Stokes problem.

$$\begin{aligned} a_0(u, v) + b(u, v) &= g(v) \\ b(v, u) &= 0 \qquad \forall v \in V \quad . \end{aligned}$$

By the regularity of Stokes problem and Sobolev imbedded theorem, we have

<u>Lemma 5.1</u> 1) Mapping $T \in L(W_0, V)$ is compact;

2) For each $u \in V$, $G(u) \in W_0$ and $G: \to V \to W_0$ is a C^∞ quadratic mapping;

3) $\| D_u G(u) \|_{L(v, w_0)} \le k_0 \| u \|_v$

$$\| G(u) \|_{w_0} \le k_0 \| u \|_v^2 \quad , \quad \forall u \in V \tag{5.4}$$

where k_0 is a constant.

Naturally, an operator form of (2.1) is the following

$$F(\lambda, u) = \lambda u - TG(u) - f = 0 \tag{5.5}$$

where $\quad f = T(g, 0) \quad .$

Introduce the discrete spaces V_h as Section 2, and define an approximate mapping T_h: $g = (g, 0) \in W_0 \to T_h g = (u_h, p_h) = u_h \in V_h$ as the solution of discrete Stokes problem:

$$\lambda a_0(u_h, v) + b(u_h, v) = g(v)$$
$$b(v, u_h) = 0 \qquad \forall v \in V_h \quad . \tag{5.6}$$

<u>Lemma 5.2</u> Under conditions (H1)-(H3) in [4], we have

$$\text{i) } T_h \in L(W_0, V_h) \quad , \quad \| T - T_h \|_{L(w_0, v)} \le C_0 h \tag{5.7}$$

and

$$\text{ii) } \| T - T_h \|_{L(H^{m-1}(\Omega)^n \times \{0\}, V)} \le C_0 h^m \quad . \tag{5.8}$$

The proof follows directly from the approximation theory of Stokes problem [9].

Hence, an approximate form of (5.5) is chosen as:

$$F_h(\lambda_h, u_h) \equiv \lambda_h u_h - T_h G(u_h) - f_h \tag{5.9}$$

where

$$f_h = T_h g \in V_h \quad .$$

In addition, we assume:

(NA1) λ_h is an eigenvalue of linearized Navier-Stokes equations:

$$\lambda_h \eta_h - T_h DuG(u_h)\eta_h = 0 \tag{5.10}$$

Its eigenfunction space is $V_2 = \text{span}\{\eta_h\}$. Let η^*h be an eigenfunction w.r.t. λ_h of adjoint problem of (5.10).

(NA2) For the decomposition $V = V_1 \oplus V_2$, $V = V_2^{\perp}$ we have

$$\| \lambda_h W - TD_u G(u_h)^V \|_V \geq d \| W \|_V \quad , \quad \forall w \in V_1 . \tag{5.11}$$

(NA3) There exist constants $q, \varepsilon \in [0, 1/5]$ independent of h such that

$$d.\ \min\{1, |\alpha_h|, |\beta_h|\} \geq qh^{\varepsilon} \tag{5.12}$$

where

$$\alpha_h = (\eta^* h, TG(\eta_h)) \quad , \quad \beta_h = (\eta^* h, u_h)$$

Theorem 5.1 Under the hypothesis (H1)-(H3), (NA1)-(NA3), if sgn $\alpha_h \beta_h = -1$, there is a constant $h_0 > 0$. For $h \leq h_0$, there exists $\lambda^* \in [\lambda_h - \Delta\lambda_h, \lambda_h + \Delta\lambda_h]$ such that

1) For all $\lambda \in [\lambda_h - \Delta\lambda_h, \lambda^*]$, there are exactly two continuous solutions of (5.5) $u^+(\lambda), u^-(\lambda) \in S(u_h, \rho)$

2) For $\lambda = \lambda^*$, there is a unique solution $u^* \in S_x(0, \rho)$ of (5.5) and

$$\lim_{\lambda \to \lambda^*} u^+(\lambda) = \lim_{\lambda \to \lambda^*} u^-(\lambda) = u^* \tag{5.13}$$

3) For $[\lambda^*, \lambda_h + \Delta\lambda_h]$, there is no solution of (5.6) in $S(u_h, \rho)$ where $\Delta\lambda_h = C_0/\alpha_h/\beta_h/h^{3\varepsilon}$, $\rho = coh^{3\varepsilon/2}$.

The proof consists of a series Lemma which can be found in [8].

Remark 5.1 Similar conclusions hold for $\text{Sgn}\alpha_h\beta_h = 1$.

Remark 5.2 In fact, case (3) indicates that solution $u(\lambda)$ is far away from u_h since there is at least one solution of (5.5) for each $\lambda > 0$.

6. Implements

Now, we consider the problem (1.1) and (1.2) again for general C^m bounded operator F, F_h. Suppose we have

(h1) For $\ell = 0, 1$,

$$\lim_{h\to 0} \sup_{(\lambda,u)\in\Lambda\times V} \| D^\ell(F - F_h)(\lambda, u)\|_{L_\ell(R\times V,\, V)} = 0 \qquad (6.1)$$

where $\Lambda\times C \in R\times V$ is an arbitrary solution set of (1.1)

(h2) For $\ell \in \{0, 1,\ldots,m\}$

$$\sup_{h} \sup_{(\lambda,u)\in\Lambda\times C} \| D^\ell(F - F_h)(\lambda, u)\|_{L(R\times V,\, V)} < +\infty \qquad (6.2)$$

where $\Lambda\times C$ is an arbitrary bounded subset of $1R\times V$.

<u>Theorem 6.1</u> Assume (λ_0, u_0) is a simple singular point of F. Conditions (h1), (h2) hold. Then there exist constants α_0, h_0, $k_0 > 0$, such that, for $h \le h_0$, there are unique C^m functions $\{(\lambda(\alpha), u(\alpha)), |\alpha| \le \alpha_0\}$ and $\{(\lambda_h(\alpha), u_h(\alpha), |\alpha| \le \alpha_0\}$ satisfying (1.1) and (1.2) respectively. Furthermore, we have

i) $$\| \frac{d^k}{d\alpha^k}(\lambda,u)(\alpha^k) - \frac{d^k}{d\alpha^k}(\lambda_h, u_h)(\alpha)\|_{R\times V}$$

$$\le K_0{}^* \left\{ \sum_{\ell=0}^{K} \| \frac{d^\ell}{d\alpha^k}(F - F_h)(\lambda(\alpha), u(\alpha))\|_V + |\alpha - \alpha^k| \right\}$$

$$\forall\alpha,\ \alpha^k \in [-\alpha_0, +\alpha_0] \quad , \quad K \in \{0, 1,\ldots,m-1\}$$

ii) $$\sup_{[-\alpha_0,\, \alpha_0]} \max(\| u_h^{(m)}(\alpha)\|_V, |\lambda_h^{(m)}(\alpha)|) \le K_0 \quad .$$

So, we can take properly F for the Navier-Stokes equations (2.1) and choose the corresponding approximate mapping F_h by various methods, such as, FDM, FEM, and spectral approximation method. By this way, bifurcation phenomena of Navier-Stokes equations can be investigated in more detail. For more information refer to [4].

References

[1] Brezzi, F., Rappaz, J., Raviart, P.A., "Finite dimensional approximation of nonlinear problem", Part I. Branches of non-singular solutions, Numer. Math. 36, 1-25 (1980).

[2] Brezzi, F., Rappaz, J., Raviart, P.A., "Finite dimensional approximation of nonlinear problems", Part II. Limit points, Numer. Math. 37, 1-28 (1981).

[3] Mei Zhen, Bifurcation problem and its numerical analysis of non-linear equations, to appear.

[4] Mei Zhen, Li Kaitai: "Numerical determination of turning point bifurcation", Scientific reports of Xi'an Jiaotong University, No. 486, (1984).

[5] Li Kaitai, *et al.*, "Bifurcation problems of the Navier-Stokes equations", Scientific reports of Xi'an Jiaotong University, No. 195, (1983).

[6] Scholz, R., "Computation of turning points of the stationary Navier-Stokes equations using mixed finite element", ISUM 54, Bifurcation problems and their numerical solution, (1980).

[7] Kikuchi, F., "Finite element of approximation of bifurcation problems of turning point type", I.N.R.I.A. Meeting, "Methodes de calcul scientifique et technique III", (1977).

[8] Ciarlet, P.G., The finite element method for elliptic problems, Amsterdam: North-Holland, (1978).

[9] Girault, V., Raviart, P.A., Finite element approximation of the Navier-Stokes equations, Lecture notes in Math. Vol. 749, (1979).

[10] Temam, R., Navier-Stokes equations. Amsterdam: North-Holland. (1977).

[11] Guan Zhaozhi, *et al.*, Linear functional analysis (in Chinese), (1980).

A LINEAR MIXED FINITE ELEMENT APPROXIMATION OF THE NAVIER-STOKES EQUATION *

Li Likang

(Fudan University Shanghai China)

§1 Introduction

Let Ω be an open set of R^2, whose boundary Γ is sufficiently smooth. In this paper we discuss mixed finite element approximation of the following Navier-Stokes equations

$$\begin{cases} -\nu\Delta\vec{u}+(\vec{u}\cdot\nabla)\vec{u}+\nabla p=\vec{f}, & \text{in } \Omega, \\ \nabla\cdot\vec{u}=0, & \text{in } \Omega, \\ \vec{u}=0, & \text{on } \Gamma . \end{cases} \tag{1.1}$$

Glowinski and Pironneau [1] introduce a mixed finite element approximation of the Stokes equation based on a new variational principle. Le Tallec [2] extended the above approximation to Navier-Stokes equations. They use linear Lagrange elements (or quadratic Lagrange elements) for the velocity and linear Lagrange elements for the pressure only over convex polygon. We give a new finite element for Navier-Stokes equations based on the variational principle by Glowinski and Pironneau. In this paper we use linear Lagrange element for the velocity and linear Lagrange element for the pressure over sufficiently smooth domain. Moreover, degrees freedom of finite element of velocity in this paper is less than those in [1] or [2].

Certainly, the result in this paper will be true for the convex polygon Ω .

We shall not give definition of Sobolev space $H^m(\Omega)$, $H_0^1(\Omega)$, $\cdots$, which are well known. We denote norm and seminorm of sobolev space $H^m(\Omega)$, $H_0^1(\Omega)$, $\cdots$, by $\|\cdot\|_m$, $|\cdot|_m$, $\cdots$. We denote

*Project supported by science fund of the Chinese Academy of Science

$H^0(\Omega)=L^2(\Omega)$ and $L_0^2(\Omega)=L^2(\Omega)\setminus R$. We denote norm and seminorm of $H^m(\Omega)\cap L_0^2(\Omega)$ by $\|\cdot\|_{m\backslash 1}$ and $|\cdot|_{m\backslash 1}$ respectively.

We define

$$W_0=\{(\vec{v},\phi)\mid(\vec{v},\phi)\in(H_0^1(\Omega))^2\times H_0^1(\Omega),\ (\nabla\phi,\nabla w)=(\nabla\cdot\vec{v},w),\forall w\in H^1(\Omega)\},$$

$$W=\{(\vec{v},\phi)\mid(\vec{v},\phi)\in(H_0^1(\Omega))^2\times H_0^1(\Omega),\ (\nabla\phi,\nabla w)=(\nabla\cdot\vec{v},w),\forall w\in H_0^1(\Omega)\}.$$

Variational problem P: Find $(\vec{u},\psi)\in W_0$ satisfying

$$\nu(\nabla\vec{u},\nabla\vec{v})=(\vec{f}-(\vec{u}\cdot\nabla)\vec{u},\vec{v}+\nabla\phi),\quad\forall(\vec{v},\phi)\in W_0. \tag{1.2}$$

Variational problem P_1 : Find $(\vec{u},\psi,p)\in(H_0^1(\Omega))^2\times H_0^1(\Omega)\times H^1(\Omega)$ satisfying

$$\begin{cases}\nu(\nabla\vec{u},\nabla\vec{v})+(\nabla p,\nabla\phi)-(p,\nabla\cdot\vec{v})\\ \qquad=(\vec{f}-(\vec{u}\cdot\nabla)\vec{u},\vec{v}+\nabla\phi),\quad\forall(\vec{v},\phi)\in(H_0^1(\Omega))^2\times H_0^1(\Omega).\\ (\nabla\psi,\nabla w)=(\nabla\cdot\vec{u},w),\quad\forall\ w\in H^1(\Omega).\end{cases} \tag{1.3}$$

Variational problem P_2 : Find $(\vec{u},\psi,p)\in(H_0^1(\Omega))^2\times H_0^1(\Omega)\times H^1(\Omega)$ satisfying

$$\begin{cases}\int_\Omega\nabla p\cdot\nabla\phi\,dx=\int_\Omega\vec{f}\cdot\nabla\phi\,dx,\quad\forall\ \phi\in H_0^1(\Omega), & (1.4)\\ \nu\int_\Omega\nabla\vec{u}\cdot\nabla\vec{v}dx-\int_\Omega p\nabla\cdot\vec{v}dx+\int_\Omega(\vec{u}\cdot\nabla)\vec{u}\cdot\vec{v}dx=\int_\Omega\vec{f}\cdot\vec{v}\ dx,\ \forall\ \vec{v}\in(H_0^1(\Omega))^2,\\ \int_\Omega\nabla\psi\cdot\nabla w dx=\int_\Omega\vec{u}\cdot\nabla w\ dx,\quad\forall\ w\in H^1(\Omega).\end{cases}$$

Boundary value problem P_3 : $(\vec{u},\psi,p)$ is the solution in $(H_0^1(\Omega))^2\times H_0^1(\Omega)\times H^1(\Omega)$ of the following boundary value problem

$$\begin{cases}\Delta p=\nabla\cdot(\vec{f}-(\vec{u}\cdot\nabla)\vec{u}),\quad\text{in }\Omega,\\ -\nu\Delta\vec{u}+\nabla p=\vec{f}-(\vec{u}\cdot\nabla)\vec{u},\quad\text{in }\Omega,\\ -\Delta\psi=\nabla\cdot\vec{u},\quad\text{in }\Omega,\\ \vec{u}=0,\qquad\psi=\frac{\partial\psi}{\partial n}=0,\quad\text{on }\Gamma.\end{cases} \tag{1.5}$$

If $p\in H^1(\Omega)$, then the above problems are equivalent. Moreover, $(\vec{u},p)$ in the above problem is the solution of (1.1).

The mixed finite element discretization we shall study is based upon (1.4).

Let $\{K_i\}$ $(i=1,\cdots,m_0+m_1)$ be a triangulation of Ω, $\Omega=\bigcup_{i=1}^{m_0+m_1}K_i$. Triangulations satisfy the conditions in [3].

We denote all boundary elements by K_i $(i=1,\cdots, m_0)$ and all interior elements by K_i $(i=m_0+1,\cdots, m_0+m_1)$, where K_i $(i=1,\cdots,m_0)$ are curved triangles. We denote all boundary nodes by A_i $(i=1,\cdots, m_0)$ and all interior nodes by A_i $(i=m_0+1, \cdots,m_0+m_1)$ (see Fig 1).

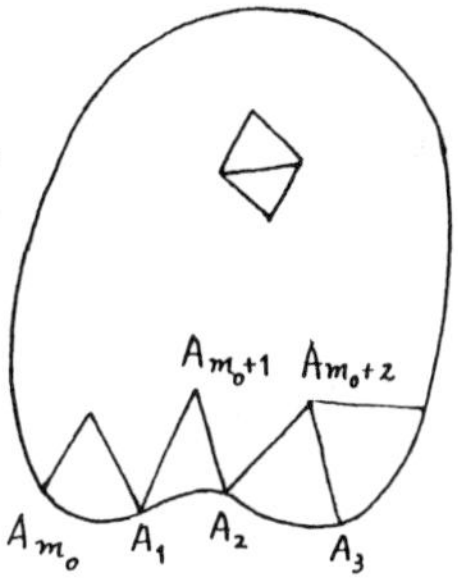

Fig 1

Let $\tilde{K}_i$ $(i=1,\cdots, m_0+m_1)$ be triangles, three vertices of $\tilde{K}_i$ $(i=1,\cdots, m_0+m_1)$ coincide with three vertices of K_i $(i=1, \cdots, m_0+m_1)$. We denote $\Omega_h = \bigcup_{i=1}^{m_0+m_1} \tilde{K}_i$. Obviously, $K_i=\tilde{K}_i$ $(i=m_0+1, \cdots, m_0+m_1)$. We denote barycenter of $\tilde{K}_i$ by G_i . Let K_i be an arbitrary element, whose three vertices are D_{i1} , D_{i2} and D_{i3} (if K_i is a boundary element, then D_{i1} , $D_{i2} \in \Gamma$). We denote triangle (or curved triangle) by K_{i3} =triangle (or curved triangle) $G_i D_{i1} D_{i2}$ and

$$K_{i2} =\Delta G_i D_{i3} D_{i1} , \quad K_{i1} =\Delta G_i D_{i2} D_{i3}$$

(see Fig 2).

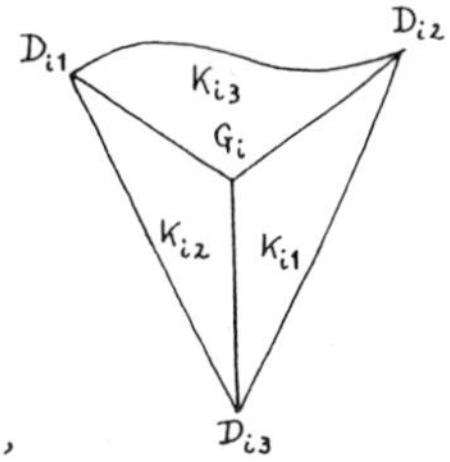

Fig 2

We define

$V_h =\{ v_h | v_h \in C^0(\Omega)$, v_h on each K_{ij} is a linear function $(i=1,\cdots, m_0+m_1, j=1,2,3)\}$,

$V_h^0=\{ v_h | v_h \in V_h$, $v_h(A_i)=0$ $(i=1,\cdots ,m_0)\}$,

$X_h =\{ v_h | v_h \in C^0(\Omega)$, v_h on each K_i is a linear function $(i=1,\cdots, m_0+m_1)\}$,

$X_h^0 =\{ v_h | v_h \in X_h$, $v_h(A_i)=0$ $(i=1,\cdots ,m_0)\}$,

$$W_{oh} = \{ (\vec{v}_h , \phi_h) | (\vec{v}_h , \phi_h) \in (V_h^0)^2 \times X_h^0 , \ (\nabla\phi_h,\nabla z_h)=-(\vec{v}_h ,\nabla z_h), \ \forall z_h \in X_h \backslash R \},$$

$$W_h = \{ (\vec{v}_h , \phi_h) | (\vec{v}_h , \phi_h) \in (V_h^0)^2 \times X_h^0 , \ (\nabla\phi_h,\nabla z_h)=-(\vec{v}_h ,\nabla z_h), \ \forall z_h \in X_h^0 \} .$$

Obviously W_{oh} and W_h are nonempty.

Variational problem P_h : Find $(\vec{u}_h, \psi_h) \in W_{oh}$ satisfying

$$\nu(\nabla\vec{u}_h, \nabla\vec{v}_h) = (\vec{f} - (\vec{u}_h\cdot\nabla)\vec{u}_h, \vec{v}_h + \nabla\phi_h), \quad \forall\, (\vec{v}_h, \phi_h) \in W_{oh} .$$

Variational problem P_{h1} : Find $(\vec{u}_h, \psi_h, p_h) \in (V_h^o)^2 \times X_h^o \times (X_h \setminus R)$ satisfying

$$\begin{cases} (\nabla p_h, \nabla\phi_h) = (\vec{f} - (\vec{u}_h\cdot\nabla)\vec{u}_h, \nabla\phi_h), & \forall\, \phi_h \in X_h^o, \\ \nu(\nabla\vec{u}_h, \nabla\vec{v}_h) = (\vec{f} - \nabla p_h - (\vec{u}_h\cdot\nabla)\vec{u}_h, \vec{v}_h), & \forall\, \vec{v}_h \in (V_h^o)^2, \\ (\nabla\psi_h, \nabla q_h) = -(\vec{u}_h, \nabla q_h), & \forall\, q_h \in X_h \setminus R . \end{cases}$$

If we cancel the trems $(\vec{u}\cdot\nabla)\vec{u}$ and $(\vec{u}_h\cdot\nabla)\vec{u}_h$ in all above formulas, then we obtain Stokes problems and its approximate problems.

In §2 we shall give some inequalities and approximation. In §3 we shall prove Brezzi's inequality. In §4 and §5 we shall give the error estimates for Stokes equations and Navier-Stokes equations respectively.

In this paper C denotes a generic constant with possibly different values in different contexts. We denote $h_i = \text{diam}\,(K_i)$, $h = \max_{1\le i\le m_0+m_1} h_i$. We suppose

$$h_i/h_j \le C \quad , \; i,j = 1,\cdots, m_0 + m_1 \quad . \qquad (1.6)$$

§2 Some inequalities and approximation

Lemma 2.1 We assume $\Gamma \in C^2$. Let K_i be an arbitrary boundary element, $\gamma_i = \Gamma \cap K_i$, $E_i = (K_i - \tilde{K}_i) \cup (\tilde{K}_i - K_i)$. If $v_h \in V_h^o$ (X_h^o), $v_{hi} = v_h|_{K_i}$ then

$$\int_{\gamma_i} v_{hi}^2\, ds \le Ch^3 \int_{K_i} \Big[\Big(\frac{\partial v_{hi}}{\partial x_1}\Big)^2 + \Big(\frac{\partial v_{hi}}{\partial x_2}\Big)^2\Big] dx , \qquad (2.1)$$

$$\int_{\gamma_i} \Big(\frac{\partial v_{hi}}{\partial s}\Big)^2 ds \le Ch \int_{K_i} \Big[\Big(\frac{\partial v_{hi}}{\partial x_1}\Big)^2 + \Big(\frac{\partial v_{hi}}{\partial x_2}\Big)^2\Big] dx , \qquad (2.2)$$

$$\int_{E_i} (v_{hi})^2 dx \le Ch^3 \int_{K_i} v_{hi}^2\, dx \;\Big(\text{or } Ch^3 \int_{\tilde{K}_i} v_{hi}^2\, dx\Big), \qquad (2.3)$$

$$\int_{E_i} \Big[\Big(\frac{\partial v_{hi}}{\partial x_1}\Big)^2 + \Big(\frac{\partial v_{hi}}{\partial x_2}\Big)^2\Big] dx \le Ch \int_{K_i\,(\text{or } \tilde{K}_i)} \Big[\Big(\frac{\partial v_{hi}}{\partial x_1}\Big)^2 + \Big(\frac{\partial v_{hi}}{\partial x_2}\Big)^2\Big] dx , \qquad (2.4)$$

where s denotes length of arc along Γ .

Proof. We only prove (2.1)-(2.4) for $v_h \in X_h^o$. We know that the values of integrations in (2.1)-(2.4) are independent of the choice of coordinates. We put K_i on the coordinates x_1-x_2, which is described in Fig 3.

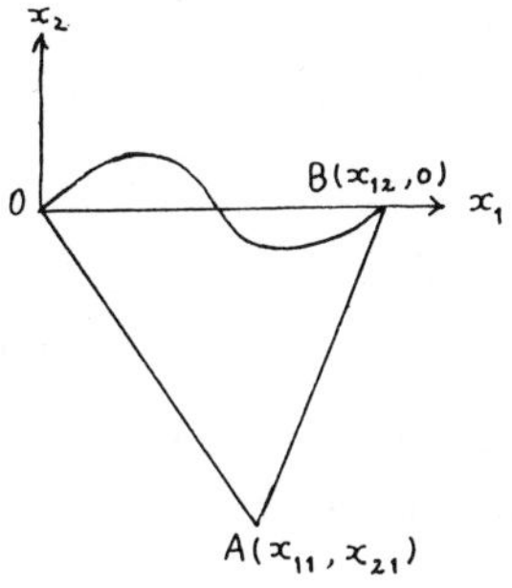

Fig 3.

We have

$$v_{hi} = \frac{a}{x_{21}} x_2 \quad , \qquad (2.5)$$

where a is a value of v_h at point A. We denote arc oB on Γ by $x_2 = f(x_1)$, then

$$S = \int_0^{x_1} \sqrt{1+f'^2}\, dx_1 \quad . \qquad (2.6)$$

From $\Gamma \in C^2$, we obtain

$$|f(x_1)| \leq ch_i^2 \quad , \quad |f'(x_1)| \leq ch_i \, , \quad \forall \; x_1 \in (0, x_{12}) . \qquad (2.7)$$

In addition, from (1.6) we have

$$c_1 h^2 \leq \text{meas } (K_i) \leq c_2 h^2 \quad . \qquad (2.8)$$

Using (2.5)-(2.8), we can easily prove (2.1)-(2.4).

Lemma 2.2 Let $\Gamma \in C^2$. We have

$$\| v_h \|_{\frac{1}{2},\Gamma} \leq c h | v_h |_{1,\Omega} \, , \quad \forall \; v_h \in V_h^o \; (\text{or } X_h^o \,) . \qquad (2.9)$$

Proof. From (2.1) and (2.2) we obtain

$$\| v_h \|_{0,\Gamma} \leq c h^{\frac{1}{2}} | v_h |_{1,\Omega} \, , \qquad (2.10)$$

$$| v_h |_{1,\Gamma} \leq c h^{\frac{1}{2}} | v_h |_{1,\Omega} \, . \qquad (2.11)$$

Using (2.10), (2.11) and property of interpolation space, we obtain (2.9).

Lemma 2.3 Let $p \in H^1(\Omega) \backslash R$. There exists $q_h \in X_h \backslash R$ satisfying

$$| p - q_h |_1 \leq c | p |_{1 \backslash 0} \quad , \qquad | p - q |_0 \leq ch | p |_{1 \backslash 0} \, . \qquad (2.12)$$

Proof. Combining the method in [4] and the method in [5], we may prove that there exists $\tilde{q}_h \in X_h$ satisfying

$$|p-\tilde{q}_h|_1 \leqslant c|p|_{1,0} \quad , \quad |p-\tilde{q}_h|_0 \leqslant ch|p|_{1,0} \ .$$

We define $a = \int_\Omega \tilde{q}_h dx /$ meas (Ω) and have

$$|a| = |\int_\Omega (\tilde{q}_h - p) dx / meas(\Omega)| \leqslant C[\int_\Omega (\tilde{q}_h - p)^2 dx]^{\frac{1}{2}} \leqslant Ch|p|_{1,0} .$$

We take $q_h = \tilde{q}_h - a (\in X_h \setminus R)$, then q_h satisfies (2.12).

Lemma 2.4 Let $\Gamma \in C^2$ and $u \in H_0^1(\Omega) \cap H^2(\Omega)$. There exists $v_h \in V_h^0$ satisfying

$$|u-v_h|_0 \leqslant ch^2 \| u \|_2 \ , \quad |u-v_h|_1 \leqslant ch \| u \|_2 \ .$$

The proof of this Lemma is easy and is omitted. We may also see [5].

Lemma 2.5 We have

$$\| v_h \|_{1,\Omega} \leqslant C|v_h|_{1,\Omega} \ , \quad \forall \ v_h \in X_h^0 \quad \text{or} \quad v_h \in V_h^0 \ .$$

Proof. By Lemma 2.1 the proof of above inequality is easy.

Lemma 2.6 Let $(\vec{v}_{h1}, \phi_{h1}) \in W_{oh}$ and $(\vec{v}_{h2}, \phi_{h2}) \in W_{oh}$. If $\vec{v}_{h1} = \vec{v}_{h2}$, then $\phi_{h1} = \phi_{h2}$.

Lemma 2.7 Let $(\vec{v}_h, \phi_h) \in W_{oh}$. $[|\vec{v}_h|_1^2 + |\phi_h|_1^2]^{\frac{1}{2}}$ and $|\vec{v}_h|_1$ are norm in W_{oh}. Moreover, they are equivalence.

The proof of these Lemma is easy and is omitted.

§3 Brezzi inequality

In this section we prove the following important inequality

$$\sup_{\vec{v}_h \in (V_h^0)^2 - \{\vec{0}\}} \frac{(q_h, div\, \vec{v}_h)}{|\vec{v}_h|_{1,\Omega}} \geqslant \beta^* \| q_h \|_{0,\Omega} , \ \forall q_h \in X_h \setminus R . \quad (3.1)$$

From Lemma 3.1 of chapter 1 in [6] we know that there exists $\vec{v} \in (H_0^1(\Omega))^2$ for $q_h \in X_h \setminus R$ satisfying

$$div\ \vec{v} = q_h \ , \quad |\vec{v}|_1 \leqslant c \| q_h \|_{0,1} \ . \quad (3.2)$$

By Lemma 2.5 there exists $\vec{w}_h \in (V_h^0)^2$ satisfying

$$(\nabla(\vec{w}_h - \vec{v}), \nabla \vec{y}_h) = 0 \ , \quad \forall \ \vec{y}_h \in (V_h^0)^2 .$$

Moreover, we have

$$|\vec{w}_h|_1 \leqslant |\vec{V}|_1 . \tag{3.3}$$

Now, we are in a position to define the $\vec{v}_h \in (V_h^\circ)^2$: Let K_i be an arbitrary element, we have

$$\begin{cases} \vec{v}_h(D_{ij}) = \vec{w}_h(D_{ij}), & j=1,2,3, \\ \int_{K_i} (\vec{v}_h - \vec{v})\,dx = 0, \end{cases} \tag{3.4}$$

where D_{ij} $(j=1,2,3)$ are three vertices of K_i. Thus (3.4) is a system of four linear equations with four unknowns. Therefore we must show that $\vec{w}_h=0$ and $\vec{v}=0 \Rightarrow \vec{v}_h=0$.

Let K_i be an interior element. By $\vec{v}_{hi}(D_{ij})=0$ $(j=1,2,3)$ and $\int_{K_i} \vec{v}_{hi}\,dx=0$ we obtain $\vec{v}_{hi}=0$. Let K_i be a boundary element. By $\vec{v}_{hi}(D_{ij})=0$ $(j=1,2,3)$ $\vec{v}_{hi}$ is the form

$$\vec{v}_{hi} = \vec{\alpha}_i p_i , \tag{3.5}$$

where p_i is a linear function on each $\Delta G_i D_{i2} D_{i3}$, $\Delta G_i D_{i3} D_{i1}$ and curved triangle $G_i D_{i1} D_{i2}$ and satisfies

$$p_i(D_{ij})=0, \quad j=1,2,3 \quad p(G_i)=1. \tag{3.6}$$

Moreover p_i are continuous functions on each K_i. Using Lemma 2.1, we obtain

$$\left| \int_{E_i} P_i\,dx \right| \leqslant \left(\int_{E_i} dx \right)^{\frac{1}{2}} \left(\int_{E_i} P_i^2\,dx \right)^{\frac{1}{2}} \leqslant c h^3 |P_i|_{0,\tilde{K}_i} .$$

From here we obtain that $\int_{K_i} p\,dx$ is positive for sufficiently small h and we have $\vec{v}_{hi}=0$. Thus there exists one and only one function $\vec{v}_h \in (V_h^\circ)^2$ defined by (3.4).

Lemma 3.1 $\vec{v}_h$ satisfies the following inequality

$$|\vec{v}_h|_1 \leqslant C \|q_h\|_{0\backslash 1} . \tag{3.7}$$

Proof. Set $\vec{e}_h = \vec{v}_h - \vec{w}_h$ $(\in (V_h^\circ)^2)$, $\vec{e} = \vec{v} - \vec{w}_h$. We have

$$|\vec{v}_h|_1 \leqslant |\vec{w}_h|_1 + |\vec{e}_h|_1 \leqslant C\|q_h\|_{0\backslash 1} + |\vec{e}_h|_1 . \tag{3.8}$$

On each K_i $\vec{e}_h$ is the form

$$\vec{e}_h = \vec{\alpha}_i P_i = \vec{e}_h(G_i) P_i , \tag{3.9}$$

where p_i is defined by (3.6). Obviously

$$|P_i|_{1,K_i} \leqslant C. \tag{3.10}$$

For sufficiently small h we have

$$|\vec{e}_h(G_i)| = \left|\int_{K_i} P_i\,dx\right|^{-1} \left|\int_{K_i} \vec{e}_h\,dx\right| = \left|\int_{K_i} P_i\,dx\right|^{-1} \left|\int_{K_i} (\vec{v}-\vec{w}_{hi})dx\right|$$

$$\leqslant \frac{c}{h^2}\left[\int_{K_i} dx\right]^{\frac{1}{2}} \left[\int_{K_i} (\vec{v}-\vec{w}_{hi})^2 dx\right]^{\frac{1}{2}} \leqslant \frac{c}{h}\|\vec{e}\|_{0,K_i} .$$

Thus we obtain

$$|\vec{e}_h|_{1,\Omega} \leqslant \frac{c}{h}\|\vec{e}\|_0 . \qquad (3.11)$$

Using Aubin-Nitsche method, we obtain

$$\|\vec{e}\|_0 \leqslant Ch\,|\vec{e}|_1 \leqslant Ch\,\|q_h\|_{0\backslash 1} . \qquad (3.12)$$

From (3.8)-(3.12) we obtain (3.7).

Now we prove inequality (3.1).

Let q_h be an arbitrary element in $X_h\backslash R$. $\vec{v}_h$ is defined by the discussion above. Let $s_h \in X_h\backslash R$. We have

$$|(\text{div}\,\vec{v}_h - q_h,\ s_h)| = \left|\sum_{i=1}^{m_0+m_1}\int_{K_i}(\vec{v}_{hi}-\vec{v})\cdot \text{grad}\, s_{hi}\ dx + \int_\Gamma (\vec{v}_h\cdot n)s_h\,ds\right|$$

$$= \left|\int_\Gamma (\vec{v}_h\cdot n)s_h\,ds\right| \leqslant \left[\int_\Gamma (\vec{v}_h\cdot n)^2 ds\right]^{\frac{1}{2}} \left[\int_\Gamma s_h^2\,ds\right]^{\frac{1}{2}} \leqslant Ch^{\frac{3}{2}}|\vec{v}_h|_1\,\|s_h\|_1 .$$

Using lemma 3.1 , inverse inequality and the above inequality, we have

$$(\text{div}\,\vec{v}_h, q_h) = \|q_h\|_{0\backslash 1}^2 + (\text{div}\,\vec{v}_h - q_h, q_h) \geqslant \|q_h\|_{0\backslash 1}^2 - C_1 h^{\frac{1}{2}}|\vec{v}_h|_1\,\|q_h\|_{0\backslash 1}$$

$$\geqslant \left(\frac{1}{C} - C_1 h^{\frac{1}{2}}\right)\|q_h\|_{0\backslash 1}\,|\vec{v}_h|_1 .$$

From here, we obtain (3.1) for sufficiently small h.

Lemma 3.2 We have

$$\sup_{\vec{v}_h\in(V_h^0)^2-\{\vec{0}\}} \frac{\int_\Omega \nabla q_h\cdot\vec{v}_h\,dx}{|\vec{v}_h|_1} \geqslant \beta_1^*\|q_h\|_{0,\Omega},\quad \forall\, q_h\in X_h\backslash R, \qquad (3.13)$$

where β_1^* is a positive constant independent of h.

Proof. Using Stokes formula, we have

$$\int_\Omega \nabla q_h\cdot\vec{v}_h\,dx = -\int_\Omega q_h\,\nabla\cdot\vec{v}_h\,dx + \int_\Gamma q_h(\vec{v}_h\cdot n)\,ds . \qquad (3.14)$$

By Lemma 2.1 we obtain

$$\left|\int_\Gamma q_h(\vec{v}_h\cdot n)ds\right| \leqslant Ch^{\frac{3}{2}}\|q_h\|_1\,|\vec{v}_h|_1 \leqslant Ch^{\frac{1}{2}}\|q_h\|_0\,|\vec{v}_h|_1 . \qquad (3.15)$$

From (3.14), (3.15) and (3.1) we obtain (3.13).

§ 4 Error Estimate for Stokes equation

Lemma 4.1 Suppose that $p(x)\in H^1(\Omega)\setminus R$. We define

$$X_h(p)=\{q_h \mid q_h\in X_h\setminus R,\ \int_\Omega \nabla(q_h-p)\cdot\nabla w_h\,dx=0,\ \forall\ w_h\in X_h^0\}.$$

Then there exists positive constant C satisfying

$$\inf_{q_h\in X_h(p)} \|p-q_h\|_0 \leqslant Ch\|p\|_{1\setminus 0}. \tag{4.1}$$

Proof. There exists a unique $p_h\in X_h\setminus R$ satisfying variational problem Q_h:

$$\int_\Omega \nabla(p_h-p)\cdot\nabla w_h\,dx=0,\quad \forall\ w_h\in X_h\setminus R. \tag{4.2}$$

Obviously $X_h^0\not\subset X_h\setminus R$. But we can easily prove that (4.2) hold for $w_h\in X_h$. Thus we obtain $p_h\in X_h(p)$ and

$$\inf_{q_h\in X_h(p)} \|p-q_h\|_0 \leqslant \|p-p_h\|_0 \tag{4.3}$$

From (4.2) we have

$$|p_h-p|_1^2=\int_\Omega\nabla(p_h-p)\cdot\nabla(p_h-p)\,dx=\int_\Omega\nabla(p_h-p)\cdot\nabla(q_h-p)\,dx,\ \forall\, q_h\in X_h\setminus R,$$

and

$$|p_h-p|_1^2=\inf_{q_h\in X_h\setminus R}|q_h-p|_1\leqslant\inf_{q_h\in X_h}|q_h-p|_1\leqslant|p|_1.$$

Using Aubin-Nitsche method and the above inequality, we derive

$$\|p_h-p\|_0\leqslant Ch\|p\|_{1\setminus 0}. \tag{4.4}$$

From (4.3) and (4.4) we obtain (4.1).

Lemma 4.2 Suppose $\vec{u}\in(H^2(\Omega)\cap H_0^1(\Omega))^2$. We have

$$\inf_{(\vec{v}_h,\phi_h)\in W_{0h}}|\vec{u}-\vec{v}_h|_1\leqslant Ch\|\vec{u}\|_2.$$

Proof. We define

$$Z_h=\{(\vec{v}_h,\phi_h)\mid \vec{v}_h\in(V_h^0)^2,\ \exists\phi_h\in X_h^0 \text{ such that } (\vec{v}_h,\phi_h)\in W_{0h}\}$$

Let $\vec{y}_h\in(V_h^0)^2$. Variational problem I_h: Find $(\vec{z}_h,\theta_h)$ satisfying

$$\begin{cases}\int_\Omega\nabla\vec{z}_h\cdot\nabla\vec{v}_h\,dx=\int_\Omega\nabla\vec{y}_h\cdot\nabla\vec{v}_h\,dx,\quad\forall\,(\vec{v}_h,\phi_h)\in W_{0h},\\ (\vec{z}_h,\theta_h)\in W_{0h}.\end{cases} \tag{4.5}$$

In order to prove the following problem II_h has a unique solution $(\vec{z}_h, \theta_h, \Pi_h) \in (V_h^o)^2 \times X_h^o \times (X_h \setminus R)$

$$\begin{cases} \int_\Omega \nabla \Pi_h \cdot \nabla w_h \, dx = 0, \quad \forall w_h \in X_h^o, \\ \int_\Omega \nabla \vec{z}_h \cdot \nabla \vec{v}_h \, dx + \int_\Omega \nabla \Pi_h \cdot \vec{v}_h \, dx = \int_\Omega \nabla \vec{y}_h \cdot \nabla \vec{v}_h \, dx, \quad \forall \vec{v}_h \in (V_h^o)^2, \\ \int_\Omega \nabla \theta_h \cdot \nabla q_h \, dx = -\int_\Omega \vec{z}_h \cdot \nabla q_h \, dx, \quad \forall q_h \in X_h \setminus R, \end{cases} \tag{4.6}$$

we must show that $\vec{y}_h = 0 \Rightarrow (\vec{z}_h, \theta_h, \Pi_h) = (\vec{0}, 0, 0)$. From the third equality of (4.6), we know $(\vec{z}_h, \theta_h) \in W_{oh}$. Taking $\vec{v}_h = \vec{z}_h$, $q_h = \Pi_h$, $w_h = \theta_h$, we obtain

$$\begin{cases} \int_\Omega \nabla \Pi_h \cdot \nabla \theta_h \, dx = 0, \\ \int_\Omega \nabla \vec{z}_h \cdot \nabla \vec{z}_h \, dx + \int_\Omega \nabla \Pi_h \cdot \vec{z}_h \, dx = 0, \\ \int_\Omega \nabla \theta_h \cdot \nabla \Pi_h \, dx = -\int_\Omega \vec{z}_h \cdot \nabla \Pi_h \, dx. \end{cases} \tag{4.7}$$

Form here we have $\int_\Omega \nabla \vec{z}_h \cdot \nabla \vec{z}_h \, dx = 0$, this implies $\vec{z}_h = 0 \Rightarrow \theta_h = 0$.

Using Lemma 3.2, from (4.6) we obtain $\Pi_h = 0$ for sufficiently small h.

Now we prove the equivalence of problems I_h and II_h. It is well known that problem I_h and II_h have a unique solution respectively.

We only prove that the solution of I_h is $(\vec{z}_h, \theta_h)$ in the solution $(\vec{z}_h, \theta_h, \Pi_h)$ of II_h. Let $(\vec{z}_h, \theta_h, \Pi_h)$ be the solution of II_h. Suppose that $(\vec{v}_h, \phi_h)$ is an arbitrary element of W_{oh}. We have

$$\int_\Omega \nabla \phi_h \cdot \nabla q_h \, dx = -\int_\Omega \vec{v}_h \cdot \nabla q_h \, dx, \quad \forall q_h \in X_h \setminus R.$$

Taking $q_h = \Pi_h$ in the above equality, $w_h = \phi_h$ in the II_h, we have

$$\int_\Omega \nabla \phi_h \cdot \nabla \Pi_h \, dx = -\int_\Omega \vec{v}_h \cdot \nabla \Pi_h \, dx,$$

$$\int_\Omega \nabla \Pi_h \cdot \nabla \phi_h \, dx = 0.$$

From here we derive $\int_\Omega \vec{v}_h \cdot \nabla \Pi_h \, dx = 0$ and

$$\int_\Omega \nabla \vec{z}_h \cdot \nabla \vec{v}_h \, dx = \int_\Omega \nabla \vec{y}_h \cdot \nabla \vec{v}_h \, dx, \quad \forall (\vec{v}_h, \phi_h) \in W_{oh}.$$

Thus we have proved the equivalence of problems I_h and II_h.

We have

$$\inf_{(\vec{v}_h,\phi_h)\in W_{oh}} |\vec{v}_h - \vec{u}|_1 \leq |\vec{z}_h - \vec{u}|_1 \leq |\vec{u}-\vec{y}_h|_1 + |\vec{y}_h - \vec{z}_h|_1 . \qquad (4.8)$$

Using second formula of II_h and lemma 3.2 , we obtain

$$|\vec{y}_h - \vec{z}_h|_1 = \sup_{\vec{v}_h \in (V_h^o)^2 - \{\vec{0}\}} \frac{|\int_\Omega \nabla(\vec{y}_h - \vec{z}_h)\cdot\nabla\vec{v}_h dx|}{|\vec{v}_h|_1}$$

$$= \sup_{\vec{v}_h \in (V_h^o)^2 - \{\vec{0}\}} \frac{|\int_\Omega \nabla\pi_h \cdot \vec{v}_h dx|}{|\vec{v}_h|_1} \geq \beta_1^* \|\pi_h\|_0 .$$

Taking $\vec{v}_h = (\vec{y}_h - \vec{z}_h)$ in II_h, we obtain

$$\int_\Omega \nabla\pi_h\cdot(\vec{y}_h - \vec{z}_h)\,dx = |\vec{y}_h - \vec{z}_h|_1^2 \geq \beta_1^* |\vec{y}_h - \vec{z}_h| \, \|\pi_h\|_0 \qquad (4.9)$$

Since $\int_\Omega \nabla\pi_h\cdot\vec{z}_h dx=0$ we have

$$\int_\Omega \nabla\pi_h\cdot(\vec{y}_h - \vec{z}_h)\,dx = -\int_\Omega \nabla\pi_h\cdot(\vec{u}-\vec{y}_h)dx = \int_\Omega \pi_h\cdot\nabla(\vec{u}-\vec{y}_h)dx + \int_\Gamma \pi_h \vec{y}_h\cdot n\,ds. \qquad (4.10)$$

Using Lemma 2.1 and inequality (see [7])

$$\left[\sum_{i=1}^{m_0} |u|_{1,K_i}^2\right]^{\frac{1}{2}} \leq ch^{\frac{1}{2}} \|u\|_2 , \quad \forall\, u \in H^2(\Omega)\cap H_0^1(\Omega), \qquad (4.11)$$

we obtain

$$|\int_\Gamma \pi_h\cdot(\vec{y}_h\cdot n)ds| \leq ch^{\frac{3}{2}} \left(\sum_{i=1}^{m_0} |\vec{y}_h|_{1,K_i}\right)\|\pi_h\|_1$$

$$\leq Ch^{\frac{1}{2}} \|\pi_h\|_0 \left(|\vec{u}-\vec{y}_h|_1 + h^{\frac{1}{2}}\|\vec{u}\|_2\right). \qquad (4.12)$$

Using (4.10) and (4.12), from (4.9) we obtain

$$\beta_1^* |\vec{y}_h - \vec{z}_h|_1 \|\pi_h\|_0 \leq \|\pi_h\|_0 |\vec{u}-\vec{y}_h|_1 + Ch^{\frac{1}{2}}\|\pi_h\|_0\left(|\vec{u}-\vec{y}_h|_1 + h^{\frac{1}{2}}\|\vec{u}\|_2\right). \qquad (4.13)$$

From the second formula of problem II_h we know that $(\vec{z}_h - \vec{y}_h)$ and π_h are simultaneously different from 0 or equal to 0. Thus we obtain from (4.13)

$$\beta_1^* |\vec{y}_h - \vec{z}_h|_1 \leq |\vec{u}-\vec{y}_h|_1 + Ch^{\frac{1}{2}}\left(|\vec{u}-\vec{y}_h|_1 + h^{\frac{1}{2}}\|\vec{u}\|_2\right). \qquad (4.14)$$

From (4.8) and (4.14) we easily prove this lemma.

Theorem 4.1 Let Ω be a domain with sufficiently smooth boundary Γ , $\vec{f}\in(L^2(\Omega))^2$, $\mathrm{div}\vec{f}\in L^2(\Omega)$. Then

(1) $(\vec{u},p)\in(H^2(\Omega)\cap H_0^1(\Omega))^2 \times (H^1(\Omega)\setminus R)$ and $\left.\frac{\partial p}{\partial n}\right|_\Gamma \in H^{-\frac{1}{2}}(\Gamma)$.

(2) We have the following error estimates

$$\| \vec{u} - \vec{u}_h \|_1 \leq Ch \left(\| \vec{f} \|_0 + \| \frac{\partial p}{\partial n} \|_{-\frac{1}{2},\Gamma} + \| \vec{f} \cdot n \|_{-\frac{1}{2},\Gamma} \right), \tag{4.15}$$

$$\| p - p_h \|_{0,1} \leq Ch \left(\| \vec{f} \|_0 + \| \frac{\partial p}{\partial n} \|_{-\frac{1}{2},\Gamma} + \| \vec{f} \cdot n \|_{-\frac{1}{2},\Gamma} \right), \tag{4.16}$$

$$\| \psi_h \|_1 \leq Ch \left(\| \vec{f} \|_0 + \| \frac{\partial p}{\partial n} \|_{-\frac{1}{2},\Gamma} + \| \vec{f} \cdot n \|_{-\frac{1}{2},\Gamma} \right). \tag{4.17}$$

Proof (1) Using the first formula of (1.5) and assumption and Theorem 1.2 of chapter 1 in [8], we obtain the proof.

(2) Let $\vec{v}_h$ be an arbitrary element in $(V_h^0)^2$. We have

$$\nu | \vec{v}_h - \vec{u}_h |_1^2 = \nu \int_\Omega \nabla (\vec{v}_h - \vec{u}_h) \cdot \nabla \vec{v}_h \, dx - \nu \int_\Omega \nabla \vec{u}_h \cdot \nabla (\vec{v}_h - \vec{u}_h) \, dx. \tag{4.18}$$

If $\vec{v}_h$ is replaced by $\vec{v}_h - \vec{u}_h$ in the second formula of variational problem P_{h1}, we obtain

$$\nu \int_\Omega \nabla \vec{u}_h \cdot \nabla (\vec{v}_h - \vec{u}_h) \, dx = \int_\Omega \vec{f} \cdot (\vec{v}_h - \vec{u}_h) \, dx - \int_\Omega \nabla p_h \cdot (\vec{v}_h - \vec{u}_h) \, dx. \tag{4.19}$$

From the second formula of (1.5) we have

$$\begin{aligned} \nu \int_\Omega \nabla \vec{u} \cdot \nabla (\vec{v}_h - \vec{u}_h) \, dx = & \int_\Omega \vec{f} \cdot (\vec{v}_h - \vec{u}_h) \, dx - \int_\Omega \nabla p \cdot (\vec{v}_h - \vec{u}_h) \, dx \\ & + \nu \int_\Gamma \frac{\partial \vec{u}}{\partial n} \cdot (\vec{v}_h - \vec{u}_h) \, ds. \end{aligned} \tag{4.20}$$

From (4.18)-(4.20) we derive

$$\begin{aligned} \nu | \vec{v}_h - \vec{u}_h |_1^2 = & \nu \int_\Omega \nabla (\vec{v}_h - \vec{u}) \cdot \nabla (\vec{v}_h - \vec{u}_h) \, dx + \int_\Omega \nabla (p_h - p) \cdot (\vec{v}_h - \vec{u}_h) \, dx \\ & + \nu \int_\Gamma \frac{\partial \vec{u}}{\partial n} \cdot (\vec{v}_h - \vec{u}_h) \, ds. \end{aligned} \tag{4.21}$$

Using triangle inequality and the above equality, we obtain

$$\begin{aligned} | \vec{u} - \vec{u}_h |_1 & \leq | \vec{u} - \vec{v}_h |_1 + | \vec{v}_h - \vec{u}_h |_1 \\ & \leq 2 | \vec{u} - \vec{v}_h |_1 + \frac{1}{\nu} \frac{\int_\Omega \nabla (p_h - p) \cdot (\vec{v}_h - \vec{u}_h) \, dx}{| \vec{v}_h - \vec{u}_h |_1} + \frac{\int_\Gamma \frac{\partial \vec{u}}{\partial n} \cdot (\vec{v}_h - \vec{u}_h) \, ds}{| \vec{v}_h - \vec{u}_h |_1}, \end{aligned} \tag{4.22}$$

$$\forall \ \vec{v}_h \in (V_h^0)^2 \text{ and } \vec{v}_h \neq \vec{u}_h.$$

Let $(\vec{v}_h, \phi_h)$ and q_h be an arbitrary element in W_{oh} and $X_h(p)$ respectively. Obviously $(\vec{v}_h - \vec{u}_h, \phi_h - \psi_h) \in W_{oh}$ and $p_h - q_h \in X_h$. We have

$$\int_\Omega \nabla (\phi_h - \psi_h) \cdot \nabla (p_h - q_h) \, dx = - \int_\Omega (\vec{v}_h - \vec{u}_h) \cdot \nabla (p_h - q_h) \, dx, \tag{4.23}$$

Let w_h be an arbitrary element in X_h^0. Using the first formula of (1.5), we obtain

$$\int_\Omega \nabla p\cdot\nabla w_h\, dx = \int_\Omega \vec{f}\cdot\nabla w_h\, dx + \int_\Gamma \frac{\partial p}{\partial n} w_h\, ds - \int_\Gamma (\vec{f}\cdot n) w_h\, ds,\ \forall\, w_h \in X_h^0 ,$$

we notice that $\int_\Gamma \frac{\partial p}{\partial n} w_h\, ds$ and $\int_\Gamma (\vec{f}\cdot n) w_h\, ds$ are values of functional defining in $H^{-\frac{1}{2}}(\Gamma)$. Using (4.24) and the first formula of variational problem P_{h1} and the definition of $X_h(p)$ we obtain

$$\int_\Omega \nabla(p_h - q_h)\cdot\nabla w_h\, dx = -\int_\Gamma \frac{\partial p}{\partial n}\cdot w_h\, ds + \int_\Gamma (\vec{f}\cdot n) w_h\, ds ,\ \forall\, w_h \in X_h^0 .$$

Taking $w_h = \phi_h - \psi_h$, we have

$$\int_\Omega \nabla(p_h - q_h)\cdot\nabla(\phi_h - \psi_h)\, dx = -\int_\Gamma \frac{\partial p}{\partial n}(\phi_h - \psi_h)\, ds + \int_\Gamma (\vec{f}\cdot n)(\phi_h - \psi_h)\, ds. \quad (4.25)$$

Using (4.23) and the above equality we derive

$$\begin{aligned} &\int_\Omega \nabla(p_h - p)\cdot(\vec{v}_h - \vec{u}_h)\, dx \\ &= -\int_\Omega (q_h - p)\,\nabla\cdot(\vec{v}_h - \vec{u}_h)\, dx + \int_\Gamma (q_h - p)\,((\vec{v}_h - \vec{u}_h)\cdot n)\, ds \\ &\quad + \int_\Gamma \frac{\partial p}{\partial n}(\phi_h - \psi_h)\, ds - \int_\Gamma (\vec{f}\cdot n)(\phi_h - \psi_h)\, ds. \end{aligned} \quad (4.26)$$

We have

$$\left|\int_\Gamma (q_h - p)((\vec{v}_h - \vec{u}_h)\cdot n)\, ds\right| \leqslant C h^{\frac{3}{2}} \| q_h - p\|_1 \, |\vec{v}_h - \vec{u}_h|_1 . \quad (4.27)$$

Using inverse inequality and Lemma 2.3 , we obtain

$$h\, \| p - q_h \|_1 \leqslant C\left(h\, \| p\|_{1\backslash 0} + \| p - q_h \|_0 \right). \quad (4.28)$$

From (4.27) and (4.28) we derive

$$\begin{aligned} &\left|\int_\Gamma (q_h - p)((\vec{v}_h - \vec{u}_h)\cdot n)\, ds\right| \\ &\quad\leqslant C\left(h^{\frac{3}{2}} \|p\|_{1\backslash 0} + h^{\frac{1}{2}} \| p - q_h\|_0 \right) |\vec{v}_h - \vec{u}_h|_1 . \end{aligned} \quad (4.29)$$

Using the same method and Lemma 2.2, we obtain

$$\left|\int_\Gamma \frac{\partial p}{\partial n}(\phi_h - \psi_h)\, ds\right| \leqslant \left\|\frac{\partial p}{\partial n}\right\|_{-\frac{1}{2},\Gamma} \|\phi_h - \psi_h\|_{\frac{1}{2},\Gamma} \leqslant Ch \left\|\frac{\partial p}{\partial n}\right\|_{-\frac{1}{2},\Gamma} |\phi_h - \psi_h|_1 , \quad (4.30)$$

$$\left|\int_\Gamma (\vec{f}\cdot n)(\phi_h - \psi_h)\, ds\right| \leqslant Ch \|\vec{f}\cdot n\|_{-\frac{1}{2},\Gamma} \, |\phi_h - \psi_h|_1 . \quad (4.31)$$

Using Lemma 2.7, from (4.26)-(4.31) we derive

$$\begin{aligned} &\left|\int_\Omega \nabla(p_h - p)\cdot(\vec{v}_h - \vec{u}_h)\, dx\right| \\ &\quad\leqslant C\left(h \left\|\frac{\partial p}{\partial n}\right\|_{-\frac{1}{2},\Gamma} + h\, \|\vec{f}\cdot n\|_{-\frac{1}{2},\Gamma} + \|p - q_h\|_0 + h^{\frac{3}{2}} \|p\|_{1\backslash 0}\right) |\vec{v}_h - \vec{u}_h|_1 . \end{aligned} \quad (4.32)$$

Using $\| \vec{u} \|_2 + \| p \|_{1\backslash 0} \leqslant c\, \| \vec{f} \|_0$, we have

$$\left|\int_\Gamma \frac{\partial \vec{u}}{\partial n}\cdot(\vec{v}_h - \vec{u}_h)\, ds\right| \leqslant C h^{\frac{3}{2}} \|\vec{f}\|_0 \, |\vec{v}_h - \vec{u}_h|_1 . \quad (4.33)$$

Using (4.32) and (4.33), from (4.22) we obtain

$$|\vec{u}-\vec{u}_h|_1 \leq 2|\vec{u}-\vec{v}_h|_1 + c_1 \left(\| p-q_h\|_0 + h^{\frac{3}{2}} \|\vec{f}\|_0 \right) + C_2 h \left(\left\|\frac{\partial p}{\partial n}\right\|_{-\frac{1}{2},\Gamma} + \|\vec{f}\cdot n\|_{-\frac{1}{2},\Gamma} \right), \ \forall (\vec{v}_h, \phi_h) \in W_{0h},\ q_h \in X_h(p).$$

Using Lemma 4.1, Lemma 4.2 and (4.11), we obtain

$$|\vec{u}-\vec{u}_h|_1 \leq Ch \left(\|\vec{f}\|_0 + \left\|\frac{\partial p}{\partial n}\right\|_{-\frac{1}{2},\Gamma} + \|\vec{f}\cdot n\|_{-\frac{1}{2},\Gamma} \right),$$

$$\|\vec{u}-\vec{u}_h\|_1 \leq C \left(|\vec{u}-\vec{u}_h|_1 + \|\vec{u}-\vec{u}_h\|_{0,\Gamma} \right) \leq Ch \left(\|\vec{f}\|_0 + \left\|\frac{\partial p}{\partial n}\right\|_{-\frac{1}{2},\Gamma} + \|\vec{f}\cdot n\|_{-\frac{1}{2},\Gamma} \right).$$

Now we prove (4.16). Let q_h be an arbitrary element of $X_h \setminus R$. From (3.1) we obtain

$$\| p_h - q_h\|_{0\backslash 1} \leqslant C \sup_{\vec{v}_h \in (V_h^0)^2 - \{\vec{0}\}} \frac{\int_\Omega (p_h - q_h)\, \nabla\cdot\vec{v}_h\, dx}{|\vec{v}_h|_1} . \tag{4.34}$$

We have

$$\begin{aligned} \int_\Omega (p_h - q_h)\, \nabla\cdot\vec{v}_h\, dx &= -\int_\Omega \nabla(p_h - q_h)\cdot\vec{v}_h\, dx + \int_\Gamma (p_h - q_h)(\vec{v}_h\cdot n)\, ds \\ &= \int_\Omega \nabla(p-q_h)\cdot\vec{v}_h\, dx + \int_\Omega \nabla(p_h - p)\cdot\vec{v}_h\, dx + \int_\Gamma (p_h-q_h)(\vec{v}_h\cdot n)\, ds . \end{aligned} \tag{4.35}$$

From the second formula of (1.5) we obtain

$$\int_\Omega \nabla p\cdot\vec{v}_h\, dx = \int_\Omega \vec{f}\cdot\vec{v}_h\, dx - \nu\int_\Omega \nabla\vec{u}\cdot\nabla\vec{v}_h\, dx + \nu\int_\Gamma \frac{\partial\vec{u}}{\partial n}\cdot\vec{v}_h\, ds .$$

Using the second formula of variational problem P_{h1}, from above equality we obtain

$$\int_\Omega \nabla(p_h - p)\cdot\vec{v}_h\, dx = \nu\int_\Omega \nabla(\vec{u}-\vec{u}_h)\cdot\nabla\vec{v}_h\, dx - \nu\int_\Gamma \frac{\partial\vec{u}}{\partial n}\cdot\vec{v}_h\, ds .$$

We have

$$\begin{aligned} \int_\Omega \nabla(p_h - q_h)\cdot\vec{v}_h\, dx &= \int_\Omega \nabla(p-q_h)\cdot\vec{v}_h\, dx + \nu\int_\Omega \nabla(\vec{u}-\vec{u}_h)\cdot\nabla\vec{v}_h\, dx \\ &\quad - \nu\int_\Gamma \frac{\partial\vec{u}}{\partial n}\cdot\vec{v}_h\, ds = -\int_\Omega (p-q_h)\nabla\cdot\vec{v}_h\, dx + \int_\Gamma (p-q_h)(\vec{v}_h\cdot n)\, ds \\ &\quad + \nu\int_\Omega \nabla(\vec{u}-\vec{u}_h)\cdot\nabla\vec{v}_h\, dx - \nu\int_\Gamma \frac{\partial\vec{u}}{\partial n}\cdot\vec{v}_h\, ds . \end{aligned} \tag{4.36}$$

$$\left| \int_\Gamma \frac{\partial\vec{u}}{\partial n}\cdot\vec{v}_h\, ds \right| \leqslant Ch^{\frac{3}{2}} \|\vec{u}\|_2\, |\vec{v}_h|_1 . \tag{4.37}$$

Using (4.36) and (4.37), we obtain

$$\left| \int_\Omega \nabla(p_h - q_h)\cdot\vec{v}_h\, dx \right| \leqslant \|p-q_h\|_0 |\vec{v}_h|_1 + \nu |\vec{u}-\vec{u}_h|_1 |\vec{v}_h|_1 + C\|p-q_h\|_1 h^{\frac{3}{2}} |\vec{v}_h|_1 + Ch^{\frac{3}{2}} \|\vec{u}\|_2 |\vec{v}_h|_1 .$$

Using Lemma 4.1, we know that there exists a $q_{oh} \in X_h \backslash R$ satisfying $|p-q_{oh}|_1 \leq c \| p \|_{1\backslash 0}$ and $\| p-q_{oh} \|_0 \leq ch \| p \|_{1\backslash 0}$. From the above equality we derive

$$\left| \int_\Omega \nabla (P_h - q_{oh}) \cdot \vec{V}_h dx \right| \leq Ch \left(\|\vec{f}\|_0 + \left\| \frac{\partial h}{\partial n} \right\|_{-\frac{1}{2},\Gamma} + \|\vec{f} \cdot n\|_{-\frac{1}{2},\Gamma} \right) |\vec{V}_h|_1 . \tag{4.38}$$

In addition, we have

$$\left| \int_\Gamma (P_h - q_{oh})(\vec{V}_h \cdot n) ds \right| \leq Ch^{\frac{1}{2}} \| P_h - q_{oh} \|_0 \, |\vec{V}_h|_1 . \tag{4.39}$$

Using (4.35), (4.38) and (4.39), from (4.34) we obtain for sufficiently small h

$$\| P_h - q_{oh} \|_{0\backslash 1} \leq Ch \left(\| \vec{f} \|_0 + \left\| \frac{\partial h}{\partial n} \right\|_{-\frac{1}{2},\Gamma} + \| \vec{f} \cdot n \|_{-\frac{1}{2},\Gamma} \right) . \tag{4.40}$$

Using triangle inequality, from (4.40) we obtain (4.16).

Now we prove (4.17). We know that the third formula of variational problem P_{h1} holds for $q_h \in X_h$. Thus we obtain

$$\begin{aligned} |\psi_h|_1^2 &= \int_\Omega |\nabla \psi_h|^2 dx = -\int_\Omega \vec{u}_h \cdot \nabla \psi_h dx \\ &= \int_\Omega (\nabla \cdot (\vec{u}_h - \vec{u})) \psi_h dx - \int_\Gamma (\vec{u}_h \cdot n) \psi_h ds , \end{aligned} \tag{4.41}$$

$$\left| \int_\Gamma (\vec{u}_h \cdot n) \psi_h ds \right| \leq \| \vec{u}_h \cdot n \|_{0,\Gamma} \| \psi_h \|_{0,\Gamma} \leq Ch^3 (|\vec{u}_h - \vec{u}|_1 + |\vec{u}|_1) |\psi_h|_1 , \tag{4.42}$$

$$\left| \int_\Omega (\nabla \cdot (\vec{u}_h - \vec{u})) \psi_h ds \right| \leq |\vec{u}_h - \vec{u}|_1 \| \psi_h \|_0 \leq C |\vec{u}_h - \vec{u}|_1 |\psi_h|_1 . \tag{4.43}$$

Using (4.42) and (4.43), from (4.41) we obtain (4.17)

§5 Error estimate for Navier-Stokes equation

We define

$$\gamma_h(\vec{f}) = \sup_{(\vec{V}_h, \phi_h) \in W_h - \{\vec{0}\}} \frac{(\vec{f}, \vec{V}_h + \nabla \phi_h)}{|\vec{V}_h|_1} , \tag{5.1}$$

$$\gamma(\vec{f}) = \sup_{(\vec{V}, \phi) \in W - \{\vec{0}\}} \frac{(\vec{f}, \vec{V} + \nabla \phi)}{|\vec{V}|_1} , \tag{5.2}$$

$$\gamma_h(\beta) = \sup_{\substack{\vec{u}_h, \vec{v}_h \in (V_h^o)^2 - \{\vec{0}\} \\ (\vec{w}_h, \eta_h) \in W_h - \{\vec{0}\}}} \frac{((\vec{u}_h \cdot \nabla)\vec{v}_h, \vec{w}_h + \nabla \eta_h)}{|\vec{u}_h|_1 \, |\vec{v}_h|_1 \, |\vec{w}_h|_1}, \tag{5.3}$$

$$\gamma(\beta) = \sup_{\substack{\vec{u}, \vec{v} \in (H_0^1)^2 - \{\vec{0}\} \\ (\vec{w}, \eta) \in W - \{\vec{0}\}}} \frac{((\vec{u} \cdot \nabla)\vec{v}, \vec{w} + \nabla \eta)}{|\vec{u}|_1 \, |\vec{v}|_1 \, |\vec{w}|_1}. \tag{5.4}$$

Lemma 5.1 Let $u \in W^{2,p}(\Omega)$ and $W^{2,p}(\Omega) \hookrightarrow W^{m,q}(\Omega)$ (where $\hookrightarrow$ denotes imbedding). There exists $v_h \in X_h$ (or V_h) satisfying

$$|u - v_h|_{m,q} \leq ch^{\frac{2}{q} - \frac{2}{p} + 2 - m} \|u\|_{2,p}. \tag{5.5}$$

Lemma 5.2 Let $u_h \in V_h$ or X_h, We have

$$|u_h|_{1,4,\Omega_h} \leq ch^{-\frac{1}{2}} |u_h|_{1,2,\Omega_h}, \tag{5.6}$$

$$|u_h|_{1,4,\Omega} \leq ch^{-\frac{1}{2}} |u_h|_{1,2,\Omega}. \tag{5.7}$$

The proof of Lemma 5.1 - Lemma 5.2 is easy and is omitted.

Lemma 5.3 Let $(\vec{w}, \eta) \in W$. There exists a sequence $(\vec{w}_h, \eta_h)$ in W_h satisfying

$$\lim_{h \to 0} (\|\vec{w}_h - \vec{w}\|_1 + \|\eta_h - \eta\|_1) = 0. \tag{5.8}$$

Proof. Using density of $C_0^\infty(\Omega)$ in $H_0^1(\Omega)$, we know that there exist $\vec{w}_h \in (V_h^o)^2$, $\bar{\eta}_h \in X_h^o$ satisfying

$$\lim_{h \to 0} (\|\vec{w}_h - \vec{w}\|_1 + \|\bar{\eta}_h - \eta\|_1) = 0. \tag{5.9}$$

Let $\eta_h \in X_h^o$, which is defined by the following system

$$(\nabla\eta_h, \nabla z_h) = -(\vec{w}_h, \nabla z_h), \quad \forall\ z_h \in X_h^o . \tag{5.10}$$

Obviously $(\vec{w}_h, \eta_h) \in W_h$. By assumption, we know $\eta \in H^2(\Omega) \cap H_o^1(\Omega)$ and

$$-\Delta\eta = \nabla\cdot\vec{w} \qquad \text{in } \Omega , \tag{5.11}$$

$$(\nabla\eta, \nabla\eta) = -(\vec{w}, \nabla\eta) , \tag{5.12}$$

$$(\nabla\eta, \nabla z_h) = -(\vec{w}, \nabla z_h) + \int_\Gamma \frac{\partial\eta}{\partial n}\cdot z_h\, ds, \quad \forall\ z_h \in X_h^o . \tag{5.13}$$

From above equality we obtain

$$(\nabla\eta, \nabla(\eta-\eta_h)) = -(\vec{w}, \nabla(\eta-\eta_h)) - \int_\Gamma \frac{\partial\eta}{\partial n}\cdot\eta_h\, ds,$$

$$(\nabla\eta_h, \nabla(\bar{\eta}_h-\eta_h)) = -(\vec{w}_h, \nabla(\bar{\eta}_h-\eta_h)),$$

and

$$\begin{aligned} |\eta_h-\eta|_1^2 &= (\nabla(\eta_h-\eta), \nabla(\eta_h-\eta)) \\ &= -(\vec{w}-\vec{w}_h, \nabla(\eta-\eta_h)) - (\nabla(\eta-\bar{\eta}_h), \vec{w}_h+\nabla\eta_h) - \int_\Gamma \frac{\partial\eta}{\partial n}\cdot\eta_h\, ds, \end{aligned} \tag{5.14}$$

$$\left|\int_\Gamma \frac{\partial\eta}{\partial n}\cdot\eta_h\, ds\right| \le C h^{\frac{3}{2}} \|\eta\|_2 \left(|\eta|_1 + |\eta-\eta_h|_1\right) . \tag{5.15}$$

From (5.14) and (5.15), we have $\lim\limits_{h\to 0}|\eta_h-\eta|_1=0$ and $\lim\limits_{h\to 0}\|\eta_h-\eta\|_1=0$.

Lemma 5.4 Let $(\vec{w}_h, \eta_h) \in W_h$ and η_{oh} satisfy

$$\begin{cases} -\Delta\eta_{oh} = \nabla\cdot\vec{w}_h , & \text{in } \Omega , \\ \eta_{oh} = 0, & \text{on } \Gamma . \end{cases} \tag{5.16}$$

Then we have: 1. $\eta_{oh} \in H_o^1(\Omega) \cap H^2(\Omega)$ and

$$\|\eta_{oh}\|_2 \le C\, |\vec{w}_h|_1 . \tag{5.17}$$

$$2. \quad |\eta_h - \eta_{oh}|_{1,4} \le C h^{\frac{1}{2}} |\vec{w}_h|_1 . \tag{5.18}$$

Proof 1. When boundary Γ of domain Ω is sufficiently smooth, it is a standard result.

2. From (5.16) and definition of W_h we have

$$(\nabla \eta_{oh}, \nabla Z) = -(\vec{w}_h, \nabla Z), \qquad \forall Z \in H_0^1(\Omega), \tag{5.19}$$

$$(\nabla \eta_h, \nabla Z_h) = -(\vec{w}_h, \nabla Z_h), \qquad \forall Z_h \in X_h^0, \tag{5.20}$$

$$(\nabla \eta_{oh}, \nabla Z_h) = -(\vec{w}_h, \nabla Z_h) + \int_\Gamma \frac{\partial \eta_{oh}}{\partial n} Z_h \, ds + \int_\Gamma (\vec{w}_h \cdot n) z_h \, ds, \qquad \forall Z_h \in X_h^0. \tag{5.21}$$

From (5.20) and (5.21) we obtain

$$(\nabla(\eta_h - \eta_{oh}), \nabla Z_h) = -\int_\Gamma \frac{\partial \eta_{oh}}{\partial n} z_h \, ds - \int_\Gamma (\vec{w}_h \cdot n) z_h \, ds, \ \forall z_h \in X_h^0. \tag{5.22}$$

Let β_h be an arbitrary element in X_h^0 . We have

$$\begin{aligned}
&(\nabla(\beta_h - \eta_{oh}), \nabla(\beta_h - \eta_{oh})) \\
&= (\nabla(\beta_h - \eta_h), \nabla(\beta_h - \eta_h)) + (\nabla(\eta_h - \eta_{oh}), \nabla(\eta_h - \eta_{oh})) \\
&\qquad + 2(\nabla(\beta_h - \eta_h), \nabla(\eta_h - \eta_{oh})),
\end{aligned} \tag{5.23}$$

From (5.22) and (5.23) we derive

$$\begin{aligned}
|\beta_h - \eta_h|_1^2 &= (\nabla(\beta_h - \eta_h), \nabla(\beta_h - \eta_h)) \\
&\leq (\nabla(\beta_h - \eta_{oh}), \nabla(\beta_h - \eta_{oh})) + 2\int_\Gamma \frac{\partial \eta_{oh}}{\partial n}(\beta_h - \eta_h) ds \\
&\qquad + 2\int_\Gamma (\vec{W}_h \cdot n)(\beta_h - \eta_h) ds \\
&\leq |\beta_h - \eta_{oh}|_1^2 + C(\|\eta_{oh}\|_2 + \|\vec{W}_h\|_1) h^{\frac{3}{2}} |\beta_h - \eta_h|_1 .
\end{aligned} \tag{5.24}$$

Using equivalent of $\|\vec{w}_h\|_1$ and $|\vec{w}_h|_1$, from (5.24) we obtain

$$|\beta_h - \eta_h|_1 \leq c_1 |\beta_h - \eta_{oh}|_1 + c_2 h^{\frac{3}{2}} |\vec{W}_h|_1 . \tag{5.25}$$

It is easy to know that there exists $\beta_{ho} \in X_h^0$ satisfying

$$\|\beta_{ho} - \eta_{oh}\|_1 \leq Ch \|\eta_{oh}\|_2 . \tag{5.26}$$

Taking $\beta_h = \beta_{ho}$ in (5.25), we derive

$$|\beta_{ho} - \eta_h|_1 \leq Ch |\vec{W}_h|_1 . \tag{5.27}$$

From triangle inequality and Lemma 5.1 and Lemma 5.2 we obtain

$$|\eta_{oh} - \eta_h|_{1,4} \leqslant |\eta_{oh} - \beta_{ho}|_{1,4} + |\beta_{ho} - \eta_h|_{1,4} \leqslant C h^{\frac{1}{2}} |\vec{w}_h|_1 . \qquad (5.28)$$

Theorem 5.1 We have

$$\lim_{h\to 0} \gamma_h(\beta) = \gamma(\beta), \quad \lim_{h\to 0} \gamma_h(\vec{f}) = \gamma(\vec{f}) . \qquad (5.29)$$

Proof The proof of Theorem 5.1 is similar to the one of the convergence of $\gamma_h(\beta)$ given in [9] and is omitted.

We define

$$B_0 = \{ (\vec{w}_h, \eta_h) \mid (\vec{w}_h, \eta_h) \in W_{oh}, \ |\vec{w}_h|_1 \leqslant \frac{2}{\nu} \gamma_h(\vec{f}) \},$$

$$B_1 = \{ (\vec{w}_h, \eta_h, Z_h) \mid (\vec{w}_h, \eta_h) \in B_0, \ Z_h \in X_h \backslash R \} .$$

Theorem 5.2 The variational problem P_h has a unique solution in B_0, if the following condition is satisfied:

$$\gamma_h(\vec{f}) \ \gamma_h(\beta) \leqslant \frac{\nu^2}{4} . \qquad (5.30)$$

The proof of theorem 5.2 is similar to Theorem 2.3 in [9] and is omitted.

Theorem 5.3 The variational problem P_{h1} has a unique solution in B_1, if the condition (5.30) is satisfied.

Proof Let $(\vec{u}_{oh}, \psi_{oh})$ be a unique solution of variational problem P_h in B_0 .

Variational problem P_{oh} : Find $(\vec{u}_h, \psi_h) \in W_{oh}$ satisfying

$$\nu(\nabla\vec{u}_h, \nabla\vec{v}_h) = (\vec{f} - (\vec{u}_{oh}\cdot\nabla)\ \vec{u}_{oh}, \ \vec{v}_h + \nabla\phi_h), \quad \forall\ (\vec{v}_h, \phi_h) \in W_{oh} .$$

By Lemma 2.7 and Lax-Milgram Theorem the variational problem P_{oh} has a unique solution. Moreover, its solution belongs to B_0 . It is easy to prove that the variational problem P_{oh} is equivalent to the following variational problem P_{oh1} :

$$\begin{cases} (\nabla p_h, \nabla \phi_h) = (\vec{f} - (\vec{u}_{oh} \cdot \nabla)\vec{u}_{oh}, \nabla \phi_h), \ \forall \phi_h \in X_h^o, \\ \nu(\nabla \vec{u}_h, \nabla \vec{v}_h) = (\vec{f} - \nabla p_h - (\vec{u}_{oh} \cdot \nabla)\vec{u}_{oh}, \vec{v}_h), \ \forall \vec{v}_h \in (V_h^o)^2, \\ (\nabla \psi_h, \nabla q_h) = -(\vec{u}_h, \nabla q_h), \ \forall q_h \in X_h \backslash R. \end{cases}$$

Moreover, $(\vec{u}_h, \psi_h)$ is the solution of variational problem P_{oh1} equals the solution $(\vec{u}_{oh}, \psi_{oh})$ of variational problem P_{oh}. Thus variational problem P_{h1} has a unique solution in B_1.

If we can prove that $(\vec{u}_h, \psi_h)$ in the solution of variational problem P_{h1} is the solution of variational problem P_h, then Theorem is true. The proof of above result is easy, so is omitted.

Theorem 5.4 If the condition (5.30) is satisfied, then we have

$$\| \vec{u}-\vec{u}_h \|_1 \leqslant ch, \qquad \| p-p_h \|_0 \leqslant ch. \qquad (5.31)$$

Proof. Combining the method of § 4 in this paper and the one of §3 in [9], we can easy prove (5.31) and we omit it.

Reference

[1] Glowinski,R., Pironneau,O. On a mixed finite element approximation of Stokes problem. Numer. Math. 33, 397-427(1979)

[2] Le Tallec, P. A mixed finite element approximation of the Navier-Stokes equations. Numer. Math. 35, 381-404(1980)

[3] Ciarlet, P.G. The finite element method for elliptic problems, North-Holland Publishing Company, 1978

[4] Clement, P . Approximation by finite element functions using local regularization. RAIRO, Série Analyse Numérique, R-2, p.77-84 (1975)

[5] Li Likang, A finite element method for the stationary Stokes equations. Numerical Mathematics, A Journal of Chinese Universities 6 (1984) 104-115.

[6] Girault, V., Raviart, P.A., Finite element approximation of the Navier-Stokes equations. Lecture Notes in Mathematics, Springer-Verlag, Berlin, 1979.

[7] Blair J.J., High order approximations to the boundary conditions for the F.E.M., Math. Comp. 30, 250-262 (1976)

[8] Temam, R., Navier-Stokes equations. Amsterdam, North-Holland, 1979.

[9] Li Likang, On a mixed isoparametric finite element method of the Navier-Stokes problem. (to appear)

VACCUM STATES AND EQUIDISTRIBUTION OF THE RANDOM SEQUENCE FOR GLIMM'S SCHEME

by

LONG - WEI LIN

HUA CHIAO UNIVERSITY

(I)

Glimm's scheme is a main tool for studying hyperbolic conservation laws. It has been shown that in order for Glimm's difference approximations approach to the generalized solution, the random sequence $a \equiv \{a_n\}$ should be equidistributed in the interval (-1, 1) [1,2]. However, in the standard application for Glimm's scheme, the compactness is established for all sequences and not only for the equidistributed ones. This is a source of essential difficulties for the standard application for Glimm's scheme.

In this paper, we only consider the Glimm's schemes generated by equidistributed sequences. We expect that this is the proper application for Glimm's scheme.

Consider the initial value problem for isentropic gas dynamics equations in Lagrangian coordinates

$$v_t - u_x = 0 \ , \ u_t + p(v)_x = 0, \qquad (-\infty, \infty) \times (0, \infty) \tag{E}$$

$$(u(x,0), v(x,0)) = (u_0(x), v_0(x)), \quad (-\infty, \infty) \tag{I}$$

Where the pressure $p=p(v)>0$ is a C^2 function of the specific volume $v > 0$ and u is the velocity of the gas. We assume that $p'(v)<0, p''(v) > 0, \quad \lim_{v\to\infty} p(v) = 0$, and

$$\int_1^\infty \sqrt{-p'(v)} \ dv < \infty \tag{C}$$

Condition (C), which is dictated by physics, is a source of

essential difficulties in the mathematical treatment of this problem. For instance, it is well-known that vaccum may occur, that is, v may be infinite, in a solution of the Riemann problem consisting of rarefaction waves.

Our main purpose is to show that the flow and its Glimm's difference approximations contain no vaccum, if they consist of rarefaction waves and the initial data are Lipschitz continuous. Here we assume that the Glimm's schemes are generated by an equidistributed sequence with sufficient small mesh lengths.

It is well-known that a necessary condition for the existence of a global bounded continuous solution is

$$r_0(x_1) \leq r_0(x_2), \quad s_0(x_1) \leq s_0(x_2), \qquad x_1 < x_2 \quad , \tag{M}$$

Where r and s are the Riemann invariants,

$$r(u,v) = u + \int_1^v \sqrt{-p'(s)}\, ds \quad , \; s(u,v) = u - \int_1^v \sqrt{-p'(s)}\, ds \tag{R}$$

and $r_0(x) = r(u_0(x), v_0(x))$, $s_0(x) = s(u_0(x), v_0(x))$, (see 3 , 4). On the other hand, it has also been shown (see 4 - 7) that, conversely, when (M) holds, a continuous global solution of (E) (I) exists, provided

$$\{(r,s) \mid r_0(-\infty) \lesssim r \leq r_0(\infty), s_0(-\infty) \leq s \leq s_0(\infty)\} \subset \Omega \tag{V}$$

where $\Omega \equiv \{(r,s) \mid r-s < 2\int_1^\infty \sqrt{-p'(v)}dv\}$. The condition (V) assures that the solution obtained is globally bounded, i.e. strictly away from vaccum. When (V) fails, the methods that have been used so far are not necessarily effective. For instance, if one attempts to solve (E), (I) using Glimm's scheme, he may end up with a difference approximation in which vaccum appears. However, condition (V) is not necessarily justified by physics. For example, suppose $p(V) = \frac{1}{8v^2}$, then $\int_1^\infty \sqrt{-p'(v)}\, dv = 1$,

We consider the initial values

$$\left.\begin{aligned} v_o(x) &= 1 \\ u_o(x) &= \begin{cases} 1+\varepsilon, & \text{as } x > 1+\varepsilon \\ x, & \text{as } |x| < 1+\varepsilon \\ -(1+\varepsilon), & \text{as } x < -(1+\varepsilon), \end{cases} \end{aligned}\right\} (I)_\varepsilon$$

$\varepsilon > 0$. There initial data just violate slightly condition (V). We now briefly discuss Glimm's difference scheme for this case (see [1], [2], [8]). We select as our mesh length h=l=1/n, n=1,2,...... , in the x and t directions, respectively, (Note: $h/l=1 \geqslant \sqrt{-p'(v)} = \lambda(v)$, for $v \geqslant 1$). We now construct a Glimm's difference solution $(u_h(x,t), v_h(x,t))$ Corresponding to the sequence a = (0,0, ...), which yields mesh points $a_{m,n} = (mh, nl)$. It is easy to see that vaccum appears in the difference solution after n steps in t direction, i.e. $v(0,1) = \infty$. However, the solution of the initial value problem (E), $(I)_\varepsilon$ does not contain vaccum for any $\varepsilon > 0$. Even for $\varepsilon = \infty$, i.e.

$$v_o(x) = 1 \; , \; u_o(x) = x \; , \qquad (I)_\infty$$

the solution of the initial value problem (E), $(I)_\infty$ is $v(x,t) = t+1$, $u(x,t) = x$. There is no vaccum. What is the matter? The sequence a = (0,0, ...) is not equidistributed. Condition (V) is necessary in order to establish an uniform bound on all difference solutions of Glimm's scheme.

(II)

Consider the initial value problem (E), (I). Following [2], we now describe briefly the Glimm's scheme. Let l and h be the mesh lengths in the x and t direction respectively, ratio $\delta \equiv l/h$ satisfies the Courant-Friedrichs-Lewy condition: $\delta \equiv l/h \geqslant \max \lambda(x)$ for all v under consideration. The approximate solution $(u_h(x,t,a),$ $v_h(x,t,a))$ is constructed inductively according to a prechosen sequence $a \equiv \{a_n\}$, $-1 < a_n < 1$, in the following way. Set $(u_h(x,o,a),$ $v_h(x,o,a)) = (u_o(kl), v_o(kl))$, k even, for $(k-1)l < x < (k+1)l$. Suppose that $(u_h(x,t,a), v_h(x,t,a))$ is defined on $0 \leqslant t < nh$, then we set

$$(u_h(x,nh,a), v_h(x,nh,a))$$
$$= (u_h((k+a_n)l-0,\ nh-0),\ v_h(k+a_n)l-0,\ nh-0)),$$
$$(k-l)l < x < (k+l)l$$

for any integer k with k+n even, thus $(u_h(x,kh,a), v_h(x,kh,a))$ is a step function of x with possible discontinuity at $((k+1)l,\ nh)$, n+k even. We then define $(u_h(x,t,a), v_h(x,t,a))$, $nh \leqslant t < (n+1)h$, by resolving these discontinuities, so that in the zone $nh \leqslant t < (n+1)h$, the approximate solution is exact and consists of elementary waves issued from $((k+1)l,\ nh)$, n+k even. Here we now solve each Riemann problem and it is sufficient to extend the solution forward by one time step, because of the Courant - Friedrichs - Lewy condition. The scheme piecewise together solutions of Riemann problems to obtain an approximate solution $(u_h(x,t,a),\ v_h(x,t,a))$ for general class of data.

<u>Definition</u> We say a sequence $a \equiv \{a_n\}$ is uniformly equidistributed on the interval (-l,l), if for each $\varepsilon > 0$, there is a $N(\varepsilon) > 0$, such that, as $n > N(\varepsilon)$,

$$|B(j,n,I)/n - \mu(I)/2| < \varepsilon \qquad \text{(D)}$$

holds for any integer $j \geqslant 0$ and any subinterval I in the interval $(-1,1)$, where $B(j,n,I)$ denotes the number of m, $j \leqslant m < j+n-1$, with $a_m \in I$, and $\mu(I)$ is the length of I. The integer $N(\varepsilon) > 0$ only depends on ε, but is independent of j and I.

<u>Theorem</u> Let $u_o(x)$, $v_o(x)$ be bounded, $0 < v_* \leqslant v_o(x) \leqslant V_o$. Lipshitz continuous, and satisfy

$$|\Phi(v_o(x_2)) - \Phi(v_o(x_1))| \leqslant u_o(x_2) - u_o(x_1),\ x_1 < x_2, \qquad (M)_1$$

i.e.

$$r_o(x_1) \leqslant r_o(x_2),\ s_o(x_1) \leqslant s_o(x_2),\quad x_1 \leqslant x_2, \qquad (M)_2$$

where $\Phi(v) \equiv \int_1^v \sqrt{-p'(s)}\, ds$. Suppose that the random sequence is uniformly equidistributed on the interval $(-1,1)$. For given $T > 0$, if the mesh lengths $h > 0$ are sufficient small, the ratio $\delta \equiv l/h > 0$, $\lambda_* \equiv \sqrt{-p'(v_*)}$, is a constant, then the Glimm's approximate solutions $(u_h(x,t,a), v_h(x,t,a))$ of (E), (I) are uniformly bounded respect to h in the zone $(-\infty, \infty) \times [0,T)$.

First of all, we prove the following lemma.

Denote $f_h^{(n)} \equiv f_h(kl, nh, a)$, here $f = u,v,r,s$, $n+k$ = even.

<u>Lemma</u> Suppose that the initial data (I) satisfy the condition (M). For given integer $n > 0$, if

$$0 \leqslant r_{k+2}^{(n)} - r_k^{(n)} \leqslant b,\ 0 \leqslant s_{k+2}^{(n)} - s_k^{(n)} \leqslant b,$$

holds for all k, k+n = even, then

$$0 \leqslant r_{k+1}^{(n+1)} - r_{k-1}^{(n+1)} \leqslant b, \qquad (1)$$

$$0 \leqslant s_{k+1}^{(n+1)} - s_{k-1}^{(n+1)} \leqslant b, \qquad (2)$$

$$\Phi_{n+1} \leq \Phi_n + \frac{b}{2} \qquad (3)$$

where $\Phi_n \equiv \sup_k \Phi_k^{(n)}$, $\Phi_k^{(n)} \equiv \frac{1}{2}(r_k^{(n)} - s_k^{(n)})$ *.

*) Obviously, $\Phi_k^{(n)} = \Phi(v_k^{(n)})$, if and only if

Proof

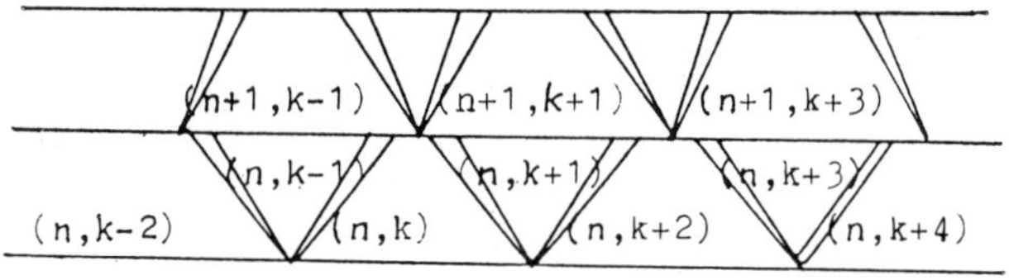

Without loss of generality, we assume $a_n \in (0,1)$. Let $w^{(n)} > 0$ satisfy

$$a_n = \sqrt{-p'(w^{(n)})}\ h/l,$$

According to the condition (M), the Glimm's difference solution consists of rarefaction waves . Denote

$$(r_{k+1}^{(n)}, s_{k+1}^{(n)}) = (r_h((k+1)l,(n+1)h-0),\ s_h((k+1)l,(n+1)h-0)),\ n+k = \text{even}.$$

Obviously, $r_{k+1}^{(n)} = r_{k+2}^{(n)}$, $s_{k+1}^{(n)} = s_k^{(n)}$. Recalling the construction of Glimm's scheme,

$$r_{k-1}^{(n+1)} = r_k^{(n)}, \quad s_{k-2}^{(n)} \leqslant s_{k-1}^{(n+1)} \leqslant s_k^{(n)} \quad ,$$

$$r_{k+1}^{(n+1)} = r_{k+2}^{(n)}, \quad s_k^{(n)} \leqslant s_{k+1}^{(n+1)} \leqslant s_{k+2}^{(n)} \quad ,$$

then we get (1),

$$0 \leqslant r_{k+1}^{(n+1)} - r_{k-1}^{(n+1)} = r_{k+2}^{(n)} - r_k^{(n)} \leqslant b \ .$$

We now prove (2) only for the most complicated case. Suppose that

$$\sqrt{-p'(v_{k+2}^{(n)})}\, h/l \leq a_n \leqslant \sqrt{-p'(v_{k-1}^{(n)})}\, h/l \qquad (4)_1$$

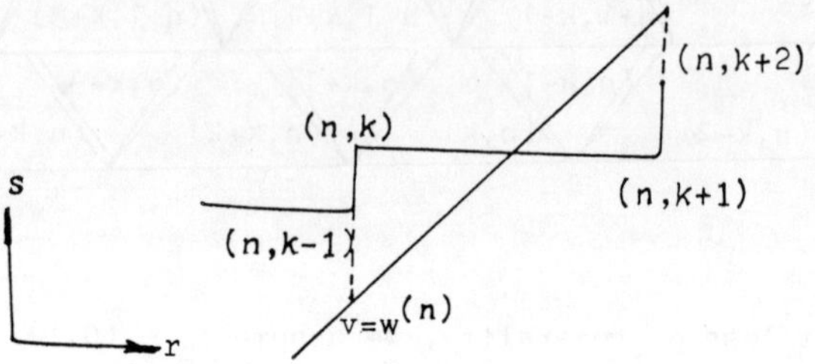

i.e.

$$\sqrt{-p'(v_{k+2}^{(n)})} \leqslant \sqrt{-p'(w^{(n)})} \leqslant \sqrt{-p'(v_{k-1}^{(n)})}\ ,$$

$$v_k^{(n)} \leqslant v_{k-1}^{(n)} \leqslant w^{(n)} \leqslant v_{k+2}^{(n)} \leqslant v_{k+1}^{(n)}\ , \qquad (4)_2$$

Hence

$$r_{k+2}^{(n)} - s_{k+2}^{(n)} \geqslant 2\Phi(w^{(n)}) \geqslant r_{k-1}^{(n)} - s_{k-1}^{(n)} = r_k^{(n)} - s_{k-2}^{(n)}\ ,$$

$$0 \leqslant s_{k+2}^{(n)} - s_{k-2}^{(n)} \leqslant r_{k+2}^{(n)} - r_k^{(n)} \leqslant b. \qquad (5)$$

Recalling the construction of Glimm's scheme, from (4),

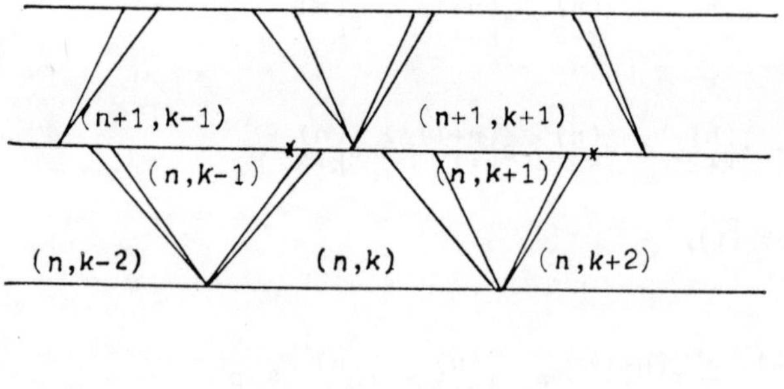

$$s_{k-1}^{(n+1)} = s_{k-1}^{(n)} = s_{k-2}^{(n)}\ , \quad s_{k+1}^{(n+1)} = s_{k+2}^{(n)}\ ,$$

then

$$s_{k+1}^{(n+1)}-s_{k-1}^{(n+1)} = s_{k+2}^{(n)} - s_{k-2}^{(n)} . \tag{6}$$

Combining (5) and (6), we get (2).

We now prove (3). According to $(4)_2$

$$\Phi_{k+1}^{(n+1)} \leqslant \Phi_{k+1}^{(n)} = ½(r_{k+1}^{(n)}-s_{k+1}^{(n)}) = ½(r_{k+2}^{(n)}-s_{k}^{(n)}),$$

then

$$\Phi_{k+1}^{(n+1)}-\Phi_{k}^{(n)} \leqslant ½(r_{k+2}^{(n)} -r_{k}^{(n)}) \leqslant b/2,$$

$$\Phi_{k+1}^{(n+1)}-\Phi_{k+2}^{(n)} \leqslant ½(s_{k+2}^{(n)} -s_{k}^{(n)}) \leqslant b/2.$$

Hence

$$\Phi_{k+1}^{(n+1)} - \Phi_n \leqslant b/2$$

holds for all k, k+n = even. Consequently we get (3). Q.E.D.

<u>Proof of the theorem</u> For given $0 < \varepsilon < ¼$, choose sufficient large natural numbers N,M,

$$N > N(\varepsilon_1), \text{ where } \quad \varepsilon_1 \equiv \varepsilon\delta^{-1}\Phi'(V_0 + 2LT), \tag{7}$$

$$\frac{M}{LT} > \max \left\{ 4C, \max_{V_0 \leqslant V \leqslant V_0 + \frac{3}{2}LT} \frac{\Phi'(V)+\Phi'(V_0+2LT)}{\Phi(V_0+2LT)-\Phi(V)} \right\}, \tag{8$_1$}$$

where

$$C = \frac{\Phi'(V_0)+\Phi'(V_0+2LT)}{[\Phi'(V_0+2LT)]^2} \max_{V_0 \leqslant V \leqslant V_0+2LT} [-\Phi''(V)], \tag{8$_2$}$$

L is a Lipschitz constant of $r_0(x)$ and $s_0(x)$, $N(\varepsilon_1)$ is defined in Definition 1. Let $\tau \equiv T/M$, $h \equiv \tau/N = T/MN$,

$$\lambda_0 \equiv \sqrt{-p'(V_0)} = \Phi'(V_0), \quad I_0 \equiv (-\lambda_0\delta^{-1}, \lambda_0\delta^{-1}).$$

From (D),

$$0 \leqslant N_1 \equiv B(0,N,I_0) \leqslant N(\lambda_0\delta^{-1} + \varepsilon_1), \tag{9}$$

We now prove that

$$0 \leqslant \Phi(V_1) - \Phi(V_0) \leqslant N_1 Ll, \tag{10}$$

here $V_1 \equiv \sup_{D_N} v_h(x,t,a)$,

$$D_n \equiv \{ (x,t) | -\infty < x < \infty, \ 0 \leqslant t \leqslant nh \}.$$

In fact, the Glimm's approximation under consideration consists of rarefaction waves. $v_h(n,a) \equiv \sup_{D_n} v_h(x,t,a)$ is non-decreasing respect to n. It is strictly increasing, if and only if there is an interaction of rarefaction waves. Notice that

$\Phi_n = \Phi(\tilde{v}_h(n,a))$, $0 \leqslant r^{(0)}_{k+2} - r^{(0)}_k \equiv r_0((k+2)l) - r_0((k)l) \leqslant 2Ll$,
$0 \leqslant s^{(0)}_{k+2} - s^{(0)}_k \leqslant 2Ll$, k:even. According to the Lemma, we get

$$0 \leqslant \Phi_{j+1} - \Phi_j \leqslant Ll, \qquad (0 \leqq j, \ j+1 \leqslant N) \tag{11}$$

In order to estimate the upper bound of $V_1 \equiv \sup_{D_N} v_h(x,t,a)$ without loss of generality we may assume $V_h \geqslant V_0$. Consequently, if there is an interaction of rarefaction waves at $t = kh$, then $a_k \in I_0 \equiv (-\lambda_0\delta^{-1}, \lambda_0\delta^{-1})$, $k=1,2,\ldots$. Recalling the definition of $N_1 \equiv B(0,N,I_0)$, from (11), we can get (10).

Combining (7), $(8)_1$, (9) and (10),

$$0 \leqslant \Phi(V_1) - \Phi(V_0) \leqslant Ll N\delta^{-1}(\Phi'(V_0) + \varepsilon\Phi'(V_0+2LT))$$
$$= \frac{LT}{M}(\Phi'(V_0) + \varepsilon\Phi'(V_0+2LT)) \leqslant \Phi(V_0+2LT) - \Phi(V_0), \tag{12}$$

Since $\Phi'(v) = \sqrt{-p'(v)} > 0$, then

$$V_0 \leqslant V_1 \leqslant V_0 + 2LT,$$

Moreover,

$$\Phi(V_1) - \Phi(V_0) \geqslant (V_1 - V_0)\Phi'(V_1) \tag{13}$$
$$\geqslant (V_1 - V_0)\Phi'(V_0+2LT) \tag{14}$$

Substituting (13) into (12), we get

$$0 \leqslant V_1 - V_0 \leqslant \frac{LT}{M}\left(\frac{\Phi'(V_0)}{\Phi'(V_0+2LT)} + \varepsilon\right)$$

Hence

$$0 \leqslant \Phi'(V_0) - \Phi'(V_1) = (V_1 - V_0)(-\Phi''(\theta)) \leqslant \frac{LT}{M}\left(\frac{\Phi'(V_0)}{\Phi'(V_0+2LT)} + \varepsilon\right) \max_{V_0 \leqslant V \leqslant V_0+2LT}(-\Phi''(V))$$

here $V_0 < \theta < V_1$, Notice $0 < \varepsilon < \frac{1}{4}$ and (8), then

$$0 \leqslant \frac{\Phi'(V_0)}{\Phi'(V_1)} - 1$$

$$\leqslant \frac{LT}{M}\left(\frac{\Phi'(V_0)+\Phi'(V_0+2LT)}{\Phi'(V_0+2LT)^2}\right) \max_{V_0<V<V_0+2LT}\left(-\Phi''(V)\right)$$

$$= \frac{LTC}{M}.$$

From (12), (13) and (15)

$$0 \leqslant V_1 - V_0 \leqslant \frac{LT}{M}\left(\frac{\Phi'(V_0)}{\Phi'(V_1)} + \varepsilon\frac{\Phi'(V_0+2LT)}{\Phi'(V_1)}\right)$$

$$\leqslant \frac{LT}{M}\left(1 + \frac{LTC}{M} + \varepsilon\right).$$

Since $0<\varepsilon<\frac{1}{4}$, $M > 4LTC$, then

$$0 \leqslant V_1 - V_0 \leqslant LT/(1+1/4+1/4) \leqslant 3LT/2M,$$

$$V_0 \leqslant V_1 \leqslant V_0 + \frac{3LT}{2}.$$

Hence, from (8),

$$\frac{LT}{M} \leqslant \frac{\Phi(V_0+2LT)-\Phi(V_1)}{\Phi'(V_1)+\Phi'(V_0+2LT)}.$$

Because the sequence $a \equiv \{a_n\}$ is uniformly equidistributed, we get, similarly,

$$V_k \equiv \sup_{D_{kN}} v_h\ (x,t,a), \quad k = 1,2, \dots, M$$

$$\lambda_k \equiv \sqrt{-p'(V_k)} = \Phi'(V_k),$$

$$I_k \equiv (-\lambda_k\delta^{-1}, \lambda_k\delta^{-1}),$$

$$0 \leqslant N_k \equiv B((k-1)N, N, I_{k-1}) \leqslant N(\lambda_{k-1}\delta^{-1} + \varepsilon_1),$$

$$0 \leqslant \Phi(V_k) - \Phi(V_{k-1}) \leqslant N_k L1 \tag{10$_k$}$$

$$\leqslant \frac{LT}{M}\left(\Phi'(V_{k-1}) + \varepsilon\Phi'(V_0+2LT)\right) \tag{12$_k$}$$

$$\leqslant \Phi(V_0 + 2LT) - \Phi(V_{k-1})$$

$$V_{k-1} \leqslant V_k \leqslant V_o + 2LT,$$

$$\Phi(V_k) - \Phi(V_{k-1}) \geqslant (V_k - V_{k-1})\,\Phi'(V_k) \qquad (13)_k$$

$$\geqslant (V_k - V_{k-1})\,\Phi'(V_o+2LT), \qquad (14)_k$$

$$0 \leqslant V_k - V_{k-1} \leqslant \frac{LT}{M}\left(\frac{\Phi'(V_{k-1})}{\Phi'(V_o+2LT)} + \varepsilon \right).$$

$$0 \leqslant \frac{\Phi'(V_{k-1})}{\Phi'(V_k)} - 1 \leqslant \frac{LTC}{M} \qquad (15)_k$$

$$0 \leqslant V_k - V_{k-1} \leqslant \frac{LT}{M}\left(1 + \frac{LTC}{M} + \varepsilon \right)$$

$$V_k - V_{k-1} \leqslant \frac{LT}{M}\left(\frac{\Phi'(V_{k-1})}{\Phi'(V_k)} + \varepsilon \right]$$

$$\leqslant \frac{LT}{M}\left(1 + \frac{LTC}{M} + \varepsilon \right] \leqslant \frac{3LT}{2M} \qquad (16)_k$$

$$0 \leqslant V_k - V_o \leqslant \frac{3LT}{2},$$

$$V_o \leqslant V_k \leqslant V_o + \tfrac{3}{2}LT$$

$$\frac{LT}{M} \leqslant \frac{\Phi(V_o+2LT) - \Phi(V_k)}{\Phi'(V_k) + \Phi'(V_o+2LT)}$$

Adding (16) from k=1 to k=M, we get

$$0 \leqslant V_M - V_o \leqslant LT(1 + LTC/M + \varepsilon) \qquad (17)$$

So $V_h(x,t,a)$ is uniformly bounded respect to h in the zone $(-\infty, \infty) \times (0,T)$. And the uniform boundness for $u_h(x,t,a)$ is easy to be proved. Q.E.D.

<u>Corollary</u> If initial data (I) satisfy the conditions of the theorem, then there is a Lipschitz continuous solution $(u(x,t), v(x,t))$ of the initial value problem (E), (I), and

$$V_o \leqslant \sup_{D_T} v(x,t) \leqslant V_o + LT, \qquad (18)$$

here $D_T = \{ (x,t) \mid -\infty < x < \infty,\ 0 \leqslant t \leqslant T \}$

Proof According to the theorem , it is easy to prove that the set of Glimm's approximations is compact. Following [2], one can prove that the limit function is a Lipschitz continuous solution of (E), (I). And (18) can be obtained from (17).

Note 1 The author obtained the corollary by polygonal approximation (9) in [10].

Note 2 The equality of (18) holds for the initial function $(u_0(x), v_0(x)) = (u_0 + Lx, v_0)$, where $L > 0$, u_0, $v_0 > 0$ are constants.

Acknowledgement This paper was completed at Maryland University. I am grateful to Professor Tai-Ping Liu for intimate discussions. I want to thank Professor Peter Lax for his encouragement.

Reference

[1] J.Glimm, Solutions in the large for nonlinear hyperbolic systems of equations, Comm. Pure Appl. Math. 18(1965) 697-715.

[2] T.P Liu, The deterministic version of the Glimm scheme. Comm. Math. Phys. 57(1977), 135-148.

[3] P.D.Lax, Development of singularities of solutions on nonlinear hyperbolic partial differential equations. J.Math. Phy. 5.(1964) 611-613.

[4] L.W.Lin, On the global existence of the continuous solutions of the reducible quasi-linear hyperbolic systems. Acta Scien. Nat. Jilin Univ. 4(1963)83-96.

[5] J.L.Johnson. Global continuous solutions of hyperbolic systems of quasi-linear equations. Bull. Amer. Math. Soc. 73 (1967) 639-641.

[6.] M.Yamaguti and T, Nishida, On the global solution for quasi-linear hyperbolic equations. Funkcial Evac. 11(1968) 51-57.

[7.] D.Hoff, A constructive theory for shock - free isentropic flow. J. Diff. Equ. (1980)1-31.

[8] T.Nishida and J.A. Smoller, Solutions in the large of some non-linear hyperbolic conservation laws. Comm. Pure Appl. Math. 26 (1973) 193-200.

[9] C.M.Dafermos, Polygonal approximations of solutions of the initial value problem for a conservation law. J. Math. Anal. and Appl. 38(1972) 33-41.

[10] L.W. Lin, On the vaccum state for the equations of isentropics gas dynamics. (to appear)

Global Classical Solutions of Nonlinear Generalized Vorticity Equations And Its Applications

Mu Mu

Mathematical Institute of Fudan Univercity,

Shanghai, China

1. Introduction

In atmospheric dynamics, baroclinic quasi-geostrophic quasi-nondivergent model is a fundamental one. It is described by the following vorticity equation (See [1], Chapter 12)

$$\frac{\partial}{\partial t} A_0 \psi + J(\psi, A_0 \psi) + 2a^2 J(\psi, \omega \cos\theta) = 0 . \qquad (1.1)$$

Where

$\psi = \psi(\lambda, \theta, s, t)$ is unknown function (stream function).

λ is longitude, $0 \le \lambda \le 2\pi$.

θ is colatitude(i.e. the North Pole corresponds to the point $\theta = 0$).

s belongs to interval $[s_0, 1]$ and s_0 is a given constant satisfying $0 < s_0 < 1$.

t is time variable. a is radius of the earth.

ω is angular speed of the rotation of the earth.

$$\Delta = \frac{1}{\sin\theta}\left(\frac{\partial}{\partial\theta}\sin\theta\frac{\partial}{\partial\theta} + \frac{1}{\sin\theta}\frac{\partial^2}{\partial\lambda^2}\right) ,$$

$$A_0 = \frac{\partial}{\partial s}\left(\tilde{C}\frac{\partial}{\partial s}\right) + \Delta ,$$

$$J(F, G) = \frac{1}{\sin\theta}\left(\frac{\partial F}{\partial\lambda}\frac{\partial G}{\partial\theta} - \frac{\partial F}{\partial\theta}\frac{\partial G}{\partial\lambda}\right) .$$

$\tilde{C}$ is a given positive function.

Many authors have worked for numerical solutions and particular solutions of various initial-boundary value problems for equation (1.1). In the other side, perhaps because of the difficulties, there is a little work about the well-posedness of these problems. In [1], Zeng Qing-cun considered the following problem

$$\begin{cases} \frac{\partial}{\partial t} A_0\psi + J(\psi, A_0\psi) + 2a^2 J(\psi, \omega\cos\theta) = 0, \quad \text{in } M_0\times(0,+\infty) \\ \frac{\partial\psi}{\partial s}\Big|_{s=s_0} = 0, \quad \left(\frac{\partial}{\partial s} + \alpha_s\right)\psi\Big|_{s=1} = 0. \\ \psi|_{t=0} = \psi_0. \end{cases} \qquad (V)$$

where α_s is a known positive function, $M_0=S^2\times[s_0,1]$, and $S^2 = \{(\lambda,\theta) | 0\le\lambda\le2\pi, 0\le\theta\le\pi\}$ is the unit sphere. He has proved that there exists a local weak solution of problem (V) and the classical solution of problem (V) is unique.

In [2], J.Dutton made the β-plane approximation to simplify equation (1.1), under the same initial-boundary conditions as problem (V), he considered well-posedness of the corresponding initial-boundary value problem for the equation simplified. Unfortunately, Zen Qing-cun has pointed out that there is an essential mistake in the proof, so the problem has not been solved even in this simple case.

Because equation (1.1) and problem (V) play important roles in atmospheric dynamics, we are very interested in the well-posedness of problem (V). In this paper, we consider initial-boundary value problem for generalized vorticity equations on compact Riemannian manifolds. Utiliing the peculiarity of composite type of the equation, we give some accurate estimates and obtain the uniqueness and existence of global classical solution of initial-boundary value problem for a generalized vorticity equation on a Riemannian manifold. As its applications, we obtain the uniqueness and existence of global classical solution of problem (V).

2. Generalized Vorticity Equations On Riemannian Manifolds

In equation (1.1), the point $\theta = 0$ and $\theta = \pi$ are two singular points caused by the choice of coordinate system. To overcome this difficulty, we shall introduce the definition of a generalized voticity equation on a Riemannian manifold. Equation (1.1) is one of this type. Let M be a n-dimensional, compact, oriented C^∞-Riemannian manifold with C^∞-boundary ∂M. $\{\Omega_j, \varphi_j\}_{j=1}^{J}$ is a C^∞-atlas on M. $\tilde{\Omega}_j = \varphi_j(\Omega_j) \subset \mathbb{R}^n$. We denote a point $x^j \in \tilde{\Omega}_j$ by $x^j = (x_1^j, \ldots, x_n^j)$, $\{\Omega_j|_{\partial M}, \varphi_j|_{\partial M}\}_{j=J_1}^{J}$ is a C^∞-atlas of boundary manifold ∂M. We also assume that $\varphi_j(\Omega_j \cap \partial M) = \{x^j \in \tilde{\Omega}_j \,|\, x_n^j = 0\}$ for $J_1 \leq j \leq J$.

In following, we shall denote by (P, t) a point in $M \times [0, +\infty)$, We will write $F_j = F \circ \varphi_j^{-1}$ defined in $\tilde{\Omega}_j \times [0, +\infty)$ if F is a function defined in $M \times [0, +\infty)$.

Let T(P, t) be a given 2-order contravariant tensor field on M which depends on parameter t smoothly. In each $\tilde{\Omega}_j \times [0, +\infty)$, T(P, t) can be designated by functions $\{\gamma_{\alpha\beta}^j(x^j, t)\}_{\alpha,\beta=1}^{n}$. We also assume that $\vec{R}(P, t)$ $\vec{\tilde{R}}(P, t)$, $\vec{R}_0(P, t)$ are three given C^∞-vector fields on M which also depend on parameter t smoothly, $\varphi_{j*}(\vec{R}) = \{\gamma_\alpha^j(x^j, t)\}_{\alpha=1}^{n}$, $\varphi_{j*}(\vec{\tilde{R}}) = \{\tilde{\gamma}_\alpha^j(x^j, t)\}_{\alpha=1}^{n}$, $\varphi_{j*}(\vec{R}_0) = \{\tilde{\gamma}_{0\alpha}(x^j, t)\}_{\alpha=1}^{n}$.

Where φ_{j*} is the mapping induced by mapping φ_j.

For any differentiable function $V(P,t)$, we define on M a family of vec tor fields $\vec{W}(T,\vec{R},\vec{\tilde{R}},V)$ varying with parameter t as follows.

$$\vec{W}(T,\vec{R},\vec{\tilde{R}},V)(P,t)$$

$$=(\varphi_j^{-1})_*\left(\left\{\sum_{\alpha=1}^{n}\gamma_{\alpha\beta}^{j}(x^j,t)\frac{\partial V_j}{\partial x_\alpha^j}+\gamma_\beta^j(x^j,t)V_j+\tilde{\gamma}_\beta^j(x^j,t)\right\}_{\beta=1}^{n}\right),$$

$$\text{when } (P,t)=\varphi_j^{-1}(x^j,t)\in\tilde{\Omega}_j\times[0,+\infty).$$

We shall always abbreviate $\vec{W}(T,R,R,V)$ to $\vec{W}(V)$. Let b,d be two given smooth functions on $M\times[0,+\infty)$. By well-known methods (see [3]), we define two first-order linear partial differential operators $L_{(\vec{W}(V),b)}$ and $L_{(\vec{R}_0,d)}$ respectively.

We assume that $A(t)$ is a family of second-order linear elliptic operators on M. which varies smoothly with parameter t. For any differentiable function u on $M\times[0,+\infty)$, we define

$$L(V,u)=\left(\frac{\partial}{\partial t}+L_{(\vec{W}(V),b)}\right)A(t)u+L_{(\vec{R}_0,d)}u,$$

and call $L(u,u)=f$ a generalized voticity equation.

In this paper, we shall study the following initial-boundary value problem for a generalized vorticity equation

$$\begin{cases}\left(\frac{\partial}{\partial t}+L_{(\vec{W}(u),b)}\right)A(t)u+L_{(\vec{R}_0,d)}u=f, \quad \text{in } M\times(0,+\infty) & (2.1)\\ B(t)u\big|_{\partial M\times(0,+\infty)}=0. & (2.2)\\ u\big|_{t=0}=u_0. & (2.3)\end{cases}$$

where $B(t)$ is a family of linear differential operators on ∂M depending on parameter t smoothly. u_0 and f are given functions.

We introduce Sobolev space $W^{k,p}(M)$, $H^k(M)$ according to usual definition. Let $\{\alpha_j\}_{j=1}^{J}$ be a C^∞-partition of unity which is subordinate to $\{\Omega_j,\varphi_j\}_{j=1}^{J}$. For $u \in C^{k+1}(M)$, we define

$$|||u(x)|||_{C^k(M)} = \sum_{j=1}^{J} |||(\alpha_j u)_j|||_{C^k(\tilde{\Omega}_j)}, \quad \text{integer } k \geq 0.$$

Where

$$|||(\alpha_j u)_j|||_{C^k(\tilde{\Omega}_j)} = \|(\alpha_j u)_j\|_{C^k(\tilde{\Omega}_j)} + \sup_{\substack{x,y \in \tilde{\Omega}_j \\ |x-y|<1}} \sum_{|\alpha|=k} \frac{|D^\alpha u(x) - D^\alpha u(y)|}{\chi(|x-y|)},$$

$$\chi(s) \equiv s(1-\ln s), \quad 0<s<1.$$

Let $\vec{R}(t)$ be a family of C^∞-vector fields on M with parameter t, $\varphi_{j*}(\vec{R}) = \{\gamma_\alpha^j(x^j,t)\}_{\alpha=1}^{n}$. We define $|||L_{(\vec{R})}|||_{C^0([0,T];M)} = \sup_{0\le t\le T} \sum_{1\le\alpha\le n} |||\gamma_\alpha^j(x^j,t)|||_{C^0(M)}$.

If for $\forall t \geq 0$, in each boundary map Ω_j $(J_1 \leq j \leq J)$,

$$\gamma_n^j(x^j,t)\big|_{x_n^j=0} = 0.$$

the manifold $\partial M \times (0,+\infty)$ is then called the characteristic manifold of the operator $L_{(R)}$.

In this paper, we make the following fundamental assumptions.

(A1). In each boundary map Ω_j $(J_1 \leq j \leq J)$, we have

$$\gamma_{\alpha n}^j(x^j,t)\big|_{x_n^j=0} = 0, \quad 1\le\alpha\le n; \quad \gamma_n^j(x^j,t)\big|_{x_n^j=0} = \tilde{\gamma}_n^j(x^j,t)\big|_{x_n^j=0} = 0.$$

(A2). For $\forall\, t \geq 0$, the elliptic boundary value problem $\{A(t)u = 0, \text{ in } M;\ B(t)u|_{\partial M} = 0\}$ satisfies Lopatinski condition. (See [6]).

(A3). Suppose $u \in C^\infty(M\times[0,+\infty))$, $Bu|_{\partial M\times(0,+\infty)} = 0$. For every $T>0$, there exists a constant C_T which does not depend on u and t such that

$$\|u(t)\|_{W^{k+2,p}(M)} \leq C_T \|A(t)u(t)\|_{W^{k,p}(M)}, \quad 0\le t\le T. \quad 1<p<+\infty.$$

$$|||U(t)|||_{C^1(M)} \le C_T \|A(t)U(t)\|_{C^0(M)}, \quad 0 \le t \le T.$$

It is easy to verify that assumption (A1) is equivalent to the following assumption

(A1). For any differentiable function V, $\partial M \times (0,+\infty)$ is a characteristic manifold of operator $L_{(\vec{W}(V))}$.

In this paper, all functions are real functions and all functional spaces are real functional spaces.

3. Main Results.

In following, we only state the essential lemmas and the main results of this paper. The details can be found in [7], [8].

Lemma 3.1. Suppose the assumptions (A1)-(A3) are fulfilled. T is a positive constant chosen arbitrarily. V, U, u and f belong to space $C^\infty(M\times[0,T])$, $U_0 \in C^\infty(M)$ such that

$$\left\{\begin{array}{ll} (\frac{\partial}{\partial t} + L_{(\vec{W}(V),b)})A(t)U + L_{(\vec{R}_0,d)}u = f, \quad \text{in } M\times(0,T) & (3.1) \\ B(t)U|_{\partial M\times(0,T)} = 0. & (3.2) \\ U|_{t=0} = U_0 & (3.3) \end{array}\right.$$

then, for any integer $k \geq 4$, real p satisfying $n < p < +\infty$, we have the following estimates

$$\|U(t)\|^p_{W^{k,p}(M)} \le C\Big(\|U_0\|^p_{W^{k,p}(M)} + \int_0^t \big(\|u(\tau)\|^p_{W^{k-1,p}(M)} + \|f(\tau)\|_{W^{k-2,p}(M)}\big)d\tau\Big)\cdot$$

$$\cdot \exp\Big\{C\int_0^t \big(\|V(\tau)\|_{W^{k-1,p}(M)} + 1\big)d\tau\Big\}, \quad 0 \le t \le T. \qquad (3.4)$$

particularly, if U=u in (3.1), then, for every integer $k \geq 4$, real p satisfying $n < p < +\infty$,

$$\|U(t)\|^p_{W^{k,p}(M)} \le C\Big(\|U_0\|^p_{W^{k,p}(M)} + \int_0^t \|f(\tau)\|^p_{W^{k-2,p}(M)} d\tau \Big) \cdot$$

$$\cdot \exp\Big\{ C\int_0^t \big(\|V(\tau)\|_{W^{k+1,p}(M)} + 1\big) d\tau \Big\}, \quad 0 \le t \le T. \tag{3.5}$$

$$\|U(t)\|^p_{W^{3,p}(M)} \le C\Big(\|U_0\|^p_{W^{k,p}(M)} + \int_0^t \|f(\tau)\|^p_{W^{1,p}(M)} d\tau \Big) \cdot$$

$$\cdot \exp\Big\{ C\int_0^t \big(\|V(\tau)\|_{C^2(M)} + 1\big) d\tau \Big\}, \quad 0 \le t \le T. \tag{3.6}$$

Furthermore, if V=U=u in (3.1), then for any integer $k \geqslant 1$, we have inequalities

$$\|u(t)\|^2_{H^{k+2}(M)} \le C\Big(\|U_0\|^2_{H^{k+2}(M)} + \int_0^t \|f(\tau)\|^2_{H^k(M)} d\tau \Big) \cdot$$

$$\cdot \exp\Big\{ C\int_0^t \big(\|u(\tau)\|_{H^{k+1}(M)} + \|u(\tau)\|_{C^2(M)} + 1\big) d\tau \Big\}, \quad 0 \le t \le T. \tag{3.7}$$

In above inequalities, we have denoted by C the different constants which depend on k, p, T and M only.

According to assumption (A1), $\partial M \times (0,+\infty)$ is a characteristic manifold of operator $L_{(\vec{W}(V))}$ for any $V \in C^\infty(M \times [0,+\infty))$, hence the characteristic curve of hyperbolic operator $\left(\frac{\partial}{\partial t} + L_{(\vec{W}(V))}\right)$ exists on $[0,+\infty)$ globally. Integrating along the characteristic curve, we can prove

Lemma 3.2. Suppose assumptions (A1)-(A3) hold. If $u, v, f \in C^\infty(M \times [0,T])$ and $u_0 \in C^\infty(M)$ such that

$$\begin{cases} \left(\frac{\partial}{\partial t} + L_{(\vec{W}(V),\, b)}\right) Au + L_{(\vec{R}_0,\, d)} u = f, & \text{in } M \times (0, T) \\ B(t)u|_{\partial M \times (0,T)} = 0. \\ u|_{t=0} = u_0. \end{cases}$$

then

$$|||u(s)|||_{C^1(M)} \le C_1(T, \|u_0\|_{C^2(M)}, \|f(t)\|_{C^0(M\times[0,T])}), \quad 0 \le s \le T. \tag{3.8}$$

Here and afterwards, we denot by T a constant chosen arbitrarily and by $C_i(F_1, \ldots, F_k)$ the constants which only depend on $(F_1, \ldots, F_k)$.

Remark. The right side of (3.8) does not depend on function V, we denote the constant by

$$\tilde{D} = C_1(T, \|u_0\|_{C^2(M)}, \|f(t)\|_{C^0(M\times[0,T])}) . \qquad (3.9)$$

Suppose $\sup_{0\le t\le T} |||V(t)|||_{C^1(M)} \le E_1$ .Then we can verify easily that $|||L(\vec{W}(V))|||_{C^0(M\times(0,T))} \le C_2(E_1)$. We define the constant D_1 by

$$D_1 = C_2(\tilde{D}) \qquad (3.10)$$

Where $\tilde{D}$ has been given by (3.9).

Utilizing the second inequality in assumption (A3) and Lemma 2.6 in [4], intergrating along the characteristic curve, we can prove

Lemma3.3. Suppose all assumptions in Lemma 3.2 are fulfilled. constant D_1 is given by (3.10). Letting $J_0=1+[2T\sqrt{n}D_1]$ ([x] denotes the largest integer less than x) and assuming $|||L(\vec{W}(V))|||_{C^0(M\times(0,T))} \le D_1$. Let α and β be two real numbers satisfying $\alpha = \beta e^{-2TD_1J_0}$ .Then, we have

$$\|u(s)\|_{C^{2,\alpha}(M)} \le C_3(T, D_1, \|u_0\|_{C^{2,\beta}(M)}, \|f(t)\|_{C^0([0,T]; C^{0,\beta}(M))}), \quad 0\le s\le T. \qquad (3.11)$$

Utilizing prior estimate (3.4) and contraction mapping theorem, we can prove the existence and uniqueness theorem for the linearized problem of (2.1)-(2.3). Then, combining estimates (3.5)-(3.8),(3.11),we get the main theorem of this paper by Schauder s fixed point theorem as follows.

Theorem 3.1. Suppose assumptions (A1)-(A3) hold and the initial date u_0 satisfies condition $B(0)u_0|_{\partial M} = 0$.(1). If $u_0 \in W^{k,p}(M)$, $f \in C^0([0,+\infty); W^{k-2,p}(M))$, $n < p < +\infty$,$k \geq 5$, then there exists a unique global classical solution of problem (2.1)-(2.3) in spaces $C^0([0,+\infty);$

$W^{k,p}(M)) \cap C^1([0,+\infty); W^{k-1,p}(M))$.(2). If $u_0 \in H^{k+2}(M)$, $f \in C^0([0,+\infty), H^k(M))$, $k > \frac{n}{2}+2$, then there exists a unique global classical solution of problem (2.1)-(2.3) in the space $C^0([0,+\infty); W^{k,p}(M)) \cap C^1([0,+\infty); W^{k-1,p}(M))$.

(3). If $u_0 \in C^\infty(M)$, $f \in C^\infty(M \times [0,+\infty))$, then the solution of problem (2.1)-(2.3) is in $C^\infty(M \times [0,+\infty))$.

Using the techniques in §6.7 of [6] and Lemma 1.4 of [4], we can prove

Lemma3.4. Suppose $\psi \in C^\infty(M_0 \times [0,+\infty))$, $\frac{\partial \psi}{\partial s}\big|_{s=s_0} = (\frac{\partial}{\partial s}+\alpha_s)\psi\big|_{s=1} = 0$,

Then, for every $T > 0$, there exists a constant C_T which does not depend on ψ such that

$$\|\psi(t)\|_{W^{k+2,p}(M)} \leq C_T \|A_0(\psi)\|_{W^{k,p}(M_0)},$$

$$\|\psi(t)\|_{C^1(M)} \leq C_T \|A_0(\psi)\|_{C^0(M)}. \qquad 0 \leq t \leq T.$$

Where $1 < p < +\infty$, integer $k \geq 0$.

From Theorem 3.2 and Lemma 3.4, we have proved the following theorem

Theorem 3.2. Suppose $\psi_0(\lambda,\theta,s)$ satisfies compatibility condition $\frac{\partial \psi_0}{\partial s}\big|_{s=s_0} = (\frac{\partial}{\partial s}+\alpha_s)\psi_0\big|_{s=1} = 0$.(1). If $\psi_0 \in W^{k,p}(M_0)$, $3 < p < +\infty$, $k \geq 5$, then there exists a unique global classical solution of problem (V) in the space $C^0([0,+\infty); W^{k,p}(M_0)) \cap C^1([0,+\infty); W^{k-1,p}(M_0))$.(2). If $\psi_0 \in H^{k+2}(M_0)$, $k \geq 4$,

then there exists a global classical solution of problem (V) in the space $C^0([0,+\infty); H^{k+2}(M)) \cap C^1([0,+\infty); H^{k+1}(M))$. (3). If $\psi_0 \in C^\infty(M_0)$, then the solution of problem (V) is in the spaces $C^\infty(M_0 \times [0,+\infty))$.

This work is completed by author under the guidance of professor

Gu Chao-hao and Professor Li Ta-qian. Professor Zeng Qing-cun at **Atmospheric Physics Institute of the Chinese Academy of Sciences gave author great encouragements and supports. I would like to express my heartfelt thanks to them.**

References

1 Zen Qingcun, Physical-Mathematical Basis of Numerical Weather Prediction, Vol.1, Science Press, Beijing, 1979, Chinese edition. (The English edition is in press).

[2] Dutton, J.A. The nonlinear quasi-geostropic equations: existence and uniqueness of solutions on bounded domain, J.Atmos.Sci, Vol.31, 2, (1974).

[3] Dieudonné, J. Treatise On Analysis, Vol.IV. Academic Press, 1974.

[4] Kato, T. Arch.Rational Mech Anal. Vol.25, 1967, 188-200.

[5] Gilbarg, D., Trudinger, N.S. Elliptic Partial Differential Equations of Second Order, Springer-Varlar, 1977.

[6] Triebel, H. Interpolation Theory, Function Spaces, Differential Operators. North-Holland Publishing Company, 1978.

[7] Mu Mu, Global Classical Solutions of nonlinear vorticity equations and its applications, to contribute to J.Mathe Phys, (Chinese Edition.)

[8] Mu Mu, Global classical solutions of Cauchy problems for nonlinear vorticity equations and its applications, to contribute to Ann. Math.(Chinese Edition).

Nonoscillatory Solution of Second Order Linear Equation and Periodic Solution of Periodic Riccati Equation

PU FUQUAN

(Tsinghua University, Beijing, China)

The interchangeability of Riccati equation and second order linear differential equation gives us a convenient way of investigating Riccati equation

$$z^1 = z^2 + p(x) \qquad (*)$$

Adamov (1948) [1] and Massera (1950) [2] gave the same result: Riccati equation with periodic coefficients has a periodic solution iff the corresponding linear differential equation possesses a nonoscillatory solution. So the investigation of existence of nonoscillatory solution of second order linear equation is equivalent to the investigation of existence of periodic solution of periodic Riccati equation.

To study the existence of nonoscillatory solution of

$$y'' + p(x)\, y = 0 \qquad (**)$$

is what we are interested in.

When p(x) is a periodic pulse function with one furning point x_0 (Fig. 1), we get conditions on (**) which garantee the existence of nonoscillatory solution of (**).

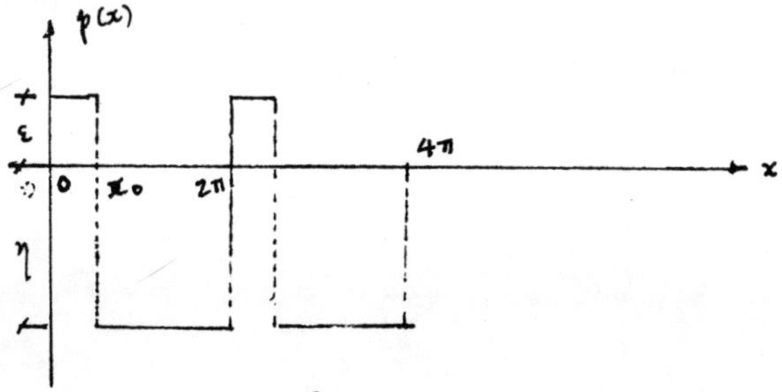

(Fig. 1)

Let $\varepsilon = 1$, and

$$x_0 < (2\pi - x_0)\eta \quad ,$$

then (**) has nonoscillatory solution if

$$0 \le x_0 < \frac{\pi}{2} \; , \quad \tan x_0 < \sqrt{\eta}$$

The cases p(x) is periodic pulse function with multiple turning points are also investigated [3] .

By application of Adamov's theorem, we study two concrete examples: a pulse function potential and the equation

$$y'' + \frac{\sin x}{1+\sin x} y = 0 \; .$$

The existence of nonoscillatory solutions are checked [4] .

References

1. N. V. Adamov, On some transformations which do not change the integral curves of second order differential equations. Mathematical sbornik Vol. 23 (65) No. 2 (1948) pp216-218

2. J. L. Massera, The existence of periodic solutions of systems of differential equations, Duke Math. J. 17 (1950) pp451-476

3. Pu Fuquan, A special kind of nonoscillatory second order linear differential equations (to appear)

4. Pu Fuqnan, Applications of Adamov's theorem (to appear)

5. D. Willet, classification of second order linear differential equations with respect to oscillations. Advomces in Math. 1 (1967) pp599-623

Global Solutions to the Cauchy Problem of a Class of Quasilinear Hyperbolic Parabolic Coupled Systems

Zheng Songmu — Institute of Mathematics
Shen Weixi — Department of Mathematics
Fudan University
Shanghai, China

In this paper we are concerned with the global existence, uniqueness and the asymptotic behavior of smooth solutions to the following Cauchy Problem of a class of quasilinear hyperbolic parabolic coupled systems

$$\begin{cases} u_{1t}-u_{2x}=0, \\ u_{2t}-p(u_1,v)_x=0, \\ c_0(u_1,v)v_t-p_v(u_1,v)u_{2x}-(K(u_1,v)v_x)_x+d(u_1,v)=0, \\ t=0: \quad u_1=u_1^0(x), \quad u_2=u_2^0(x), \quad v=v^0(x). \end{cases} \tag{1}$$

which arise in thermoelasticity, radiation hydrodynamics and other mechanical problems.

Such kinds of Cauchy problems for quasilinear systems in one space dimension are more difficult than the corresponding initial boundary problem and the Cauchy problem in higher dimensions.

First, using the spectral analysis, we derive the $L^p(p=1, 2,\infty)$ decay estimates of solutions to the linearized problem of (1)

$$\begin{cases} u_{1t}-\alpha u_{2x}=0, \\ u_{2t}-\alpha u_{1x}+\beta v_x=0, \\ v_t+\beta u_{2x}-kv_{xx}=0, \\ t=0: \quad u_1=u_1^0(x), \quad u_2=u_2^0(x), \quad v=v^0(x). \end{cases} \tag{2}$$

where α, β are nonzero constants, k is a positive constant, $U_0=(u_1^0,u_2^0,v^0)\in W^{1,1}(R)$. More precisely we prove

Theorem 1. Suppose $u_o \in W^{1,1}$. Then for i,j=1,2,3 the following estimates hold:

$$\|G_{ij}*u_o\|_{L^1} \leqslant C\|u_o\|_{L^1}$$
$$\left\|\frac{\partial}{\partial x}G_{ij}*u_o\right\|_{L^1} \leqslant C(1+t)^{-\frac{1}{2}}\|u_o\|_{W^{1,1}} \qquad \forall t \geqslant 0, \qquad (3)$$

Moreover, when $t \geqslant 1$,

$$\|G_{ij}*u_o - V_{ij}(u_o)\|_{L^1} \leqslant C(1+t)^{-\frac{1}{4}}\|u_o\|_{L^1}$$
$$\left\|\frac{\partial}{\partial x}G_{ij}*u_o - \frac{\partial}{\partial x}V_{ij}(u_o)\right\|_{L^1} \leqslant C(1+t)^{-\frac{3}{4}}\|u_o\|_{W^{1,1}} \qquad (4)$$

where G_{ij} (i,j=1,2,3) are elements of resolvent matrix of ordinary differential equations obtained by making Fourier transformation of (2), V_{ij} are the operators similar to the heat potential operator.

Theorem 2. Suppose $u_o \in W^{2,1}$. Then the following estimates hold (i,j=1,2,3):

$$\left\|\frac{\partial^2}{\partial x^2}G_{ij}*u_o\right\|_{L^1} \leqslant C(1+t)^{-1}\|u_o\|_{W^{2,1}} \qquad \forall t \geqslant 0,$$
$$\left\|\frac{\partial}{\partial t}G_{ij}*u_o\right\|_{L^1} \leqslant C(1+t)^{-1}\|u_o\|_{W^{2,1}} \qquad \forall t \geqslant 0,$$
$$\left\|\frac{\partial^2}{\partial x^2}G_{ij}*u_o - \frac{\partial^2}{\partial x^2}V_{ij}(u_o)\right\|_{L^1} \leqslant C(1+t)^{-\frac{5}{4}}\|u_o\|_{W^{2,1}} \qquad \forall t \geqslant 1, \qquad (5)$$
$$\left\|\frac{\partial}{\partial t}G_{ij}*u_o - \frac{\partial}{\partial t}V_{ij}(u_o)\right\|_{L^1} \leqslant C(1+t)^{-\frac{5}{4}}\|u_o\|_{W^{2,1}} \qquad \forall t \geqslant 1,$$

Theorem 3. Let $U(x,t)=(u_1,u_2,v)^T$ be the solution of (2). Then the following estimates hold:

$$\|D_xU(t)\|_{L^2} \leqslant C(1+t)^{-\frac{3}{4}}\|U_o\|_{L^1} + Ce^{-ct}\|U_o\|_{H^1}, \quad \forall t \geqslant 1,$$
$$\|D_xU(t)\|_{L^2} \leqslant C(1+t)^{-\frac{1}{2}}\|U_o\|_{H^2} \qquad \forall t \geqslant 0, \qquad (6)$$
$$\|D_xU(t)\|_{L^\infty} \leqslant C(1+t)^{-1}\|U_o\|_{W^{1,1}} + Ce^{-ct}(\|D_xU_o\|_{H^1} + \|D_xU_o\|_{L^1}) \qquad \forall t \geqslant 1,$$

$$\| D_x U(t) \|_{L^\infty} \leqslant C(1+t)^{-\frac{3}{4}} \| U_o \|_{H^2} \qquad \forall\ t \geqslant 0.$$

Secondly, based on the above L^p decay estimates, by delicately deriving the uniform a priori estimates of solutions for (1) we establish the global existence, uniqueness of small smooth solutions to(1) and obtain the decay rates as $t \to +\infty$ as follows.

Theorem 4. Under some suitable assumptions which are reasonable in mechanics there exists an appropriately small positive constant ε such that when $\| U_o \|_{H^3} + \| U_o \|_{W^{1,1}} \leqslant \varepsilon$, Cauchy problem (1) admits a unique global smooth solution $u_1, u_2 \in C([0,+\infty);H^3) \cap C([0,+\infty);W^{1,1}) \cap C^1([0,+\infty);H^2)$, $v \in C([0,+\infty);H^3) \cap C([0,+\infty);W^{1,1}) \cap C^1([0,+\infty);H^1)$. Moreover, as $t \to +\infty$, the solution has the decay rates:

$$\| U(t) \|_{L^\infty} = O((1+t)^{-\frac{1}{2}}),$$

$$\| U(t) \|_{L^2} = O((1+t)^{-\frac{1}{4}}),$$

$$\| DU(t) \|_{L^1} = O((1+t)^{-\frac{1}{2}}),$$

which are the same as the heat equations. The higher order derivatives of solution have the corresponding decay rates.

Finally, as the applications, we get the global existence, uniqueness and decay rates of small smooth solutions to the Cauchy problems for the system of thermoelasticity and the system of radiation hydrodynamics.

References

1. Zheng Sôngmu, Shen Weixi, L^p decay estimates of solutions to the Cauchy problem of hyperbolic-parabolic coupled system , submitted to Scientia Sinica.

2. Zheng Songmu, Shen Weixi, Global solutions to the Cauchy

problem of a class of quasilinear hyperbolic parabolic coupled system, submitted to Scientia Sinica.

3. Zheng Songmu, Global solutions and applications to a class of quasilinear hyperbolic parabolic coupled systems, Scientia Sinica, VOL XXVII No.12 1274-1286(1984).

4. S.Kawashima, T.Nishida, Global solutions to the initial value problem for the equations of one-dimensional motion of viscous polytropic gases, J.Math. Kyoto Univ., 21-4(1981),825-837.

5. S.Klainerman, Long-time behavior of solutions to nonlinear evolution equations, Arch. Rat. Mech. Anal., 78(1982),73-98.

6. A.Matsumura, An energy method for the equations of motion of compressible viscous and heat-conductive fluids, Univ. of Wisconsin-Madison MRC Tech. Sum. Report # 2194(1981)

7. Jong Uhn Kim, Solutions to the equations of one-dimensional viscoelasticity in BV, SIAM, J.Math. Anal., 14(1983), 684-694.

8. Jong Uhn Kim, Global existence of solutions of the equations of one-dimensional thermoviscoelasticity with initial data in BV and L^1, Ann. Scoula Norm. Sup. Pisa, X-3(1983), 357-427.

9. C.M.Dafermos, Conservation laws with dissipation, Nonlinear Phenomena in Math. Sci.V, Academic Press, 1981.

10. Li Tatsien, Yü Wentz, Shen Weixi, Boundary value problems and free boundary problems for quasilinear hyperbolic parabolic coupled systems, Univ. of Wisconsin-Madison, MRC Tech. Sum. Report #2273(1981).

An Elliptic-Parabolic Coupled System Arising From the Filtration with Transport of Solute

Su Ning

Introduction

In this paper we consider the flow of soil water with solute such as fertilizer or salt. The model is a nonlinear elliptic-parabolic coupled system.

As a general case of the filtration of soil water, the flow through porous media has been studied extensively, cf.[2,6,10-12] and cited references. However, except numerical computation and experiment researches, there are a few mathematical researches in the field of the solute flow through a porous medium which have been appeared, cf.[5,8]. Even in these works, the flow and the solute movement are dealt seperately, e.g. under the assumption that the soil water flow is independent of the dissolved salt. This is available to low concentration case. In general, however, the increase of the solute concentration will give rise to the increase of the water viscosity. It is the aim of the present work to study the general case, viz. the viscosity is an increasing function of the concentration. We deduce the control equation system in this case and consider its first initial boundary value problem.

1. Formulation of the Problem

Assume that the soil is a homogeneous, rigid porousmedium with imcompressible water filled the pore space. By continuity equation and Darcy's law, we have

$$\operatorname{div} \bar{q} = 0 \tag{1}$$

$$\bar{q} = -\frac{\bar{k}\gamma}{\mu} \nabla \varphi \tag{2}$$

where $\bar{q}$ is Darcy velocity, φ the piezometric head, $\bar{k}=(k_{ij})$ the soil permeability, a third-order symmetrical matrix, μ the water viscosity and γ the specific weight of water.

Suppose that there is no recharge into, or discharge from the soil, then the conservation law for the solute indicates

$$C_t + \operatorname{div} \bar{Q} + \operatorname{div}\left(\frac{C}{n}\bar{q}\right) = 0 \tag{3}$$

$$\bar{Q} = -\bar{D} \nabla C \tag{4}$$

where C denotes the concentration of the solute in the water, $\bar{Q}$ the solute flux, n the soil porisity and $\bar{D}=(D_{ij})$ the solute dispersive coefficient, a third-order symmetrical matrix.

For isotropic soil, $k_{ij} = k\,\delta_{ij}$, $D_{ij} = D\,\delta_{ij}$.

In (1) - (4), both the permeability $\bar{k}$ and the porisity n are constants, determined by the soil structure, and independent of the water and solute. According to the preceding statement, we can assume that the viscosity μ is an increasing function of the concentration , viz. $\mu=\mu(C)$. In addition, let $\bar{D}$ and γ be constants for simplicity. Clearly, both $\bar{k}$ and $\bar{D}$ are positive symmetrical matrices.

Under these assumptions, we deduce (1) - (4) to the following system

$$\begin{cases} \operatorname{div}\left(\frac{\bar{k}\gamma}{\mu(C)}\nabla\varphi\right)=0 \\ C_t=\operatorname{div}\left(\bar{D}\nabla C+\frac{\gamma\bar{k}C}{n\mu(C)}\nabla\varphi\right) \end{cases}$$

This is an elliptic-parabolic coupled system. Without loss of generality, let $k_{ij} = D_{ij} = \delta_{ij}$, γ=n=1, then we obtain

$$\begin{cases} \operatorname{div}\left(\frac{1}{\mu(C)}\nabla\varphi\right)=0 \\ C_t=\Delta C+\frac{1}{\mu(C)}\nabla\varphi\cdot\nabla C \end{cases}$$

We consider the following first initial boundary value problem: Find functions u and v defined on $\bar{Q}_T$, satisfying

$$(\mathrm{EP})\quad \begin{cases} \operatorname{div}\left(a(v)\nabla u\right)=0, & x\in\Omega,\ t\in[0,T] \\ u(x,t)=u_1(x,t), & x\in\partial\Omega,\ t\in[0,T] \\ v_t=\Delta v+a(v)\nabla u\cdot\nabla v, & (x,t)\in Q_T \\ v(x,t)=v_1(x,t), & (x,t)\in\Gamma \end{cases}$$

where $\Omega\subset\mathbb{R}^N$ is a bounded domain with sufficiently smooth boundary $\partial\Omega$, $Q_T=\Omega\times(0,T)$ with parabolic boundary Γ, known function $a(s)\in C^2(\mathbb{R})$ is positive and bounded, u_1 and v_1 are sufficiently smooth functions defined on $\partial\Omega\times[0,T]$ and Γ , respectively, and so can be extended to $\bar{Q}_T$, preserving their smoothness.

2. Gerenalized Solution and Uniqueness

For convenience, we first introduce the generalized solution for (EP) as follows:

A pair of functions (u,v) is called a generalized solution

of problem (EP) if

1° $u-u_1 \in \overset{\circ}{W}{}^{1,0}_{2q,2r}(Q_T)$, $v-v_1 \in \overset{\circ}{W}{}^{1,0}_{2q,2r}(Q_T) \cap L_{2,\infty}(Q_T)$

where q, r satisfy

$$\frac{N}{2q}+\frac{1}{r}=1 \quad \begin{cases} \frac{N}{2}<q\le\infty, & 1\le r<\infty \text{ for } N\ge 2 \\ 1\le q\le\infty, & 1\le r\le 2 \text{ for } N=1 \end{cases} \tag{5}$$

2° for almost all $t\in(0,T)$, (u,v) satisfies

$$\int_0^t\int_\Omega a(v)\nabla u\cdot\nabla\phi\,dxdt=0 \tag{6}$$

$$\int_0^t\int_\Omega(\nabla v\cdot\nabla\psi-v\psi_t-a(v)\nabla u\cdot\nabla v\psi)\,dxdt = \int_\Omega(v(x,t)\psi(x,t)-v_1(x,0)\psi(x,0))\,dx \tag{7}$$

where $\phi\in\overset{\circ}{W}{}^{1,0}_{\bar q,\bar r}(Q_T)$, $\psi\in\overset{\circ}{W}{}^{1,1}_{\bar q,\bar r}(Q_T)$ with $\bar q=\frac{2q}{2q-1}$, $\bar r=\frac{2r}{2r-1}$.

Noticing that condition (5) assures embedding

$$\overset{\circ}{W}{}^{1,1}_{\bar q,\bar r}(Q_T) \hookrightarrow L_{q,r}(Q_T)\ [9],$$

we can see that each integral in (6) and (7) is finite.

Now we consider the uniqueness problem for generalized solutions.

Let (u,v) and $(\tilde u,\tilde v)$ be the generalized solutions of problem (EP), then, by denoting $V=v-\tilde v$, $U=u-\tilde u$, we obtain from (6) and (7)

$$\iint_{Q_{t_1t_2}} a(v)\nabla U\cdot\nabla\phi\,dxdt=-\iint_{Q_{t_1t_2}}(a(v)-a(\tilde v))\nabla\tilde u\cdot\nabla\phi\,dxdt \tag{8}$$

$$\int_\Omega(V(x,t_2)\psi(x,t_2)-V(x,t_1)\psi(x,t_1))\,dx+\iint_{Q_{t_1t_2}}(\nabla V\cdot\nabla\psi-V\psi_t)\,dxdt = \iint_{Q_{t_1t_2}}(a(v)\nabla u\cdot\nabla V+a(v)\nabla U\cdot\nabla\tilde v+(a(v)-a(\tilde v))\nabla\tilde u\cdot\nabla\tilde v)\psi\,dxdt \tag{9}$$

where $Q_{t_1t_2}=\Omega\times(t_1,t_2)$ and $0\le t_1<t_2\le T$.

By L_∞- estimates [9], $v,\tilde v\in L_\infty(Q_T)$, then it follows from (8) that

$$\|\nabla U\|_{L_2(Q_{t_1t_2})}\le C\|\nabla\tilde u\|_{L_{2q,2r}(Q_{t_1t_2})}\cdot\|V\|_{L_{2q',2r'}(Q_{t_1t_2})} \tag{10}$$

where $q'=\frac{q}{q-1}$, $r'=\frac{r}{r-1}$. And (9) implies

$$\begin{aligned}\|V\|^2_{L_{2,\infty}(Q_{t_1t_2})}+\|\nabla V\|^2_{L_2(Q_{t_1t_2})} &\le \|V(\cdot,t_1)\|^2_{L_2(\Omega)} \\ &+C\|\nabla\tilde v\|_{L_{2q,2r}(Q_{t_1t_2})}\cdot\|V\|_{L_{2q',2r'}(Q_{t_1t_2})}\cdot\|\nabla U\|_{L_2(Q_{t_1t_2})} \\ &+C\|\nabla\tilde v\|_{L_{2q,2r}(Q_{t_1t_2})}\cdot\|\nabla\tilde u\|_{L_{2q,2r}(Q_{t_1t_2})}\cdot\|V\|^2_{L_{2q',2r'}(Q_{t_1t_2})}\end{aligned} \tag{11}$$

in which (6) is used. Substituting (10) into (11), we have

$$\|V\|^2_{L_{2,\infty}(Q_{t_1t_2})}+\|\nabla V\|^2_{L_2(Q_{t_1t_2})}\leq\|V(\cdot,t_1)\|^2_{L_2(\Omega)}+C\|\nabla\tilde{v}\|_{L_{2q,2r}(Q_{t_1t_2})}\cdot\|\nabla\tilde{u}\|_{L_{2q,2r}(Q_{t_1t_2})}\cdot\|V\|^2_{L_{2q',2r'}(Q_{t_1t_2})} \qquad (12)$$

Define space $\overset{\circ}{V}_2(Q_{t_1t_2}) = \overset{\circ}{W}{}^{1,0}_2(Q_{t_1t_2}) \cap L_{2,\infty}(Q_{t_1t_2})$ with norm

$$\|\cdot\|^2_{\overset{\circ}{V}_2} = \|\cdot\|^2_{L_{2,\infty}} + \|\cdot\|^2_{\overset{\circ}{W}{}^{1,0}_2}$$

Condition (5) satisfied by q, r assured imbedding

$$\overset{\circ}{V}_2(Q_{t_1t_2}) \hookrightarrow L_{2q',2r'}(Q_{t_1t_2}) \text{ [9]}$$

Hence (12) deduces to

$$\|V\|^2_{\overset{\circ}{V}_2(Q_{t_1t_2})} \leq \|V(\cdot,t_1)\|^2_{L_2(\Omega)} + \omega(t_1,t_2)\|V\|^2_{\overset{\circ}{V}_2(Q_{t_1t_2})} \qquad (13)$$

where $\omega(t_1,t_2) = C\|\nabla\tilde{v}\|_{L_{2q,2r}(Q_{t_1t_2})}\cdot\|\nabla\tilde{u}\|_{L_{2q,2r}(Q_{t_1t_2})}$.

Choosing $0=\tau_0<\tau_1<\tau_2<\cdots<\tau_m=T$ appropriately, we can let

$$\omega(\tau_{i-1},\tau_i) \leq \frac{1}{2}, \quad i=1,2,\cdots,m.$$

Therefore (13) indicates

$$\|V\|^2_{\overset{\circ}{V}_2(Q_{\tau_{i-1}\tau_i})} \leq 2\|V(\cdot,\tau_{i-1})\|^2_{L_2(\Omega)}, \quad i=1,2,\cdots,m. \qquad (14)$$

From $\|V(\cdot,\tau_0)\|_{L_2(\Omega)}=0$ step by step, we obtain $\|V\|_{\overset{\circ}{V}_2(Q_T)}=0$. In turn (10) implies $\|\nabla U\|_{L_2(Q_T)}=0$ and thus $\|U\|_{L_2(Q_T)}=0$. This completes the proof of

<u>Theorem 1</u>. There exists at most one generalized solution for (EP).

As for the uniqueness of the classical solutions, we only need to assert that a classical solution is a generalized one. This is not obvious because we have not required any integrable condition on classical solution (u,v), especially on u with respect to t. It is not hard to show, however, that u and its first derivatives in X are bounded on $\bar{Q}_T$ if we ask that the first derivatives in X of v are bounded on $\bar{Q}_T$, see (15) or (17) in section 3. In this case, a classical solution is a generalized one, and then the classical solution is unique.

3. Existence

First of all, for $v \in W^{1,0}_\infty(Q_T)$ consider the following problem

$$(E_v) \qquad \begin{cases} \operatorname{div}(a(v)\nabla u) = 0 & x\in\Omega,\ t\in[0,T] \\ u(x,t) = u_1(x,t) & x\in\partial\Omega,\ t\in[0,T] \end{cases}$$

According to the L_p- theory of elliptic equations [1], there exists a unique solution $u \in W^{2,0}_{p,\infty}(Q_T)$ and for almost all $t\in[0,T]$

$$\|u(\cdot,t)\|_{W^2_p(\Omega)} \leq C_1(\|u_1(\cdot,t)\|_{W^2_p(\Omega)} + \|v(\cdot,t)\|_{W^1_\infty(\Omega)}\cdot\|u(\cdot,t)\|_{W^1_p(\Omega)})$$

and thus

$$\|u\|_{W^{2,0}_{p,\infty}(Q_\tau)} \le C_1\left(\|u_1\|_{W^{2,0}_{p,\infty}(Q_\tau)} + \|v\|_{W^{1,0}_{\infty}(Q_\tau)} \cdot \|u\|_{W^{1,0}_{p,\infty}(Q_\tau)}\right) \tag{15}$$

where $1\le p<\infty$, $0\le\tau\le T$ are arbitrary and C_1 is independent of u, v and τ.

Continue to consider the problem

$$(P_u) \qquad \begin{cases} \tilde{v}_t = \Delta\tilde{v} + a(v)\nabla u\cdot\nabla v & (x,t)\in Q_T \\ \tilde{v}(x,t) = v_1(x,t) & (x,t)\in \Gamma \end{cases}$$

where u is the solution of problem (E_v). By the L_p approach to parabolic equations[9], problem (P_u) has a unique solution $\tilde{v}\in W^{2,1}_p(Q_T)$ and

$$\|\tilde{v}\|_{W^{2,1}_p(Q_\tau)} \le C_2\left(\|v_1\|_{W^{2,1}_p(Q_\tau)} + \|v\|_{W^{1,0}_{\infty}(Q_\tau)} \cdot \|u\|_{W^{1,0}_p(Q_\tau)}\right) \tag{16}$$

where constant C_2 does not depend on u, v, $\tilde{v}$ and τ.

Applying interpolation inequality [9]

$$\|u(\cdot,t)\|_{W^1_p(\Omega)} \le C_3 \|u(\cdot,t)\|^{\frac12}_{W^2_p(\Omega)} \cdot \|u(\cdot,t)\|^{\frac12}_{L_p(\Omega)}$$

and L_∞-estimates for elliptic equations

$$\|u(\cdot,t)\|_{L_\infty(\Omega)} \le \|u_1(\cdot,t)\|_{L_\infty(\Omega)}$$

to (15) and (16), we get

$$\|u\|_{W^{2,0}_{p,\infty}(Q_\tau)} \le C_4\left(1+\|v\|^2_{W^{1,0}_\infty(Q_\tau)}\right) \tag{17}$$

and

$$\|\tilde{v}\|_{W^{2,1}_p(Q_\tau)} \le C_5\left(1+\tau^{\frac1p}\left(\|u\|_{W^{2,0}_{p,\infty}(Q_\tau)} + \|v\|^2_{W^{1,0}_\infty(Q_\tau)}\right)\right) \tag{18}$$

Combining (17) and (18), we have

$$\|\tilde{v}\|_{W^{2,1}_p(Q_\tau)} \le C_6\left(1+\tau^{\frac1p}\left(1+\|v\|^2_{W^{1,0}_\infty(Q_\tau)}\right)\right) \tag{19}$$

where C_6 is independent of u, v, $\tilde{v}$ and τ.

Define operator $\mathcal{F}(v) = \tilde{v}$.

In the same way to show the uniqueness of generalized solution, we can varify that $\mathcal{F}: W^{1,0}_\infty(Q_\tau)\to W^{2,1}_p(Q_\tau)$ is continuous. Fix $p>N+2$, then imbedding $W^{2,1}_p(Q_\tau)\hookrightarrow W^{1,0}_\infty(Q_\tau)$ is compact[9]. This indicates $\mathcal{F}: W^{1,0}_\infty(Q_\tau)\to W^{1,0}_\infty(Q_\tau)$ is completely continuous.

Denote $B_{\tau,R} = \{w\in W^{1,0}_\infty(Q_\tau) \mid \|w\|_{W^{1,0}_\infty(Q_\tau)}\le R\}$.

If there exist τ, $R>0$ such that $\mathcal{F}$ maps $B_{\tau,R}$ into $B_{\tau,R}$, then, by Schauder's fixed point theorem, there exists a $v\in B_{\tau,R}$ such that $\mathcal{F}(v) = v$. Let u solve (E_v), (u,v) is then a generalized Solution of problem (EP).

Therefore, it suffices to find such , R. To this end, we need that $\tau>0$ is small enough, which means we get only the local existence.

To begin with, by imbedding $W_p^{2,1}(Q_\tau) \hookrightarrow W_\infty^{1,0}(Q)$ and (19) we have

$$\|\tilde{v}\|_{W_\infty^{1,0}(Q_\tau)} \leq C_7\left(1+\tau^{\frac{1}{p}}\left(1+\|v\|^2_{W_\infty^{1,0}(Q_\tau)}\right)\right) \tag{20}$$

where C_7 is still independent of u, v, $\tilde{v}$ and τ.

Now, take $R = 4C_7$ and $\tau = \min\{1, T, (\frac{2}{R^2})^p\}$ then

$$\|\tilde{v}\|_{W_\infty^{1,0}(Q_\tau)} \leq 2C_7 + \tau^{\frac{1}{p}} C_7 R^2 \leq \frac{R}{2} + \frac{2}{R^2}\cdot\frac{R}{4}\cdot R^2 = R$$

which means $\mathcal{F}(v) = \tilde{v} \in B_{\tau,R}$, providing $v \in B_{\tau,R}$

Thus we obtain

Theorem 2. There exists a $\tau > 0$ such that problem (EP) possesses a unique generalized solution in Q_τ.

To improve the regularity of the generalized solution (u,v). we first notice that (17) and (19) indicate $u \in W_{p,\infty}^{2,0}(Q_\tau)$ and $v \in W_p^{2,1}(Q_\tau)$, respectively. Next, by arbifrariness of $p \in (\frac{N}{2}, \infty)$ and imbeddings $W_p^2(\Omega) \hookrightarrow C^{1+\alpha}(\bar{\Omega})$, $W_p^{2,1}(Q_\tau) \hookrightarrow C^{1+\beta,\frac{1+\beta}{2}}(\bar{Q}_\tau)$ with $0 \leq \alpha < 1-\frac{N}{p}$ $0 \leq \beta < 1-\frac{N+2}{p}$, we can see that, for almost all $t \in (0,T)$, $u(\cdot,t) \in C^{1+\alpha}(\bar{\Omega})$ and $v \in C^{1+\beta,\frac{1+\beta}{2}}(\bar{Q}_\tau)$. Finally, by interior estimates of Schauder type and a result from [7], we get

Theorem 3. There exists $\tau > 0$ such that, in Q_τ, (EP) has a unique classical solution (u,v). Furthermore,

$$u \in L_\infty(0,\tau; C^{2+\alpha}(\Omega) \cap C^{1+\alpha}(\bar{\Omega})),\ v \in C^{2+\beta,\frac{2+\beta}{2}}(Q_\tau) \cap C^{1+\beta,\frac{1+\beta}{2}}(\bar{Q}_\tau)$$

in which $0 < \alpha, \beta < 1$ are arbitrary.

Remark 1. For simplicity and clearness, we have not striven the greatest genarality for hypertheses in this work. In fact, it is evident that the conditions on a(s), u, v and Ω can be much weakened.

Remark 2. The method used here can be extended, without any difficulty, to deal with other initial boundary value problems, such as the second or third initial boundary value problems.

Acknowledgement

The author of this paper would like to express his deep appreciation to Professor Xiao Shutie for suggesting this research topic and for his many valuable criticisms.

References

1. S. Agmon, The L_p approach to the Dirichlet problem. I. Regularity theorems. Ann. Scuola Norm. Sup. Pisa (3) 13, 405-448 (1959).

2. D.G. Aronson, Regularity properties of flows through porous media. SIAM J. Appl. Math. 17, 461-467 (1969).

3. J. Bear, Dynamics of Fluids in Porous Media. New York: American Elzevier Publishing Company Inc. 1972.

4. D. Gilbarg, N.S. Trudinger, Elliptic Partial Differential Equations of Second Order. Springer-Verlag, 1977.

5. P. Knabner, A free boundary problem arising from the leaching of the saline soil. Preprint (1984).

6. B.F. Knerr, The porous medium equation in one dimension. Trans. Amer. Math. Soc. 234, 381-415 (1977).

7. S.N. Kruzhkov, A. Kastro, M. Lopes, Estimates of Schauder type and existence theorems of solutions of basic problems for linear and nonlinear parabolic equations. Dokl. Akad. Nauk SSSR 220, 277-280 (1975).

8. V.L. Kulagin, On a parabolic problem with unknown boundary. Dokl. Akad. Nauk SSSR 252, 76-79 (1980).

9. O.A. Ladyzenskaja, V.A. Solonnikov, N.N. Ural' ceva, Linear and Quasilinear Equations of Parabolic Type. Translations of Mathematical Monographs Volume 23, Providence, R.I.: American Mathematical Society. 1968.

10. O.A. Oleinik, A.C. Kalashnikov, Czou Yui-lin, Cauchy problem and boundary value problems for equations of nonstationary filtration. Izv. Akad. Nauk SSSR, Ser. Mat. 22, 667-704 (1958).

11. C.J. Van Duyn, L.A. Peletier, Nonstationary Filtration in partially saturated porous media. Arch. Rat. Mech. Anal. 78, 173-198 (1982).

12. Xiao Shutie, Huang Zhou Chuanzhong, The infiltration problem with constant rate in partially saturated porous media. Acta Math. Appl. Sinica (English Series) 1, 108-126 (1984).

STRUCTURE IDENTIFICATION OF DISTRIBUTED PARAMETER SYSTEMS AND TRANSFORMATION OPTIMIZATION

Ne-Zheng Sun

Department of Mathematics
Shandong University, People's Republic of China

This paper presents a new kind of optimization problems, the transformation optimization, that can be used to identify the parameter structure as well as its values in distributed parameter systems.

1. INTRODUCTION

The distributed system parameter identification problem (DSPIP) has been studied extensively during the last decade. An important area of its application is the modeling of groundwater flow. Generally, the identification of unknown parameter, such as transmissivity or storage coefficient, should include the determination of both the parameter structure and its values (Sun and Yeh, 1985).

Many papers assume that the parameter structure is known as *a priori* information and only consider how to determine the values of its components. The available methods for solving this problem can be divided into two categories based upon the criterion used: the Output Least Square Error criterion and the Equation Error criterion (Chavent, 1979). However, in practice, the known geological information is generally not sufficient for predetermining the unknown parameter structure. For example, in the zonation case, it is difficult to specify the number and shape of zones.

A number of researchers assumed that the dimension of parametrization is the same as the number of nodes used in the finite element model or the number of grids used in the finite difference model (Carter, 1974; Chavent, 1975; Neuman, 1980). This approach circumvents the problem of identifying the parameter structure, but is intrinsically unstable, particularly when observation noise exists or observations are insensitive to parameter change in certain subregions (Yeh and Yoon, 1976; Shah *et al.* 1978). The instability phenomenon is caused by the fact that the structure complexity of the identified parameter does not match the quantity and quality of observations.

Emsellem and de Marsily (1971) were the first to consider the problem of the optimal zoning pattern. Yeh and Yoon (1976) presented a systematical process for this problem based on a statistical criterion. The necessity to limit the dimension of parameterization is clarified by Shah *et al.* (1978) and Yeh and Yoon (1981). However, the problem of how to optimally determine the shape of zones in the zonation case or how to optimally determine the locations of nodes in the finite element interpolation case is still unresolved. This is called the problem of parameter structure identification.

Sun and Yeh (1985) denoted that the optimum values of parameter cannot be separated from the optimum parameter structure. That is, if the parameter structure is not correct, the identified values of parameter will also be incorrect. In that paper, the problem of parameter structure identification is formulated as a problem of combinatorial optimization.

In this paper the problem of parameter structure identification is presented in a clearer form that leads to a kind of new optimization problem, the transformation optimization. An approximate solution of this problem is given. The numerical examples include those of finding the locations of faults and/or fractures in an aquifer.

2. DSPIP AND LINEAR PARAMETERIZATION

Assume that the governing equation of a distributed system is

$$L(x, u, a) = 0 \tag{1}$$

where L is a PDE operator, $x \in X$ is the independent variable, $u \in U$ is the state of the system, $a(x) \in A$ is an unknown parameter. The problem of parameter identification or the inverse problem is the determination of the unknown parameter $a(x)$ from a set of observations of the state variable u.

Let the observations be

$$u_{ob} = u(\hat{a}) + \varepsilon \tag{2}$$

where u_{ob}, $u(a)$ and ε are all K-dimensional vectors with components u_{ob}^k, $u_k(\hat{a})$ and ε_k respectively, $k = 1,2,\ldots,K$; K is the total number of observations; $\hat{a}$ is the true parameter; ε_k is the observation noise at sensor k.

The Output Least Square Error is usually taken as the criterion of performance, that is, we have the cost index:

$$E(a) = \| \tilde{u}(a) - u_{ob} \|_K^2 \tag{3}$$

where $\tilde{u}(a)$ is a K-dimensional vector with components $\tilde{u}_k(a)$ which are the computed values of state variable u at K sensors according to the parameter $a(x)$. If we use FDM or FEM to solve Eq. (1), the parameter $a(x)$ must be replaced by a vector

$$a_N = (a_{N,1}, a_{N,2}, \ldots, a_{N,N}) \tag{4}$$

where N is the total number of nodes or grids, $a_{N,i}$ is the node value of parameter $a(x)$ at node i.

The problem of parameter identification is then formulated as follows:

To find $a_N^* \in A_N$, where A_N is an admissible set of a_N, such that

$$E(a_N^*) = \min E(a_N) \quad , \quad a_N \in A_N \tag{5}$$

The dimension of unknown vector a_N is usually called the dimension of parameterization. As mentioned above, if observation noise exists, some authors proved that when the dimension of parameterization is too large, the results obtained by (5) are not reliable. Therefore, we have to limit the dimension of parameterization according to the quantity and quality of observations, i.e., we have to select a K-dimensional vector a_M to replace N-dimension vector a_N, where $M \ll N$. Several methods can be used to decrease the dimension of parameterization, for example, the piecewise constant or zonation method, the finite element interpolation methods. The problem is how to determine the shape of M zones in the zonation method or how to locate the M basis points in an interpolation method. That means we have to deal with the problem of parameter structure.

Assume that the relation between a_N and a_M can be represented as

$$a_N = \lambda a_M \tag{6}$$

where λ is an $N \times M$ matrix which is called "structure matrix". It is dependent on the method of parameterization. In the zonation case, the elements of λ are 1 or 0 (Shah *et al.*, 1978). In the finite element interpolation case, λ depends upon the location of nodes and the form of basis functions (Yeh and Yoon, 1981). When M is close to N, λ is approximately the unit matrix. In this case, the method of parameterization is of little consequence. However, if $M \ll N$, the generated parameter distribution by an identification procedure will to a great extent depend upon the method of parameterization. Therefore, the method of parameterization should be regarded as a part of the identification procedure.

3. STRUCTURE IDENTIFICATION AND TRANSFORMATION OPTIMIZATION

Definition I. For a given structure matrix λ, an M-dimensional vector $a_{M,\lambda}$ is called the associated parameter with λ, if it satisfies:

$$E(\lambda a_{M,\lambda}) = \min E(\lambda a_M) \quad , \quad a_M \in A_{M,\lambda} \tag{7}$$

where $A_{M,\lambda}$ is such a set of a_M that if $a_M \in A_{M,\lambda}$, then $a_N \in A_N$.

Definition II. A structure matrix $\lambda^* \in \Lambda$ is called the optimal structure matrix, if it satisfies:

$$E(\lambda^* a_{M,\lambda^*}) = \min E(\lambda a_{M,\lambda}) \quad , \quad \lambda \in \Lambda \tag{8}$$

where Λ is an admissible set of λ.

The problem of parameter identification in M-dimension means that we have to find an optimal structure matrix λ^* and its associated parameter a_{M,λ^*}. The estimated optimal parameter is

$$a_N^* = \lambda^* a_{M,\lambda^*} \quad . \tag{9}$$

Now we introduce a new kind of optimization problem as follows: Let $f(x_N)(R^N \to R^1)$ be an objective function, $x_N \in X_N$, transformation $\lambda(R^M \to R^N)$, $\lambda \in \Lambda$, transforms x_M to x_N, $x_N = \lambda x_M$, and, from $x_M \in X_{M,\lambda} \to x_N \in X_N$.

Problem. For a given transformation $\lambda \in \Lambda$, find $x_{M,\lambda}$ such that

$$\phi(\lambda) = f(\lambda x_{M,\lambda}) = \min f(\lambda x_M) \quad , \quad x_M \in X_{M,\lambda} \tag{10}$$

and further, to find a transformation $\lambda^* \in \Lambda$, such that

$$\phi(\lambda^*) = \min \phi(\lambda) \quad , \quad \lambda \in \Lambda \quad . \tag{11}$$

This kind of optimization problem is called transformation optimization.

Obviously, the first part of the problem is a nonlinear program problem, therefore, the whole problem includes a set of nonlinear program problems. Each of them depends upon a transformation λ. Of course, it is interesting to discuss the transformation optimization problem from the viewpoint of theory, but, in this paper, we will focus our attention on a numerical solution and its application in the identification of parameter structure.

4. NUMERICAL SOLUTION

A numerical method for solving the problem of parameter structure identification mentioned above is presented in this section. For simplicity, we only consider the zonation case.

Assume that a finite element method with N nodes is used to solve Eq. (1) in a given region (Ω). First, we put M moveable basis points in (Ω). The distance between node i and basis point k is denoted by $d_{i,k}$. Then we divided the N nodes of the finite element into M groups S_k $(k=1,2,\ldots,M)$ according to the relative distances, i.e., we define node i to be in group S_k, if

$$d_{i,k} = \min(d_{i,1}, d_{i,2}, \ldots, d_{i,M}) \quad . \tag{12}$$

In other words, the flow region is divided into M subregions with these basis points as their centers. These subregions deform themselves with the movement of the basis points and approximately represent different structure of the unknown parameter. In this case, the structure matrix λ is determined only by the locations of M basis points. The elements λ_{ik} of matrix λ are

$$\lambda_{ik} = \begin{cases} 1 & \text{if node } i \text{ belongs to } S_k \\ 0 & \text{otherwise} \quad . \end{cases}$$

Let J be the sensitivity coefficient matrix,

$$J_{KXN} = \left(\frac{\partial \tilde{u}_i}{\partial a_{N,j}} \bigg|_{a_N^0} \right) \tag{13}$$

where $a_N^0 = \lambda^0 a_M^0$; λ^0, a_M^0 are the initial estimation of structure and value of unknown parameter respectively.

Using the Gauss-Newton method, we can obtain the approximate solution of Eq. (7):

$$a_{M,\lambda} = (\lambda^T J^T J \lambda)^{-1} \lambda^T J^T e_0 \tag{14}$$

where

$$e_0 = \tilde{u}(a_N^0) - u_{ob} - Ja_N^0 \quad . \tag{15}$$

Therefore, we can easily obtain the value of $E(\lambda a_{M,\lambda})$ in Eq. (8) as an objective function. It is only dependent on structure matrix λ, or the coordinates of M basis points. That means problem (8) is also translated into a nonlinear program problem with coordinates of M basis points as optimum variables that minimize $E(\lambda a_{M,\lambda})$. The Powell method that does not require the calculation of derivatives can be used to solve this problem.

In order to obtain the optimum dimension of parameterization, we use the method presented by Yeh and Yoon (1981). First, the unknown parameter is assumed to be a constant by setting one basis point anywhere in the region, we obtain an optimum parameter a_1^*. Then we add a second basis point into the region to obtain λ_2^* and a_2^*, and then we add a third basis point and so on. At the same time, we calculate the covariance matrix of the estimated parameters by

$$\mathrm{cov}(a_M^*) = \frac{E(a_M^*)}{K - M} \lambda \hat{A}^{-1} \lambda^T \tag{16}$$

where

$$\hat{A} = \lambda^T J^T J \lambda \quad . \tag{17}$$

Since the inverse matrix was computed in the solution of Eq. (14), the trace of the covariance matrix (16) is easily obtained. The trace is used as a measure of parameter uncertainty, therefore, the optimum parameter dimension is chosen where the trace is minimum.

5. APPLICATIONS

As an example, we consider the identification of aquifer parameters. The governing equation of a flow in an inhomogeneous, isotropic, and confined aquifer is

$$\frac{\partial}{\partial x}\left(T \frac{\partial h}{\partial x}\right) + \frac{\partial}{\partial y}\left(T \frac{\partial h}{\partial y}\right) = Q + S \frac{\partial h}{\partial t} \tag{18}$$

subject to given additional conditions, where $h(x, y, t)$ — head; $T(x, y)$ — transmissivity; S — storage coefficient; Q — source function.

In the flow region (Ω), the distributed parameter $T(x, y)$ is replaced by vector

$$T_N = (T_{N,1}, T_{N,2}, \ldots, T_{N,N}) \tag{19}$$

where N is the number of nodes in the finite element discretization of (Ω), $T_{N,i}$ is the value of $T(x, y)$ at node i $(i = 1,2,\ldots,N)$. We assume that N is large enough so that the numerical error of finite element discretization can be ignored.

Following Carter (1974), the elements of sensitivity coefficient matrix J in Eq. (13) can be described as

$$\frac{\partial h_k}{\partial T_{N,i}} = \iint_{(\Omega_i)} \int_0^t \nabla q'(x, y, t-\tau) \cdot \nabla h(x, y, \tau) d\tau dx dy$$

$$(k = 1,2,\ldots,K; \quad i = 1,2,\ldots,N) \tag{20}$$

where K is the number of observations, (Ω_i) is the exclusive sub-domain of node i, ∇ is the gradient operator, $h(x, y, t)$ is the numerical solution of Eq. (18), $q'(x, y, t)$ is the time derivative of $q(x, y, t)$, and $q(x, y, t)$ is the solution of the adjoint equation of Eq. (18). For the numerical form of Eq. (20) see Sun and Yeh (1985).

By using Eq. (20), in order to obtain all elements of matrix J, the required number of simulation runs must be equal to the number of observation wells plus one. Because matrix J is independent of parameter structure matrix λ, it is computed only once in each iterative run of solving the optimization problem (8). Therefore, this method saves computer time significantly.

A numerical example was given in Sun and Yeh (1985). The flow region is divided into two zones with different transmissivity by a fault. There are two pumping wells and five observation wells in the region. The problem of structure identification is to find the location of the fault. The true values of the parameter are $T_1^t = 500\ m^2/day$, $T_2^t = 1000\ m^2/day$. The initial estimation of the parameter values is $T_1^0 = T_2^0 = 700\ m^2/day$. The true location and the estimated location of the fault are shown in Fig. 1. If we do not add noise to observations, then the correct location of the fault and the correct values of parameter are obtained only after two iteration runs.

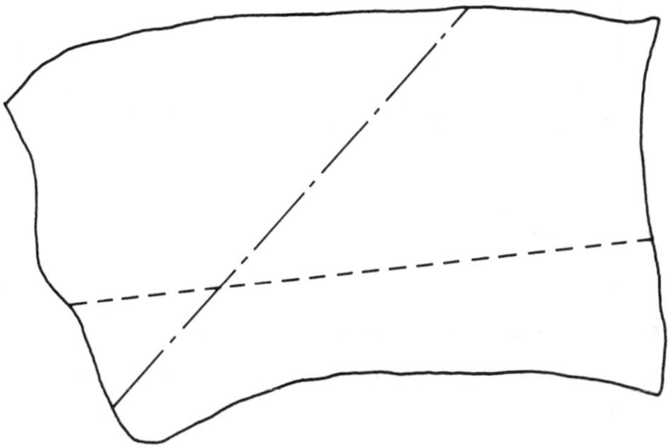

------ True location of the fault

—— · —— Initial estimation of the fault

Fig. 1 The identification of a fault.

Now we present a method of identifying locations of fractures in an aquifer. A fracture has very large transmissivity. We put M pair of points in the flow region. The link line of each pair of points represent a fracture. In this case, the structure matrix is determined by the coordinates of these points. The components of vector a_M are the transmissivities of these fractures. As before, the optimum locations of these points that minimize the objective function $E(\lambda a_{M,\lambda})$ can be obtained by an optimization procedure, see Fig. 2.

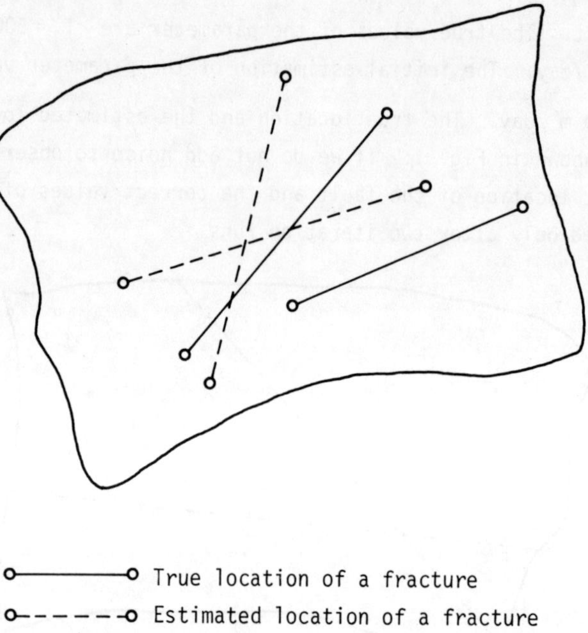

Fig. 2 The identification of fractures.

References

Carter, R.D., L.F. Kemp, A.C. Pierce and D.L. Williams, "Performance matching with constraints", Soc. Pet. Engr. J., 14 (2), 1974.

Chavent, G., M. Dupuy, and P. Lemonnier, "History matching by use of optimal control theory", Soc. Pet. Engr. J., 15 (1), 1975.

Chavent, G., "Identification of distributed parameter system: About the output least square method, its implementation and identifiability", in Identification and System Parameter Estimation Vol. 1, edited by R. Igermann, Proceedings of the Fifth IFAC Symposium, Darmstadt, Federal Republic of Germany, 24-28 September 1979, Pergamon Press.

Emsellem, Y. and G. de Marsily, "An automatic solution for the inverse problem", Water Resources Research, 7 (5), 1264-1283 (1971).

Neuman, S.P., "A statistical approach to the inverse problem of aquifer hydrology, 3. Improved solution method and added perspective",

Water Resources Research, 16 (2), 331-346 (1980).

Shah, P.C., G.R. Gavalas, and J.H. Seinfeld, "Error analysis in history matching: The optimum level of parameterization", Soc. Pet. Engr. J., 18 (3), 219-228 (1978).

Sun, N-Z., W. W-G. Yeh, "Identification of parameter structure in groundwater inverse problem", Water Resources Research, (in press), 1985.

Yeh, W. W-G. and Y.S. Yoon, "A systematic optimization procedure for the identification of inhomogeneous aquifer parameters", in Advances in Groundwater Hydrology, edited by Z.A. Saleem, pp. 72-82, American Water Resources Association, Minneapolis, Minnesota, 1976.

Yeh, W. W-G. and Y.S. Yoon, "Aquifer parameter identification with optimum dimension in parameterization, Water Resources Research, 17 (3), 664-672 (1981).

THE STUDY OF THE GLOBAL SOLUTIONS FOR A NONLINEAR PARABOLIC SYSTEM

Wang Jinghua
Institute of Systems Science
Academia Sinica

We consider the initial value problem of the nonlinear parabolic system

$$\left.\begin{aligned} \frac{\partial u}{\partial t} + \frac{\partial p(v)}{\partial x} + \alpha u &= \varepsilon \frac{\partial^2 u}{\partial x^2} \\ \frac{\partial v}{\partial t} - \frac{\partial u}{\partial x} &= \varepsilon \frac{\partial^2 u}{\partial x^2} \end{aligned}\right\} \qquad (x, t) \in R \quad , \quad x \times R^+ \quad , \tag{1}$$

with the initial value

$$(u, v)(x, 0) = (u_0(x), v_0(x)) \quad , \quad x \in R \quad . \tag{2}$$

Here $\varepsilon > 0$, $\alpha \geq 0$ are constants and

$$p(v) < 0 \quad , \quad p''(v) > 0 \quad ; \quad v > 0 \quad ; \tag{3}$$

$$\int_v^c \sqrt{-p'(\xi)}\, d\xi < +\infty \quad , \quad v > 0 \quad \text{and}$$

$$\lim_{v \to +0} \int_v^c \sqrt{-p'(\xi)}\, d\xi = +\infty \tag{4}$$

where c is a positive constant.

The system (1) can be considered as a model of compressible

Navier-Stokes equations or the system of isotropic gas dynamics equations with "artificial" viscosity with the damping effect for $\alpha > 0$ (resp. without the damping effect for $\alpha = 0$). When $\alpha = 0$ the initial value problem (1), (2) was studied by Kanel'. Ya [2]. The full compressible Navier-Stokes equations, in tree space dimensions, was studied by Matsumura and Nishida [3]. These authors have successfully applied the "energy method" to obtain the existence of a small solution defined for $t > 0$. However, in order for their method to work, it is necessary to assume that the initial values belong to some Sobolev space H^s with $s > 0$. This forces the initial values to be small at infinity and as a result it excludes some very natural choices. Thus one cannot even consider simple "Riemann problem" data, i.e. piece-wise constant data of the form

$$(u_0(x),\ v_0(x)) = \begin{cases} (u^-,\ v^-) & , \quad x < 0 \\ (u^+,\ v^+) & , \quad x > 0 \end{cases} .$$

Furthermore membership in H^s excludes taking data near a travelling wave which approximates a shock wave. Recently Nishida and Smoller [4] proved the existence of the global solution for the initial value problem (1), (2) with $\alpha = 0$ via the method of finite differences provided that the initial values have bounded total variation. Ding Xiaxi and Wang Jinghua [1] obtained the global solution for the initial value problem (1), (2) with $\alpha \geq 0$. We use the method of successive approximations and the maximum principle for weakly coupled parabolic systems. The assumption for initial values is that the initial values and their first and second derivatives are bounded.

Here we give a report on the refinement and extension of the above-mentioned work by Ding Xiaxi and myself. We prove the existence of the global solution for the initial value problem (1), (2) with $\alpha \geq 0$ provided only that the initial values are bounded and measurable, i.e.,

$$|\gamma_{i,0}(x)| \leq M \quad , \quad v_0(x) \leq M \quad , \quad x \in R \quad , \tag{5}$$

where $M > 0$ is a constant and $\gamma_{i,0}(x) = \int_{v_0(x)}^{c} \sqrt{-p'(\xi)}\, d\xi + (-1)^i u_0(x)$.

From (4) it is easy to know that the assumption (5) implies

$$|u_0(x)| \le M \quad , \quad m \le v_0(x) \le M \quad , \quad x \in R \quad , \tag{6}$$

where m is a positive number satisfying the equality $\int_m^c \sqrt{-p'(\xi)}d\xi = M$.

We also study the qualitative properties for the global solution.

For simplicity of notation we first deal with the system

$$\frac{\partial u}{\partial t} + \frac{\partial p(v)}{\partial x} + \alpha u = \frac{\partial^2 u}{\partial x^2}$$

$$\frac{\partial v}{\partial t} - \frac{\partial u}{\partial x} = \frac{\partial^2 u}{\partial x^2} \quad . \tag{1)'$$

To begin with we consider the system of integral equations

$$u(x, t) = u^0(x, t) + \int_0^t d\tau \int_{-\infty}^{+\infty} (-\alpha\Gamma(x-y), t-\tau)u(y, \tau)$$

$$+ \frac{\partial\Gamma(x-y, t-\tau)}{\partial y} p(v(y, \tau)))dy \quad ,$$

$$v(x, t) = v^0(x, t) - \int_0^t d\tau \int_{-\infty}^{+\infty} \frac{\partial\Gamma(x-y, t-\tau)}{\partial y} u(y, \tau)dy \quad , \tag{7}$$

where

$$(u^0(x, t), v^0(x, t)) = \int_{-\infty}^{+\infty} \Gamma(x-y, t)(u_0(y), v_0(y))dy$$

and

$$\Gamma(x-y, t-\tau) = \frac{1}{2\sqrt{\pi(t-\tau)}} e^{-\frac{(x-y)^2}{4(t-\tau)}} \quad .$$

By the method of successive approximations, we obtain the following local existence theorem.

Lemma 1. There exists a continuous solution $(u(x, t), v(x, t))$ for the system of integral equations (7) in the strip $\pi_1 = \{(x, t) | x \in R, 0 < t \leq t_1\}$ such that

$$\frac{m}{2} \leq v(x, t) \leq 2M \quad , \quad |u(x, t)| \leq \sqrt{2}\, M \quad , \quad (x, t) \in \pi_1 \tag{8}$$

where

$$t_1 = \min\left(\frac{\pi}{4}, \frac{\pi m^2}{16M^2}, \frac{\pi M^2}{64p(\frac{m}{2})^2}, \frac{1}{4\alpha}\right) \cdot \frac{1}{2} \quad .$$

Now we are going to prove that the solution of the system of integral equations (7) is the classical solution of the initial value problem (1)', (2) in π_1.

Theorem 1. There exists a classical solution $(u(x, t), v(x, t))$ for the initial value problem (1)', (2) in π_1 such that the following estimates hold

$$\frac{m}{2} \leq v(x, t) \leq 2M \quad , \quad |u(x, t)| \leq \sqrt{2}\, M \quad , \quad (x, t) \in \pi_1 \tag{9}$$

Outline of the proof: It is sufficient to prove that $(u(x, t), v(x, t))$, the solution of the system of the integral equations (7) obtained in Lemma 1, has continuous derivatives with respect to x up to two and continuous derivative with respect to t in π_1. First we obtain the following estimates for $0 < \beta < 1$:

$$\max\left(\frac{|\Delta_h^x u(x, t)|}{|h|^\beta}, \frac{\Delta_h^x v(x, t)}{|h|^\beta}\right) \leq C(M,m,t_1,\beta)t^{-\frac{\beta}{2}} \quad , \quad 0 < t \leq t_1$$

$$\max\left(\frac{|\Delta_h^t u(x, t)|}{|h|^\beta}, \frac{\Delta_h^t v(x, t)}{|h|^\beta}\right) \leq C(M,m,t_1,\beta)t^{-\beta} \quad , \quad 0 < t \leq t_1$$

(10)

where $\Delta_h^x u(x, t) = u(x+h, t) - u(x, t)$, $\Delta_h^t u(x, t) = u(x, t+h) - u(x, t), \ldots$, $C(M,m,t_1,\beta)$ is a constant depending on M, m, t and β. Then it follows from (10) that $(u(x, t), v(x, t))$ has continuous derivative with respect to x and the estimate

$$\sup_x |u_x(x, t)| + \sup_x |v_x(x, t)| \leq C(M,m,t_1) t^{\frac{1}{2}} \quad , \qquad t \leq t_1 \quad . \tag{11}$$

By simple calculation we have

$$\max\left(\frac{|\Delta_h^x u_x(x, t)|}{|h|^\beta}, \frac{|\Delta_h^x v_x(x, t)|}{|h|^\beta}\right) \leq C(M,m,t_1,\beta) t^{-\frac{1+\beta}{2}} \quad , \quad (x, t) \in \pi_1$$

$$\max\left(\frac{\Delta_h^t u_x(x, t)}{|h|^{\beta/2}}, \frac{\Delta_h^t v_x(x, t)}{|h|^{\beta/2}}\right) \leq C(M,m,t_1,\beta) t^{-\frac{1+\beta}{2}} \tag{12}$$

By means of (11) and (12) we conclude that $(u(x, t), v(x, t))$ has second continuous derivative with respect to x and continuous derivative with respect to t.

$$\text{Set} \quad r_i(x, t) = \int_{v(x,t)}^{C} \sqrt{-p'(\xi)}\, d\xi + (-1)^i u(x, t) \quad ,$$

$i = 1, 2$ where $C > 0$

is constant. For the solution $(u(x, t), v(x, t))$ we have the following equalities,

$$L_1(r_1) + \frac{\alpha}{2} r_2 = \frac{\partial^2 r_1}{\partial x^2} - \frac{\partial r_1}{\partial t} + \sqrt{-p'(v)}\, \frac{\partial r_1}{\partial x} - \frac{\alpha}{2} r_1 + \frac{\alpha}{2} r_2$$

$$= \left(\frac{p''(v)}{2\sqrt{-p'(v)}} \frac{\partial v}{\partial x}\right)^2 \tag{13}$$

$$L_2(r_2) + \frac{\alpha}{2} r_i = \frac{\partial^2 r_2}{\partial x^2} - \frac{\partial r_2}{\partial t} - \sqrt{-p'(v)}\, \frac{\partial r_2}{\partial x} - \frac{\alpha}{2} r_2 + \frac{\alpha}{2} r_1$$

$$= \frac{p''(v)}{2\sqrt{-p'(v)}} \left(- \frac{\partial v}{\partial x}\right)^2 \tag{13}$$

$$r_1(x, 0) = r_{1,0}(x) = \int_{v_0(x)}^{c} \sqrt{-p'(\xi)}\, d\xi - u_0(x)$$

$$r_2(x, 0) = r_{2,0}(x) = \int_{v_0(x)}^{c} \sqrt{-p'(\xi)}\, d\xi + u_0(x) \tag{14}$$

In order to obtain the global solution we need the following Lemma related to maximum principle.

Lemma 2. For $r_1(x, t)$, $r_2(x, t)$ the following estimate holds

$$\lambda \leq \Lambda \quad , \tag{15}$$

where

$$\lambda = \max\left(\sup_{\Pi_1} r_1(x, t) \cdot \sup_{\Pi_1} r_2(x, t)\right)$$

and

$$\Lambda = \max\left(\sup_{x \in R} r_{1,0}(x),\ \sup_{x \in R} r_{2,0}(x)\right) \quad .$$

Outline of proof: Set $\tilde{L}_i(z) = \frac{\partial^2 z}{\partial x^2}\ \frac{\partial z}{\partial t} + (-1)^{i+1}\sqrt{-p'(v)}\ \frac{\partial z}{\partial x} - \frac{\alpha}{2} z$

where $v = v(x, t)$ is the second component of $(u(x, t), v(x, t))$. If (15) were not true, i.e. $\lambda > \Lambda$, let (x_0, t_0) be any point in Π_1. For any given $\delta > 0$, set $R = |x_0| + \frac{\sqrt{(\lambda - \Lambda)\Phi(x_0, t_0)}}{\delta}$, where $\Phi(x, t) = x^2 + kt$. We have the following equality

$$L_i(\Phi) + \frac{\alpha}{2} \Phi \leq \frac{k}{2} < 0 \quad , \quad i = 1, 2 \quad , \quad |x| \leq R \tag{16}$$

provided that the constant $k(m, \lambda-\Lambda, x_0, t_0)$ is sufficiently large. Set $Q_i(x, t) = \frac{\delta\Phi(x, t)}{\Phi(x_0, t_0)} + \Lambda - r_i(x, t)$, $i = 1, 2$, then by (13) and (16) we obtain

$$\tilde{L}_1(Q_1) + \frac{\alpha}{2} Q_2 = \frac{\delta}{\Phi(x_0, t_0)} \left(\tilde{L}_1(\Phi) + \frac{\alpha\Phi}{2}\right) - \left(L_1(v_1) + \frac{\alpha r_2}{2} \right) < 0$$

$$\tilde{L}_2(Q_2) + \frac{\alpha}{2} Q_1 = \frac{\delta}{\Phi(x_0, t_0)} \left(\tilde{L}_2(\Phi) + \frac{\alpha\Phi}{2}\right) - \left(L_2(r_2) + \frac{\alpha r_1}{2} \right) < 0 \tag{17}$$

Here the fact that $p''(v) \cdot \frac{(v_x)^2}{2\sqrt{-p'(v)}} \geq 0$ has been used. It is easy to check that

$$Q_i(x, 0) \geq \Lambda - r_{i,0}(x) \geq 0 \quad ,$$

$$Q_i(\pm R, t) \geq \frac{\delta R^2}{\Phi(x_0, t_0)} + \Lambda - r_i(\pm R, t) \geq \lambda - r(\pm R, t) \geq 0 \quad ,$$

$$i = 1, 2 \quad . \tag{18}$$

For $\alpha \geq 0$, the maximum principle for weakly coupled parabolic system, theorem 13, Chap. 3, in [5] is applicable to the system of parabolic inequalities (17) with initial boundary condition (18) and we obtain

$$Q_i(x, t) \geq 0 \quad , \quad (x, t) \in \Pi_1 \quad , \quad i = 1, 2 \quad . \tag{19}$$

Particularly we have $Q_i(x_0, t_0) \geq 0$, $i = 1, 2$, i.e. $r_i(x_0, t_0) \leq \delta + \Lambda$, $i = 1,2$. Because of the arbitrariness of the point $(x_0, t_0) \in \Pi_1$ and the positive constant δ, we conclude that

$$r_i(x, t) \leq \Lambda \quad , \quad (x, t) \in \Pi_1 \quad , \quad i = 1, 2 \quad . \tag{20}$$

There is a contradiction.

<u>Theorem 2</u>. (Global existence). For any given $t > 0$ there exists

a solution $(u(x, t), v(x, t))$ of (1)', (2) such that the following estimate holds

$$m \leq v(x, t) \quad , \quad x \in R \quad , \quad 0 \leq t \leq T \quad . \tag{21}$$

Outline of proof: It follows from the assumption (5) and Lemma 2 that

$$\sup_{\Pi_1} r_i(x, t) \leq M \quad , \quad i = 1, 2 \quad . \tag{22}$$

Therefore $\int_{v(x,t)}^{c} \sqrt{-p'(\xi)d\xi} = \frac{r_1(x, t) + r_2(x, t)}{2} \leq M$, $(x, t) \in \Pi_1$, which implies

$$m \leq v(x, t) \quad , \quad (x, t) \in \Pi_1 \quad . \tag{23}$$

In view of Theorem 1 we obtain a classical solution $(u(x, t), v(x, t))$ for the initial value problem (1)', (2) in Π_1 such that

$$m \leq v(x, t) \leq 2M \quad , \quad |u(x, t)| \leq \sqrt{2}\, M \quad , \quad (x, t) \in \Pi_1 \tag{24}$$

$$\max \left(\sup_{\Pi_1} r_1(x, t), \sup_{\Pi_2} r_2(x, t)\right) \leq M \quad . \tag{25}$$

Suppose we obtain the classical solution $(u(x, t), v(x, t))$ for the initial value problem (1)', (2) in the strip $\Pi_1 U \Pi_2 U \ldots U \Pi_{n-1}$, where $\Pi_i = (x, t) | x \in R, t_{i-1} < t \leq t_i$, $t_0 = 0$ and

$$t_i - t_{i-1} < \min\left(\frac{\pi}{4}, \frac{\pi m^2}{16M^2}, \frac{\pi M^2}{64p(\frac{m}{2})^2}, \frac{1}{4\alpha}\right) \frac{1}{(i+1)} \quad ,$$

such that

$$m \leq v(x, t) \leq nM \quad , \quad |u(x, t)| \leq \sqrt{n}\, M \quad , \quad t \leq t_{n-1} \tag{26}$$

$$\max_{i=1,2} \left(\sup_{\Pi_1 U \ldots U \Pi_{n-1}} r_i(x, t)\right) \leq M \quad . \tag{27}$$

In a similar way, to obtain the classical solution for the initial value problem (1)', (2) and estimates (24), (25) in Π_1, we can show that there exists a classical solution $(u(x, t), v(x, t))$ in Π_n for the system (1)' with initial value $(u(x, t_{n-1}), v(x, t_{n-1}))$ at $t = t_{n-1}$ such that

$$m \le v(x, t) \le (n+1)M \quad , \quad |u(x, t)| \le \sqrt{n+1}\, M \quad ,$$

$$(x, t) \in \Pi_n \quad , \tag{28}$$

$$\max_{i=1,2} \left(\sup_{\Pi_1 \cup \ldots \cup \Pi_n} r_i(x, t) \right) \le M \tag{29}$$

where

$$\Pi_n = \{(x, t) | x \quad R, t_{n-1} < t \le t_n\}$$

and

$$t_n - t_{n-1} = \min\left(\frac{\pi}{4}, \frac{\pi m^2}{16M^2}, \frac{\pi M^2}{64p(\frac{m}{2})^2}, \frac{1}{4\alpha} \right) \frac{1}{n+1} \quad .$$

Noting that $t_n = (t_n - t_{n-1}) + (t_{n-1} - t_{n-2}) + \ldots + t_1 = \min \left(\frac{\pi}{4}, \frac{\pi m^2}{16M^2}, \frac{\pi M^2}{64p(\frac{m}{2})^2}, \frac{1}{4\alpha} \right) \sum_{i=1}^{n} \frac{1}{i+1}$, we may find an integer k such that

$$T \le t_k$$

and

$$m \le v(x, t) \le (k+1)M \quad , \quad |u(x, t)| \le \sqrt{k+1}\, M \quad ,$$

$$x \in R \quad , \quad t \le T \quad . \tag{30}$$

Now we will study the asymptotic behavior of $(u(x, t), v(x, t))$, the solution for the initial value problem (1)', (2), when $(u_0(v), v_0(x))$ the initial value (2) approaches (u^-, v^-) (resp. (u^+, v^+)) as x tends to $-\infty$ (resp. $+\infty$).

<u>Theorem 3</u>. For any given $T > 0$ and given $\delta > 0$ there exists $N > 0$ such that the following estimates holds,

$$v(x, t) \geq v^- + \delta \quad , \quad x < -N \quad , \quad 0 \leq t \leq T \quad , \tag{31}$$

$$v(x, t) \geq v^+ + \delta \quad , \quad x > N \quad , \quad 0 \leq t \leq T \quad . \tag{32}$$

<u>Outline of proof</u>: Set $w_i(x, t) = r_i(x, t) - r_i^- - (Ke^{x+\gamma t} + \frac{\sqrt{-p'(v^-)}}{2}\delta)$ $i = 1, 2$, we have

$$\tilde{L}_1(w_1) + \frac{\alpha}{2} w_2 = L_1(r_1) + \frac{\alpha}{2}(r_1^- - r_2^-) - Ke^{x+\gamma t}(1 - \gamma + \sqrt{-p'(v)}) \geq \frac{\alpha}{2}$$

$$\times (r_1^- - r_2^-) - Ke^{x+\gamma t}(1 - \gamma + \sqrt{-p'(m)}) > 0$$

$$\tilde{L}_2(w_2) + \frac{\alpha}{2} w_1 > 0 \quad . \tag{33}$$

$$w_1(x, 0) = r_{1,0}(x) - r_1^- - (Ke^x + \delta \cdot \frac{\sqrt{-p'(v^-)}}{2}) \leq 0$$

$$w_2(x, 0) = r_{2,0}(x) - r_2^- - (Ke^x + \delta \cdot \frac{\sqrt{-p'(v^-)}}{2}) \leq 0 \tag{34}$$

provided that K and γ are sufficiently large.

In a similar way to Lemma 2, we obtain

$$r_i(x, t) - r_i^- - (Ke^{x+\gamma t} + \frac{\sqrt{-p'(v^-)}}{2}\delta) < 0 \quad , \quad i = 1,2 \quad ,$$

$$x \in R \quad , \quad 0 \leq t \quad . \tag{35}$$

Similarly we obtain

$$r_i(x, t) - r_i^+ - (Ke^{-x+\gamma t} + \frac{\sqrt{-p'(v^-)}}{2}\delta) < 0 \quad , \quad i = 1, 2 \quad ,$$

$$x \in R \quad , \quad 0 < t < T \quad . \tag{36}$$

It follows from (35) and (36) that there exists a constant $N > 0$ such that

$$r_i(x, t) \leq r_i^- + \sqrt{-p'(v^-)}\delta \quad , \quad i = 1,2 \quad , \quad -N > x \quad , \quad 0 \leq t \leq T \; , \tag{37}$$

$$r_i(x,t) \le r_i^+ + \sqrt{-p'(v^+)}\delta \quad , \quad i = 1,2 \quad , \quad x > N \quad , \quad 0 \le t \le T . \tag{38}$$

We obtain from (37) and (38) that

$$\int_{v(x,t)}^{v^-} \sqrt{p'(\xi)}d\xi \le \sqrt{-p'(v^-)}\delta \quad , \quad x < -N \quad , \quad 0 \le t \le T \quad , \tag{39}$$

$$\int_{v(x,t)}^{v^+} \sqrt{-p'(\xi)}d\xi \le \sqrt{-p'(v^+)}\delta \quad , \quad x > N \quad , \quad 0 \le t \le T \quad . \tag{40}$$

The inequalities (39), (40) and (30) imply (31) and (32).

One of the most interesting constitutive relation between pressure p and specific volume v is $p(v) = v^{-\gamma}$, $\gamma > 1$. In this case we have

Theorem 4. For any given $T > 0$ and $\delta > 0$ there exists $N > 0$ such that the following estimates hold

$$u^- - \frac{2\sqrt{\gamma}}{\gamma - 1}(v^-)^{\frac{1-\gamma}{2}} - \delta \le u(x,t) \le u^- + \frac{2\sqrt{\gamma}}{\gamma - 1}(v^-)^{\frac{1-\gamma}{2}} + \delta \quad ,$$

$$x < -N \quad , \quad 0 \le t \le T \tag{41}$$

$$u^+ - \frac{2\sqrt{\gamma}}{\gamma - 1}(v^-)^{\frac{1-\gamma}{2}} - \delta \le u(x,t) \le u^- + \frac{2\sqrt{\gamma}}{\gamma - 1}(v^-)^{\frac{1-\gamma}{2}} + \delta \quad ,$$

$$x > N \quad , \quad 0 \le t \le T \quad . \tag{42}$$

<u>Outline of proof</u>: Similar to Theorem 3, we can obtain

$$r_i(x,t) \le r_i^- + \delta \quad , \quad i = 1, 2 \quad , \quad -N > x \quad , \quad 0 \le t \le T \tag{43}$$

which imply

$$\int^{u(x,t)} \le u^- + \int_{v^-}^{v(x,t)} \sqrt{-p'(\xi)}\, d\xi + \delta \le u^- + \frac{2\sqrt{\gamma}}{\gamma - 1}(v^-)^{\frac{1-\gamma}{2}} + \delta \quad ,$$

$$-u(x,t) \le -u^- + \frac{2\sqrt{\gamma}}{\gamma - 1}(v^-)^{\frac{1-\gamma}{2}} + \delta \quad , \quad -N > x \quad , \quad 0 \le t \le T \quad . \tag{44}$$

This is just (41), similarly we may obtain (42).

<u>Concluding remark:</u> In a similar way, we can obtain the existence for the global solution $(u_\varepsilon(x, t), v_\varepsilon(x, t))$ to the initial value problem (1), (2) and for any given $T > 0$ the following estimate holds

$$m \le v_\varepsilon(x, t) \quad , \quad x - R \quad , \quad 0 > t > T \tag{45}$$

and for any given $\delta > 0$, there exists N such that

$$v_\varepsilon(x, t) \ge v^- + \delta \quad , \quad x < -N \quad , \quad 0 \le t \le T \tag{46}$$

$$v_\varepsilon(x, t) \ge v^+ + \delta \quad , \quad x > N \quad , \quad 0 \le t \le T \quad . \tag{47}$$

When $P(v) = v^{-\gamma}$, $\gamma > 1$ we have

$$|u_\varepsilon(x, t)| < M + \frac{2\sqrt{\gamma}}{\gamma - 1} c^{\frac{1-\gamma}{2}} \quad , \quad x \in R \quad , \quad 0 \le t \le T \tag{48}$$

$$u^- - \frac{2\sqrt{\gamma}}{\gamma - 1} (v^-)^{\frac{\gamma - 1}{2}} - \delta \le u_\varepsilon(x, t) \le u^- + \frac{2\sqrt{\gamma}}{\gamma - 1} (v^-)^{\frac{1-\gamma}{2}} + \delta \quad ,$$

$$x < -N \quad , \quad 0 \le t \le T \tag{49}$$

$$u^+ - \frac{2\sqrt{\gamma}}{\gamma - 1} (v^-)^{\frac{\gamma - 1}{2}} - \delta \le u_\varepsilon(x, t) \le u^+ + \frac{2\sqrt{\gamma}}{\gamma - 1} (v^+)^{\frac{1-\gamma}{2}} + \delta \quad ,$$

$$x > N \quad , \quad 0 \le t \le T \quad . \tag{50}$$

There is a well-known conjecture that the solution $(u_0(x, t), v_0(x, t))$ for the quasilinear hyperbolic system

$$\left.\begin{aligned} \frac{\partial u}{\partial t} + \frac{\partial p(v)}{\partial x} + \alpha u = 0 \\ \frac{\partial v}{\partial t} - \frac{\partial u}{\partial x} = 0 \end{aligned}\right\} \tag{1)"$$

with initial value (2) could be obtained as the limit of the solution $(u_\varepsilon(x, t), v_\varepsilon(x, t))$ as $\varepsilon \to +0$. If this conjecture is true, the inequalities (45)-(50) could be used to obtain the asymptotic behavior for the solution $(u_0(x, t), v_0(x, t))$ as x tends to infinity since these inequalities hold uniformly for ε.

References

[1] Ding Xiaxi and Wang Jinghua, Global solutions for a semi-linear parabolic system, Acta Mathematica Scientia, 3 (1983) 397-414.

[2] Kanel'. Ya, On some systems of quasilinear parabolic equations. Zh. Vychislit, Matem. i Matem Fiziki, 6 (1966) 446-477.

[3] Matsumura, A. and Nishida, T., The initial value problem for the equations of motion of viscous and heat conductive gases, J. Math. Kyoto Univ., 20 (1980) 67-104.

[4] Nishida, T. and Smoller, J., A class of convergent finite difference schemes for certain nonlinear parabolic systems, Comm. Pure Appl. Math., 36 (1983) 785-808.

[5] M.H. Protter and M.F. Weinberger, Maximum principles in differential equations, Prentice-Hall, Inc., 1967.

ON THE DISTRIBUTED PARAMETER LARGE-SCALE SYSTEMS IN MODERN APPLIED MECHANICS

WANG ZHAO-LIN

Abstract

By the requirements of development in the modern scientific and technological circles, the research work, relating to the large-scale systems with their complex structures and various functions, has received the greatest amount of attention all over the world; recently, the scientific research for the dynamics of distributed parameter large-scale systems is significant.

This paper dealt mainly with three problems: firstly the flexible large-scale systems and their deployment dynamics; secondly, the large-scale systems with the partially liquid-filled rigid, flexible cavities under strong capillary and weak gravity conditions; thirdly, as the extension of their theoretical research, the dynamics of distributed parameter large-scale systems in the functional space.

1. The Flexible Large-scale Systems and their Deployment Dynamics[1-5,21,22,25]

1.1 Description of the systems

The differential equations of motion of the flexible large-scale systems (for instance, a large-spacecraft with flexible appendages) can be derived from the principle of Hamilton-Ostragradskii,

$$\int_{t_0}^{t_1} (\delta T + \delta W)dt = 0 \tag{1}$$

where T is the kinetic energy of systems, δT is the first variation of T, δW is the virtual displacement work of active force, t is the time; by the variation principle (1) Lagrangian equation can be written and then utilizing the method of assumed modes, the Lagrange equations governing the discrete large-scale systems can be given by

$$\frac{d}{dt}\frac{\partial T}{\partial \dot{q}} - \frac{\partial T}{\partial q} = \frac{\partial U}{\partial q} - \frac{\partial R}{\partial \dot{q}} + S \tag{2}$$

where $\dot{q} = (\dot{q}_1, \ldots, \dot{q}_n)^\tau$ is the vector in the Euclidean coordinate space, $R = \frac{1}{2}\dot{q}^\tau D\dot{q} + \dot{q}^\tau Fq$ is the damping function, D is the positive definite or positive semi-definite damping matrix, $F = -F^\tau$ is the anti-symmetric constraint damping matrix, S is the gyroscopic force, U is the force function.

Using linear transformation, system (2) can be described by the differential equations,

$$I\ddot{x} + (B + G)\dot{x} + (K + \Lambda)x + \Phi(x) = 0 \tag{3}$$

where $x = (x_1, \ldots, x_n)^\tau$ is the n-dimensional state vector, I is the unit matrix, B is the positive definite or positive semi-definite damping matrix, $G = -G^\tau$ is the anti-symmetric gyroscopic matrix, $K = -K^\tau$ is the anti-symmetric constraint damping matrix,

$$\Lambda = \begin{bmatrix} \lambda_1 & & 0 \\ & \ddots & \\ 0 & & \lambda_n \end{bmatrix}$$

is the diagonal potential matrix, λ_i $(i = 1, \ldots, n)$ is called the stable coefficient by Poincaré, $\Phi(x)$ is the nonlinear vector function. The linear system equations can be written as

$$I\ddot{x} + (B + G)\dot{x} + (K + \Lambda)x = 0 \tag{4}$$

1.2 Method of large-scale systems with weighting V function and the stability of motion of systems

If the trivial solution $x=0$ of (3) or (4) is considered as an undisturbed motion of systems, then these equations are called the disturbed differential equations of systems. The stability of null solutions can be researched. The way to get over the difficulty of building a Lyapunov function for a large-scale system may be described as follows:

One considers the system as constituted by the interconnection into a complex whole of several smaller subsystems. The complete system will be called composite system (or large-scale system); on the base of building V functions for subsystems the stability of motion of the large-scale system is simpler to study.

By multi-transformation the large-scale system (4) is written in the composite system of two subsystems Z_1 and Z_2:

$$\begin{bmatrix} \dot{Z}_1 \\ \hline \dot{Z}_2 \end{bmatrix} = \left[\begin{array}{c|c} -C-\beta I & I \\ \hline E-\beta H & \beta H + N \end{array}\right] \begin{bmatrix} Z_1 \\ \hline Z_2 \end{bmatrix} \qquad (5)$$

where

$$E = (B+G)C - C^2 - (K+\Lambda) \quad , \qquad N = C - (B+G) \quad ,$$

$$H = N + C + \beta I \quad , \qquad C \text{ is the chosen matrix,}$$

$$\beta \quad \text{is a smaller number} \quad .$$

Suppose there exist the matrix inequalities:

$$\begin{cases} P > 0 \quad , \quad Q < 0 \quad , \quad C^{\tau}P + PC > 0 \ (C^{\tau}P + PC = 0) \\ (P+P^{\tau}) + E^{\tau}(Q+Q^{\tau}) = 0 \\ N^{\tau}Q + QN < C \end{cases} \qquad (6)$$

where P and Q are the chosen matrices, then the following theorems can be obtained:

Theorem 1. If the matrix inequalities (6) are valid for large-scale system (4), then the undisturbed motion $x = 0$ is asymptotically stable.

Theorem 2. Suppose that

$$\frac{\|\Phi(x)\|}{\|x\|} \to 0 \quad ,$$

as $\|x\| \to 0$; then the null solution $x = 0$ of large-scale system (3) is asymptotically stable.

Theorem 3. Assume that $B = 0$, $K = 0$, $\lambda_k = 0$ $(K = 1,\ldots,s < n)$ for large-scale system (3) with the following properties: (a) $F_1(0) < 0$, as s is the even number; or (b) $F_2(0) < 0$, as s is the odd number; then the trivial solution $x = 0$ of the unisolated mechanical system with gyroscopic terms is unstable, where

$$F_1(\lambda) = \frac{|I\lambda^2 + G\lambda + \Lambda|\lambda_{k=0}}{\lambda^s} \quad , \qquad F_2(0) = F_1'(0) \quad .$$

In the process of proving the above theorems the paper presents such a method that can be applied to analysis of the stability of non-linear mechanical systems, based on Chetaev's theory and the theory of the large-scale systems: method of large-scale systems with weighting V function; moreover, this method is used for the research on some problems of attitude stability of dual-spinning satellite, the large-scale flexible spacecraft and the satellite with liquid-filled cavities. It can be seen that this method may be applied effectively to study the stability of the mechanical systems with constraint damping, the gyro-systems and the complex systems of spacecraft. In this paper, therefore, the Kelvin-Tait-Chetaev theorem and the Mingor's theorem used in the satellite attitude dynamics are extended.

1.3 Deployment dynamics of the flexible large-scale systems

Consider a satellite in a circular orbit, which consists of a central rigid body deploying two panels (solar arrays) nominally perpendicular to the orbital plane. Two simple extension procedures are considered here: Uniform deployment rate, i.e. $\dot{l}_i = V_i$, clearly, $l_i = l_{i0} + V_i t$, where l_{i0} is the initial to the ith panel; exponential deployment rate, i.e., $\dot{l}_i = V_{i0}\exp(-kt)$, i = 1,2. By the variational principle of H-O the Lagrangian equations of motion for the discrete large-scale systems can be written as

$$\frac{d}{dt}\frac{\partial T}{\partial \dot{q}} - \frac{\partial T}{\partial q} + \frac{(U + V_0)}{q} = Q \tag{7}$$

where $q = (R_c, \theta, \alpha, \beta, \gamma, q\mu\nu_i)^{\tau}$, $\mu = 1,\ldots,m$, $\nu_1 = 1,\ldots,n$, $i = 1,2$; let the position vector $\bar{R}_c$ and the true anomaly θ define the location of the instantaneous center of mass C of the spacecraft with respect to the inertial coordinate system having its origin at the center of the Earth E; α, β and γ are the attitude angles of satellite; $q\mu\nu_i$ is the coefficient in the series of the vibrational displacement of the element of the plane; the kinetic energy of the large-scale system $T = T_{orb} + \frac{1}{2}\omega \cdot J \cdot \omega + T_d$, where T_{orb} is the orbital kinetic energy, ω is the instantaneous angular velocity, J is the inertial tensor of the large-scale system, T_d is the function depending on the vibrational displacement, velocities, deployment rate and mass distribution of spacecraft; U is the gravitational potential energy of the satellite, V_0 is the strain energy stored in the planes due to their vibrations, Q is the generalized force.

Theoretical and experimental studies represent that librational response and vibrations tend to increase during deployment. Compared to uniform extension, the exponential extensions considered seems to have more adverse influence on librations but a favorable effect on vibrations.

2. The Large-scale System with Liquid-filled Cavities and Its Sub-gravity Dynamics[1-10,12-20]

2.1 Description of the system

Consider the cavities and the set of liquids as a large-scale system, the Lagrange equations of the system and the Euler equations of fluid can be derived from the principle of H-O:

$$\frac{d}{dt}\frac{\partial T}{\partial \dot{q}_j} - \frac{\partial T}{\partial q_j} = Q_j \qquad (j = 1,2,\ldots,n \le 6) \tag{8}$$

$$\frac{dV}{dt} + \omega X V = F - \frac{1}{\rho}\nabla p \tag{9}$$

$$\text{div } V = 0 \tag{10}$$

where T is the kinetic energy of the large-scale system, q_j is the generalized coordinate, Q_j is the generalized force, p is the pressure of liquid, ρ is the density of liquid, F is the force acting on the element of liquid, $V(t, x, y, z)$ is the absolute velocity vector of the element of liquid, $\frac{dV}{dt} = \frac{\partial V}{\partial t} + (U \cdot \nabla)V$, U is the relative velocity of liquid to the cavities, ω is the instantaneous angular velocity of cavity. Consider the kinetic boundary conditions:

$$U_n = U \cdot n = 0 \tag{11}$$

$$\frac{\partial \Phi}{\partial t} + U \cdot \nabla\Phi = 0 \tag{12}$$

where n is the unit vector of external normal, the equation of free surface of liquid is

$$\Phi(t, x, y, z) = 0 \quad . \tag{13}$$

The Helmholtz equation of the rotating fluids can be written as

$$\frac{d\Omega}{dt} + \omega X \Omega = (\Omega \cdot \nabla)V$$

$$\Omega = \text{rot } V \quad . \tag{14}$$

The Navier-Stokes equation of the viscous fluids is

$$\frac{dV}{dt} + \omega x V = F - \frac{1}{\rho}p + \nu\Delta V \tag{15}$$

where $\nu = \frac{\mu}{\rho}$ is the viscous kinematic coefficient; and here the boundary condition (11) is transformed into

$$U = 0 \quad . \tag{16}$$

2.2 The sub-gravity dynamics of the liquid-filled large-scale system

2.2.1 The capillary force

Along the contact surface between two different mediums i and j the surface tension has been obtained from the Gauss principle: it is a potential force where the potential energy $E = \alpha_{12}S_{12} + \alpha_{13}S_{13} + \alpha_{23}S_{23}$, 1, 2, 3 indicates the rigid-body, liquid, gas, respectively; S_{ij} (i, j = 1, 2, 3, i $\neq$ j) is the area of the contact surface and α_{ij} $(\alpha_{ij} = \alpha_{ji})$ is the coefficient of the suitable surface tension.

2.2.2 Boundary conditions

If the capillary force is considered, the principle of H - O indicates that the formulation of the large-scale system has not been variated; therefore, Eqs. (8) and (9), constraint condition (10) and the kinematic boundary conditions (11) and (12) exist. The additional dynamical boundary condition, furthermore, can be derived from the H - O principle:

$$p - p_g = \alpha K \tag{17}$$

where K is the mean curvature of the free-surface of liquid, $\alpha = \alpha_{23}$, p_g denotes the gas pressure on the free-surface of liquids; by the H - O principle the boundary condition of contact angle remaining constant can also be obtained:

$$\cos\theta = \frac{\alpha_{12} - \alpha_{13}}{\alpha} \quad . \tag{18}$$

For the viscous fluid the condition (17) is transformed into

$$p_n = -(p_g + \alpha K)n \quad . \tag{19}$$

Therefore, under the weightless or subgravity conditions the large-scale system with the viscous fluid-filled cavities can be researched by Eqs. (8), (10), (12), (15), (16), (18) and (19).

2.2.3 The liquid-filled linear large-scale system under the low-gravity conditions

(a) The stability of the principal vibrations of liquid in the cavities.

In order to obtain a stable large-scale system, not only the principal vibrations of liquid remain stable, i.e. the principal vibrations are bounded, but also the possible couplings of frequencies between the control system, the structural frame, and the sloshing fluid have never been occurred.

From the linearized equations (8)-(13), (71) and (18) the operator equation can be obtained:

$$\mu^2 H\psi = A\psi \tag{20}$$

It is well known that in Hilbert space, if operator A is self-conjugate and definite-positive, operator H is self-conjugate, completely continuous and positive, then the spectrum of problem (20) is discrete and has the finitely repeated number and a singular limit-point $\mu = \infty$, i.e., the infinite set of characteristic frequencies $\mu_1, \mu_2, \ldots, \mu_n$ ($\mu_n \to \infty$, as $n \to \infty$) exists; the number of the possible forms of the principal vibrations, suitable for every characteristic frequency, is finite; all the spectrum points are situated at the positive part of the real-axis, i.e., every principal vibration is stable; the eigen-functions construct a completion system in the norm of Friedrichs; by the method of Ritz the principal vibrations and the frequencies can be obtained.

The above analysis is valid not only for the free-oscillations of liquid contained in the fixed cavity but also for the force-oscillations of liquid contained in the moving cavity, so long as the latter part has the additional boundary conditions in displacements and in rotations.

As an example, the mechanical character of sloshing of liquids in a spherical tank under low-gravitational conditions has been investigated. After determination of the equilibrium shape of the free-surface of liquids, using a method concerning a kind of eigen-functions expanded into simple polynomial expressions, the free and forced sloshing problems have been solved. The computer results for sloshing frequencies of liquids, dynamical coefficients and damping ratio have been obtained.

(b) The stability of linear large-scale system

From the Cauchy-Lagrange integral the dynamical pressure of liquids p can be derived, therefore, the principal vector and the principal moment of the reaction force can also be obtained. Utilizing the eigen-equation of the differential equations for the cavity, the stability of the linear large-scale system can be researched.

(c) The damping problem of viscous liquid

The experimental work and the theoretical research represent that the sloshing damping of viscous liquid tends to increase during decrease of the Bond number $(N_{BO} = \frac{\rho g l}{\alpha})$.

2.2.4 <u>The liquid-filled non-linear large-scale system under sub-gravity conditions</u>

(a) Large-scale system with liquid-filled rigid cavities

Utilizing the qualitative theory of Lyapunov-Rumyantsev, the stability of non-linear large-scale system can be considered.

<u>Theorem 4</u>. Suppose that $F_1 = V_1 + V_2$ holds the isolated minimum in the equilibrium state of the liquid-filled large-scale non-linear system, then the system is stable for the ideal or viscous liquid.

Where V_1 is the potential energy of external force, V_2 is the potential energy of the capillary force.

<u>Theorem 5.</u> If $F_2 = \frac{1}{2}\frac{K_0^2}{J} + V_1 + V_2$ holds the isolated minimum in the stationary rotation of the liquid-filled large-scale non-linear system, then the system is stable for the ideal or viscous liquid.

Where J is the moment of inertia, K_0 is the moment of momentum.

(b) Large-scale system with liquid-filled flexible cavities

Theorem 6. If $F_3 = \frac{1}{2}\frac{K_0^2}{J} + V_1 + V_2 + V_3(V_1 + V_2 + V_3)$ holds isolated minimum in the heat-insulating process, then the stationary rotation (equilibrium) is stable, where V_3 is the potential energy of flexible transfiguration.

For the isothermal process there exists a similar theory to the above.

As an example, the stability of the liquid-filled satellite, considered as a large-scale non-linear system under the low-gravity conditions, can be researched.

As another example, the Heartbeat system has been considered. In fact this system has such a complex structure with various functions that it constructs a distributed parameter large-scale system. In order to analyse some significant problems of the beating of the heart, the mathematical model of the heart has been investigated, i.e., the Heartbeat belongs to the type of a kind of Relaxation Oscillations and in which the free relaxation period of the system and the influence of the impressed period force on the relaxation oscillations have been researched under the non-linear conditions.

3. Large-scale Systems in Functional Space[7,11,23,24]

3.1 Linear large-scale system in functional space

Assume that a distributed parameter large-scale system consists of two subsystems containing the multi-rigid-body system and the liquid-filled system, then the integral-differential equations governing the large-scale systems are given by the H - O principle:

$$\ddot{y}_i + \rho\int_S \varphi_i\zeta_{tt}dS + \mu_i^2 y_i + \int_S \nu_i\zeta dS = 0 \quad (i = 1,\dots,n)$$

$$\rho\sum_{j=1}^{n}\varphi_j\ddot{y}_j + \rho H\zeta_{tt} + \sum_{j=1}^{n}\nu_j + y_j + \rho g\zeta = 0 \tag{21}$$

where y_i is the canonical variable of rigid cavities, ρ is the density of liquids, φ_i and ν_i are the linear functions depending on the Stokes-Rukovskii potential and the form of cavities; the equation of free-surface of liquids is $Z = \zeta(P, t)$, $P \in S$ (equilibrium free-surface of liquid), μ_i is the eigen-frequency of large-scale system, H is the integral operator.

The operator equation governing the large-scale system (21) in the functional space can be obtained. Assume that the direct product space is written as $\mathcal{E} = E_0 \times E$, where E_0 is the n-dimensional vector space with elements $y^{(1)}, y^{(2)} \in E_0$ and inner product $< y^{(1)}, y^{(2)} > = \sum_{i=1}^{n} y_i^{(1)} y_i^{(2)}$; E is the L^2-space with elements $\zeta^{(1)}(p), \zeta^{(2)}(p) \in E$ and inner product $< \zeta^{(1)}, \zeta^{(2)} > = \int_S \zeta^{(1)} \zeta^{(2)} dp$; and for elements $x^{(1)}, x^{(2)} \in \mathcal{E}$, the inner product $< x^{(1)}, x^{(2)} > = < y^{(1)}, y^{(2)} > + < \zeta^{(1)}, \zeta^{(2)} >$; consider the following operators:

(a) Acted in E_0 the unit operator $L_{00} = I$ and the operator

$$M_{00} = \begin{bmatrix} \mu_1^2 & & & 0 \\ & \cdot & & \\ & & \cdot & \\ 0 & & & \mu_n^2 \end{bmatrix}$$

(b) Operators, acted in E, $L_{11} = H\rho$, $M_{11} = \rho g$

(c) Operators, acted from E_0 to E,

$$L_{10} y = \rho < \varphi, y, > \quad ,$$

$$M_{10} y = < \nu, y, > \quad ,$$

where $y = (y_1, \ldots, y_n)$, $\varphi = (\varphi_1, \ldots, \varphi_n)$,

$\nu = (\nu_1, \ldots, \nu_n)$.

(d) Operators, acted from E to E_0,

$$L_{01}\zeta = \gamma \quad ,$$

$$M_{01}\zeta = \delta \quad ,$$

where

$$\gamma = (\rho\int_S \varphi_1 \zeta dS, \ldots, \rho\int_S \varphi_n \zeta dS) \quad ,$$

$$\delta = (\int_S \nu_1 \zeta dS, \ldots, \int_S \nu_n \zeta dS) \quad .$$

Therefore, large-scale system can be written as

$$L_{00}\ddot{y} + L_{01}\zeta_{tt} + M_{00}y = M_{01}\zeta = 0$$

$$L_{10}\ddot{y} + L_{11}\zeta_{tt} + M_{10}y + M_{11}\zeta = 0$$

or it can be simply described by the operator equation of the linear large-scale system in the functional space:

$$L\ddot{x} + Mx = 0 \tag{22}$$

where

$$X \in \mathscr{E} \quad , \qquad L = \begin{bmatrix} L_{00} & L_{01} \\ L_{10} & L_{11} \end{bmatrix} \quad , \qquad M = \begin{bmatrix} M_{00} & M_{01} \\ M_{10} & M_{11} \end{bmatrix} \quad .$$

Then the stability of linear large-scale systems can be considered.

3.2 Non-linear large-scale system in functional space

Suppose that the large-scale system is composed of m sub-systems with the control and the perturbation, then in the B-space (Banach space) E_μ $(\mu = 1,\ldots,m)$ the sub-systems are given by

$$\frac{dx_\mu}{dt} = F_\mu(t, x_\mu, U_\mu, p_\mu)$$

$$(x_\mu(t_0) = X_{\mu_0} \in H_{\mu_0} \subset E_\mu) \tag{23}$$

where time is $t \in T = [0, +\infty)$, state $x_\mu \in H_\mu \subset E_\mu$, control $U_\mu \in U_\mu$, perturbation $p_\mu \in P_\mu$, non-linear operator $F_\mu : T \times H_\mu \times U_\mu \times P_\mu \to E_\mu$.

In the B-space $E = E_1 \times \ldots \times E_m$ the non-linear large-scale system can be written as

$$\frac{dx}{dt} = F(t, x, u, p) \tag{24}$$

the output:

$$Z = (Z_1, \ldots, Z_m) \in Z = Z_1 \times \ldots \times Z_m \tag{25}$$

where

$$x = (x_1, \ldots, x_m) \in H = H_1 \times \ldots \times H_m \quad ,$$

$$U = (U_1, \ldots, U_m) \in U = U_1 \times \ldots \times U_m \quad ,$$

$$p = (p_1, \ldots, p_m) \subset P = P_1 \times \ldots \times P_m \quad ,$$

$$F = (F_1, \ldots, F_m) : T \times H \times U \times P \to E \quad ,$$

the operator F brings a natural induced operator and norm $\|x\| \overset{\Delta}{=} \|x_1\| + \ldots + \|x_m\|$.

The equation governing the non-linear large-scale system in B-space can be obtained by

$$\frac{dx}{dt} = f(t, x) \ (x(t_0) = x_0 \in H_0 \subset E) \tag{26}$$

where the perturbation p is omitted and suppose that

$$u_\mu(t, z_1, \ldots, z_m) : T \times Z \to U_\mu \quad ,$$

$$Z_\mu(t, x_1, \ldots, x_m) : T \times H \to Z_\mu \quad .$$

In order to measure the stability of the large-scale system the non-negative functions $\rho_M : T \times E — R$ and $\rho^0_{M_0} : T_0 \times H_0 \to R$, the sets

$M \subset T \times H$ and $M_0 \subseteq T_0 \times H_0$ can be considered; it is chosen that

$$\rho_M = d_t(x, M(t)) \quad ,$$

$$\rho^0_{M_0} = d^0_{t_0}(x_0, M_0(t_0)) \quad ,$$

under the following conditions:

$$d_t(x, M(t)) = 0 \quad , \quad \text{for} \quad (t, x) \in M \quad ,$$

$$d^0_{t_0}(x_0, M_0(t_0)) = 0 \quad , \quad \text{for} \quad (t_0, x_0) \in M_0 \quad .$$

By the definition of Zubov and Mavchan the function ρ_M $(\rho^0_{M_0})$ can express a distance between the point $x(t)(x(t_0))$ and the set $M(t)$ $(M_0(t_0))$; suppose that the set of the solutions of large-scale system (26), passed through the point (t_0, x_0), is represented by $G(t_0, x_0)$ and that

$$M_0(t_0) \overset{\delta\Delta}{=} \{x_0 \in H_0 : d^0_{t_0}(x_0, M_0(t_0)) < \delta\} \quad ;$$

then the definitions of stability can be considered:

(a) Stability: $\forall\varepsilon > 0,\ t_0 \in T_0 \exists\delta(\varepsilon, t_0) > 0 \rightarrow \forall x(t_0) \in M_0(t_0)^{\delta(\varepsilon, t_0)}$ $(\forall x \in G(t_0, x_0),\ t_0 \in T_0) dt(x(t), M(t)) < \varepsilon$

(b) Asymptotical stability: (a) $\&\forall t_0 \in T_0 \exists A(t_0) \in (0, +\infty] \rightarrow \forall x(t_0) \in M_0(t_0)^{A(t_0)}$ $\forall x \in G(t_0, x_0)$

$$\lim_{t \rightarrow +\infty} d_t(x(t), M(t)) = 0 (t \in T)$$

(c) Asymptotical stability in the large: (b) $\& A(t_0) = +\infty\ \forall t_0 \in T_0$ where the logical symbols can be used: $\forall$ (for all, given), $\exists$(for some, exists), $\rightarrow$ (therefore, such that), $\&$(and, together with).

In order to analyse the stability of the non-linear distributed parameter large-scale systems, the qualitative theory of Lyapunov has received the greatest amount of attention. The approach of Lyapunov can also be used successfully in the abstract space.

References

1. Wang Zhao-lin, Liu Shou-gui, Guan Ye-hui, Huang Shi-tao, "Methods of Large-scale Systems and Satellite Attitude Dynamics", Chinese J. of Space Science, Vol. 3, No. 2 (1983) 81-102.

2. Wang Zhao-lin, Huang Shi-tao, "Methods of Large-scale Systems with Weighted V Functions and Stability of Mechanical Systems", J. of Tsinghua University, Vol. 23, No. 2 (1983) 23-34.

3. Wang Zhao-lin, Huang Shi-tao, "Stability Theorem of Mechanical Systems with Constraint Damping", Acta Mechanica Sinica, No. 5 (1983) 463-468.

4. Wang Zhao-lin, "Stability of Motion and Satellite Attitude Dynamics", Advances in Mechanics, Vol. 10, No. 4 (1980) 15-29.

5. Wang Zhao-lin, *et al.*, Fundamentals of Modern Control Theory, National Defence Industry Press, Peking, 1983, 233-241.

6. Wang Zhao-lin, Deng Zhong-ping, "Sloshing of Liquids-filled Spherical Tank under Low-gravity Conditions", Proceedings of the Third China National Congress of General Mechanics, Chong Qing, 1984.

7. N.N. Moiseev, V.V. Rumynatsev, "Dynamics of Liquid-filled Body", Science Press, Moscow, 1965.

8. V.V. Rumyantsev, "On the Motion and Stability of Body with Liquid-filled Flexible Cavity", PPM J. of Appl. Mech., Vol. 33, No. 6 (1969) 946-957.

9. A.D. Meishkis, *et al.*, "Hydromechanics of Weightless State", Moscow, (1982) 53-59.

10. D.W. Fox, J.R. Kutter, "Sloshing Frequencies", ZAMP, Vol. 34, No. 5 (1983) 668-696.

11. V.M. Matrocov, S.N. Vasiliev, "Comparison Principle in Dynamics of Distributed parameter Systems", AIT, No. 1 (1973) 5-22.

12. C.W. Kitchens, "Navier-Stokes Solutions for Spin-Up in a Liquid-filled Cylinder", AIAA J., Vol. 18, No. 8 (1980) 929-934.

13. F.T. Dodge, "Further Studies of Propellant Sloshing under Low Gravity Conditions", NASA CR-119892, 1971.

14. R.F. Glaser, "Analysis of Axisymmetrical Vibration of a Partially Liquid-filled Elastic Sphere by the Method of Green's Functions", NASA-MSFC, TN-D7472, 1973.

15. J.D. Murphy, "Accuracy of Approximations to the Navier-Stokes Equations", AIAA J., Vol. 21, No. 12 (1983) 1759-1760.

16. S. Ostrach, "Low-gravity Fluid Flows", Ann. Rev. Fl. Mech., No. 14, (1982) 313-345.

17. J.R. Kuttler, V.G. Sigillito, "Sloshing of Liquids in Cylindrical Tanks", AIAA J., Vol. 22, No. 2 (1984) 309-311.

18. A.J. Baker, M.O. Soliman, "A Finite Element Algorithm for Computational Fluid Dynamics", AIAA J., Vol. 21, No. 6 (1983) 816-827.

19. J.D. Farmer, Evolution of Order and Chaos in Physics, Chemistry and Biology, Schlose Elmau 1982, ed. by H. Haken, Springer-Verlag, 1982.

20. M.R. Guevara, L. Glass, A. Shrier, "Phase Locking, Period-Doubling Bifurcations, and Irregular Dynamics in Periodically Stimulated Cardiac Cells", Science, 18 December 1981, Vol. 214, No. 4527, 1350-1353.

21. A.E. Ibrahim, A.K. Misra, "Attitude Dynamics of Satellite During Deployment of Large-Plate-Type Structures", J. of GCD, Vol. 5, No. 5 (1982) 442-447.

22. K. Tsuchiya, "Dynamics of a Spacecraft during Extension of Flexible Appendages", J. GCD, Vol. 6, No. 2 (1983) 100-103.

23. M.S. Bhat, S.K. Shrivastava, "A Procedure to Generate Lyapunov Functional for Distributed Parameter Systems", Trans. ASME, J. of Appl. Mech., Vol. 48, No. 1 (1981) 188-190.

24. V. Lakshmikantham, S. Leela, Non-linear Differential Equations in Abstract Spaces, Pergamon Press, 1981.

25. P. Fergola, C. Tenneriello, "Partial Stability of Composite Systems with Unstable Sub-systems, Large-scale Systems. Theory and Applications", Vol. 4, No. 1 (1983) 101-105.

APPLICATION OF THE CALCULATION OF TWO-DIMENSIONAL ELASTIC-PLASTIC FLOW

Wu Shengchang, Chang Qianshun, Feng Yanling, Mu Jun
Institute of Applied Mathematics, Academia Sinica

Li Luyin, Heshunlu
Beijing Institute of Technology

1. The Problem

Since the 1970's, America, Sweden, etc. are very interested in the numerical simulation of Self Forging Fragment (SFF).

The diagram of SFF is as follows

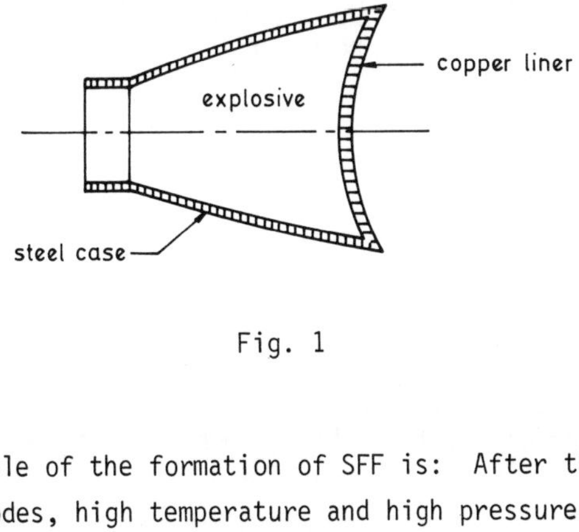

Fig. 1

The principle of the formation of SFF is: After the explosive within SFF explodes, high temperature and high pressure resulting from the explosion will impart to the copper liner a finite plastic (or fluidic) deformation in the form of a bulge as well as an acceleration. In this process, the liner forges itself into a very compact mass leaving off with high velocity.

The process of the formation of SFF is a very complicated problem of explosion. Because SFF is axisymmetric, two-dimensional elastic-plastic hydrodynamic equations are used to describe this process.

2. Differential Equations and Conditions for their Solution

Two-dimensional elastic-plastic hydrodynamic equations in xor coordinate with cylindrical symmetry about the x axis are as follows:

Conservation of mass

$$\rho \frac{\partial(\frac{1}{\rho})}{\partial t} = \frac{\partial u_x}{\partial x} + \frac{\partial u_r}{\partial r} + \frac{u_r}{r} \qquad (2.1)$$

Conservation of momentum

$$\rho \frac{\partial u_x}{\partial t} = - \frac{\partial(p+q)}{\partial x} + \frac{\partial s_x}{\partial x} + \frac{\partial \tau_{xr}}{\partial r} + \frac{\tau_{xr}}{r} \qquad (2.2)$$

$$\rho \frac{\partial u_r}{\partial t} = - \frac{\partial(p+q)}{\partial r} + \frac{\partial \tau_{xr}}{\partial x} + \frac{\partial s_r}{\partial r} + \frac{s_r - s_\theta}{r} \qquad (2.3)$$

$$u_x = \frac{\partial x}{\partial t} \qquad (2.4)$$

$$u_r = \frac{\partial r}{\partial t} \quad . \qquad (2.5)$$

Conservation of energy

$$\frac{\partial e}{\partial t} = -(p+q)\frac{\partial}{\partial t}\left(\frac{1}{\rho}\right) + \frac{1}{\rho}\left(s_x \frac{\partial \varepsilon_x}{\partial t} + s_r \frac{\partial \varepsilon_r}{\partial t} + s_\theta \frac{\partial \varepsilon_\theta}{\partial t} + \tau_{xr} \frac{\partial \varepsilon_{xr}}{\partial t}\right) \quad . \qquad (2.6)$$

Equation of state

$$p = p(\eta, e) \quad . \qquad (2.7)$$

Artificial viscosity

$$q = \begin{cases} (c_{q1}c + c_{q2}L\varphi|\dot{\varepsilon}\varphi|)\rho L\varphi|\dot{\varepsilon}\varphi| & \dot{\varepsilon}\varphi < 0 \\ 0 & \dot{\varepsilon}\varphi \ge 0 \end{cases} \quad . \qquad (2.8)$$

Velocity strains

$$\frac{\partial \varepsilon_x}{\partial t} = \frac{\partial u_x}{\partial x} \quad , \quad \frac{\partial \varepsilon_r}{\partial t} = \frac{\partial u_r}{\partial r}$$

$$\frac{\partial \varepsilon_\theta}{\partial t} = \frac{u_r}{r} \quad , \quad \frac{\partial \varepsilon_{xr}}{\partial t} = \frac{\partial u_r}{\partial x} + \frac{\partial u_x}{\partial r} \quad . \tag{2.9}$$

Stresses

When the material is within the elastic limit, stresses and strains of the material satisfy Hook's law

$$\frac{\partial s_x}{\partial t} = 2\mu\left[\frac{\partial \varepsilon_x}{\partial t} - \frac{\rho}{3}\frac{\partial}{\partial t}\left(\frac{1}{\rho}\right)\right] + \delta_x$$

$$\frac{\partial s_r}{\partial t} = 2\mu\left[\frac{\partial \varepsilon_r}{\partial t} - \frac{\rho}{3}\frac{\partial}{\partial t}\left(\frac{1}{\rho}\right)\right] + \delta_r$$

$$\frac{\partial s_\theta}{\partial t} = 2\mu\left[\frac{\partial \varepsilon_\theta}{\partial t} - \frac{\rho}{3}\frac{\partial}{\partial t}\left(\frac{1}{\rho}\right)\right]$$

$$\frac{\partial \tau_{xr}}{\partial t} = \mu\,\frac{\partial \varepsilon_{xr}}{\partial t} + \delta_{xr} \quad . \tag{2.10}$$

When the stresses of the material increase to the elastic limit, Von Mises yield condition is satisfied

$$S_1^2 + S_2^2 + S_3^2 \leq \frac{2}{3}\,(Y^0)^2 \quad . \tag{2.11}$$

That is, if the loading is continuous, stress deviators of the material will increase no longer and the material will have plastic distortions which are not restorable. If the material is unloaded again, the relation between stresses and strains satisfy Hook's law.

When the pressure of the material $p \gg Y^0$, distortion and motion of the material will show the property of a flow.

The conditions for the solution of the equations are as follows

(1) Initial condition

$$u_x = u_x^0 \ , \quad u_r = u_r^0 \ , \quad \rho = \rho^0 \ , \quad e = e^0 \ ,$$

$$s_x = s_x^0 \ , \quad s_r = s_r^0 \ , \quad s_\theta = s_\theta^0 \ , \quad \tau_{xr} = \tau_{xr}^0$$

(2) Boundary condition

In our problem, we have boundary of the following type:

A. Free surface. The normal pressure is zero on this surface.

B. Boundary condition of pressure function. We give the value normal pressure on the boundary.

C. Boundary where axisymmetry is maintained. On the axis of symmetry, the normal component of velocity is zero and the shearing stress is zero.

D. Boundary where sliding between the explosive and the steel case occurs. We assume that this interface is perfectly smooth, that is, the normal component of velocity is continuous and tangential component may be discontinuous, while normal stress is continuous and tangential stress is zero.

3. Basic Methods of Calculation

We divide the region for computation into several quadrilateral grids and then use Lagrange method for solution.

(1) We adopt the following integral definitions of the partial derivatives.

$$\left.\frac{\partial \Psi}{\partial x}\right|_{(x_0, r_0)} = \lim_{A \to 0} \frac{1}{A} \oint_L \Psi dr$$

$$\left.\frac{\partial \Psi}{\partial r}\right|_{(x_0, r_0)} = \lim_{A \to 0} \frac{1}{A} \oint_L \Psi dx \qquad (3.1)$$

where, Ψ — space function, L — some closed circle which contains (x_0, r_0) in the xor plane, A — area bounded by L.

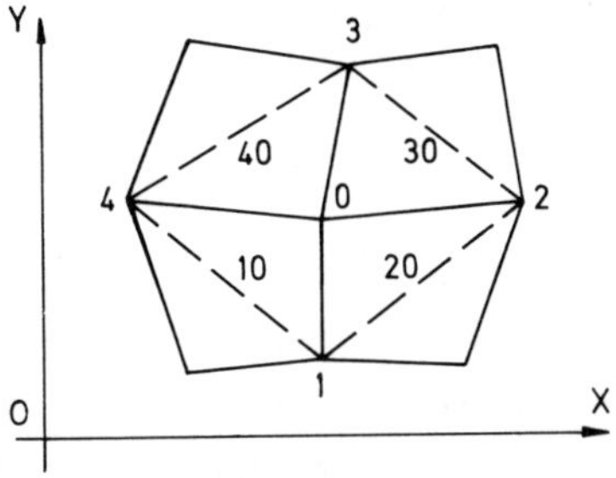

Fig. 2

Applying the above formula to the quadrilateral grid, we obtain

$$\left.\frac{\partial\Psi}{\partial x}\right|_0 \approx \frac{1}{A}\left[\Psi_{10}(r_1 - r_4) + \Psi_{20}(r_2 - r_1) + \Psi_{30}(r_3 - r_2) + \Psi_{40}(r_4 - r_3)\right]$$

$$\left.\frac{\partial\Psi}{\partial r}\right|_0 \approx \frac{-1}{A}\left[\Psi_{10}(x_1 - x_4) + \Psi_{20}(x_2 - x_1) + \Psi_{30}(x_3 - x_2) + \Psi_{40}(x_4 - x_3)\right] \quad . \tag{3.2}$$

According to (3.2), we can easily get a difference scheme of equations (2.1)-(2.11).

(2) On the interface between the two different materials sliding phenomenon will occur, owing to the action of explosion and high-speed impact. Therefore the grids in the neighbourhood of this interface will deform very much. Hence this interface will demand special treatment.

On the xor plane, sliding interface presents itself as a sliding line which changes with time. We assume that the equations of the sliding line are

$$x = \varphi_1(s, t) \quad , \quad r = \varphi_2(s, t)$$

where s — arc length, t — time. Denoting the unit tangent vector and normal vermal at some point on the sliding line by $\vec{\tau} = (\cos\alpha, \sin\alpha)$, and $\vec{n} = (-\sin\alpha, \cos\alpha)$ respectively. Then, the velocity of that point, $\vec{u}(u_x, u_r)$ can be expressed as follows

$$\vec{u} = u_\tau \vec{\tau} + u_n \vec{n}$$

where $u_\tau = u_x \cos\alpha + u_r \sin\alpha$, $u_n = -u_x \sin\alpha + u_r \cos\alpha$.

The superscript "+" and "-" are used to express the physical quantity on both sides of the sliding line, e.g. tangential velocities on both sides are u_τ^+ and u_τ^-, respectively, normal components of velocity are $u_n^+ = u_n^- = u_n$. Then by means of equations of motion and the condition of sliding line, we can obtain the equations of motion on the sliding line.

$$\begin{cases} \dfrac{\partial u_\tau^\pm}{\partial t} - u_n\left(\dfrac{\partial u_n}{\partial s} - \dfrac{u_\tau^\pm}{R}\right) = \dfrac{1}{\rho^\pm}\left[(\nabla\sigma)^\pm \cdot \vec{\tau}\right] \\ (u_\tau^+ - u_\tau^-)\left[2\dfrac{\partial u_n}{\partial s} - \dfrac{1}{R}(u_\tau^+ + u_\tau^-)\right] = \dfrac{1}{\rho^+}\left[(\nabla\sigma)^+ \cdot \vec{n}\right] - \dfrac{1}{\rho^-}\left[(\nabla\sigma)^- \cdot \vec{n}\right] \end{cases} \tag{3.3}$$

where

$$\frac{\partial}{\partial s} = \cos\alpha\frac{\partial}{\partial x} + \sin\alpha\frac{\partial}{\partial r}$$

$$(\nabla\sigma)_x = \frac{\partial\sigma_x}{\partial x} + \frac{\partial\tau_{xr}}{\partial r} + \frac{\tau_{xr}}{r}$$

$$(\nabla\sigma)_r = \frac{\partial\tau_{xr}}{\partial x} + \frac{\partial\sigma_r}{\partial r} + \frac{\sigma_r - \sigma_\theta}{r}$$

$$\sigma_i = s_i - p \quad , \qquad i = x, r, \theta \quad .$$

R is the transient local radius of curvature of the sliding line.

For the calculation of the sliding surface, we use the method of main and subordinate interfaces. The points on the subordinate interface must attach themselves to the main one, and can only slide along it.

In Fig. 3, AOC denotes a sliding line. The points on the main interface are denoted by A, 0, C while those on the subordinate interface are denoted by A^*, 0^*, C^*. The calculation of the sliding line is divided into three steps: (1) Computing the location and velocity of the point 0 on the main interface (2) Computing those of the point 0* on the subordinate interface (3) Put together the main and subordinate interfaces.

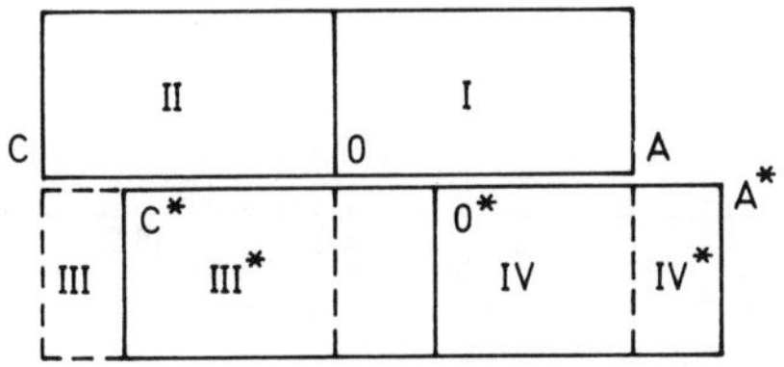

Fig. 3

4. Numerical Simulation of the Formation Process of SFF and It's Experimental Verification

After the ignition of the auxillary charge behind the explosion region, the shock wave propagates in the direction of the main charge causing a tremendous explosion.

Owing to the action of the resulting big shock wave, the liner collapses, bulges and finally is made into a bullet with highly concentrated mass. The shape of the copper liner obtained from our two-dimensional numerical simulation compares fairly with that in the high speed X-ray photograph of the actual experimental process (see Fig. 4.).

The location and velocity of the body of the bullet obtained from our calculation also compare fairly with the results of the experiment (see Fig. 5).

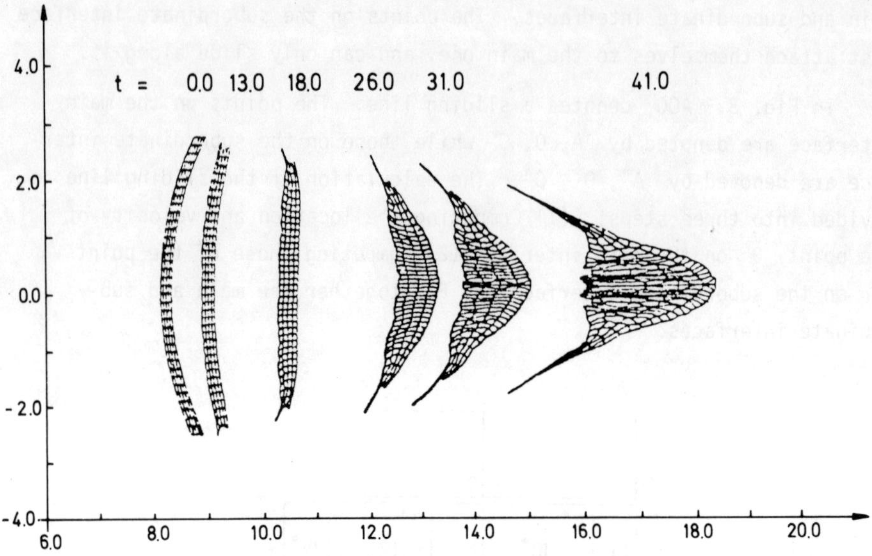
t = 0.0 13.0 18.0 26.0 31.0 41.0
4.0
2.0
0.0
-2.0
-4.0
6.0
8.0
10.0
12.0
14.0
16.0
18.0
20.0

Fig. 4

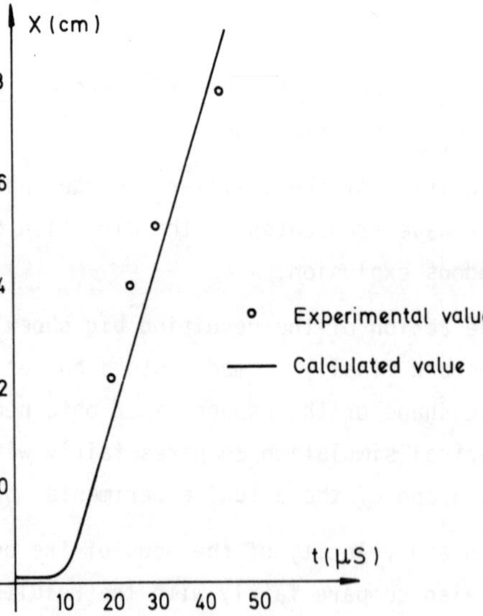
X (cm)
18
16
14
12
10
t(μS)
10 20 30 40 50
Experimental value
Calculated value

Fig. 5

References

1. J.W. Hermann, G. Randers-Pehrson, E.R. Berrus, Proc. of the Third International Symposium on Ballistics, 1977.

2. H.E. Kalsson, Proc. of the First International Symposium on Ballistics, 1974.

3. M.L. Wilkins, UCRL-7322, Rev. 1, 1969.

4. R. Grandy, Application of Finite Difference Methods to Problems in Two-Dimensional Hydrodynamics, AD-256328, 1961.

5. C.L. Mader, Numerical Modelling of Detonations, 1979.

MULTI-FIELD VARIATIONAL PROBLEMS IN LINEAR ELASTICITY

Wei-min Xue

Department of Mechanical Engineering
Tsing Hua University, Peking, China

Abstract --- A set of multi-field variational problems, which may be summarized from the general mixed-hybrid finite element methods in linear elasticity, are investigated. The existence, uniqueness, convergence, and stability properties are proved undertaking the assumptions of ellipticity and LBB-conditions. The equivalence is proved for the two abstract statements, i.e. multi-field variational problem and minimization problem with simultaneous constrains. The later are concerned in the variational principles in linear elaticity.

1. INTRODUCTION

F.Brezzi [1] considered the following variational problem: Find a pair $(u,q) \in V \times P$ such that

$$(1.1) \qquad \begin{cases} a(u,v) + b(v,q) = \langle f,v \rangle & \forall\ v \in V \\ \qquad\qquad b(u,p) = \langle g,p \rangle & \forall\ p \in P \end{cases} ,$$

where V and P are Hilbert spaces, $a(.,.)$ and $b(.,.)$ are continuous bilinear forms, f and g are given linear functionals.

L-A. Ying and S.N. Atluri [2] presended a variational statement, which is summarized from the Stokes flow, of the following type: Find $(\sigma,u,q) \in T \times V \times P$, such that

$$(1.2) \qquad \begin{cases} a(\sigma,\tau) - b(\tau,u) = \langle f,\tau \rangle & \forall\ \tau \in T \\ b(\sigma,v) - c(v,q) = \langle g,v \rangle & \forall\ v \in V \\ \qquad\qquad c(u,p) = \langle h,p \rangle & \forall\ p \in P \end{cases} .$$

For a given bilinear form $b(.,.) : V \times P \rightarrow \mathbb{R}$ we define the associated linear operator B and its dual operator B^* by

$$(1.3) \qquad \langle\ B\ v,p\ \rangle = \langle\ v,B^* p\ \rangle = b(v,p) \qquad \forall\ (v,p) \in V \times P \quad .$$

Then the variational statement (1.1) and (1.2) may be rewritten as

$$\begin{cases} A\ u + B^*\ q = f & \text{in } V^* \\ \qquad B\ u = g & \text{in } P^* \end{cases}$$

and

$$\begin{cases} A\ \sigma - B^*\ u = f & \text{in } T^* \\ B\ \sigma - C^*\ q = g & \text{in } V^* \\ \qquad C\ u = h & \text{in } P^* \end{cases}$$

Undertaking the general mixed-hybrid finite element methods [3, 4], which based on the modified Hu-Washizu principles, modified Hellinger-Reissner Principles, Modified complementary energy and potential energy principles, one may summarize the following multi-field variational problem: Find $(u_1,u_2,...,u_n)$ in space $V_1 \times V_2 \times ... \times V_n$, such that

$$(1.4)\qquad \begin{bmatrix} A & -B_1^* & & & \\ B_1 & 0 & -B_2^* & & \\ & B_2 & 0 & -B_3^* & \\ & & \cdots\cdots & & \\ & & & B_{n-1} & 0 \end{bmatrix} \begin{bmatrix} u_1 \\ u_2 \\ u_3 \\ \cdot \\ u_n \end{bmatrix} = \begin{bmatrix} f_1 \\ f_2 \\ f_3 \\ \cdot \\ f_n \end{bmatrix} \text{ in } \begin{bmatrix} V_1 \\ V_2 \\ V_3 \\ \cdot \\ V_n \end{bmatrix}^*$$

We will analyze the equivalence of variational statement (1.4) and minimization statement with simultaneous constrains.The later is known as the variational principle in linear elasticity. Then we will give the properties of existence and uniqueness of variational problem (1.4). We also give an estimate of the solution. Following the common treatment we derive the convergence and the error estimate for finite element aproximation of problem (1.4).

The method shown here is an expansion of the work created by Brezzi[1] (saddle-point problem), Ying and Atluri [2] (Stokes flow),and Raviart [5] (Navier-Stokes equation).

2. VARIATIONAL STATEMENT AND MINIMIZATION STATEMENT

We use V_i to denote the Hilbert space provided its element v_i and the norm $\|\cdot\|_{V_i}$. Let a(.,.) and b (.,.) be continuous bilinear forms. For a given bilinear form $b_i(.,.) : V_i \times V_{i+1}$ we define the associated linear operator B_i and its dual operator B_i^* by

$$(2.1)\qquad \langle B_i v_i, v_{i+1} \rangle = \langle v_i, B_i^* v_{i+1} \rangle = b_i(v_i, v_{i+1}) \quad \forall\, v_i, v_{i+1}$$

The following theorem discusses the equivalence of multi-field variational problem (1.4) and a minimization statement with simultaneous constrains.

Theorem 2.1

Suppose that the bilinear form a(.,.) is symmetric and positive definite. Then the variational problem (1.4) is equivalent to the following minimization problem : Find $(u_1,u_2,...u_n) \in V_1 \times V_2 .. \times V_n$ such that

$$(2.2)\qquad \begin{cases} L(u_1,u_2,...u_n) = \min\limits_{v_1} L(v_1,u_2,...u_n) \\ L(u_1,u_2,...u_n) = L(u_1,u_2,...v_i,...u_n) \quad \forall\, v_i \in V_i \end{cases}$$

$(i = 2,3,...,n)$

where the functional $L(u_1,u_2,\ldots,u_n)$ is given by the formula

(2.3) $L(u_1,u_2,\ldots,u_n)$

$$= \frac{1}{2}a(u_1,u_1) - \langle f_1,u_1\rangle + \sum_{i=1}^{n-1}(-1)^i [b_i(u_i,u_{i+1}) - \langle f_{i+1},u_{i+1}\rangle]$$

Proof.

Assume that $(u_1,u_2,\ldots,u_n)$ is the solution of problem (1.4). Let us investigate the diffrences

$$L(u_1+v_1,u_2,\ldots,u_n) - L(u_1,u_2,\ldots,u_n)$$

$$= \frac{1}{2}a(v_1,v_1) + [\, a(u_1,v_1) - b_1(v_1,u_2) - \langle f_1,v_1\rangle \,]$$

$$= \frac{1}{2}a(v_1,v_1) >= 0$$

and

$$L(u_1,u_2,\ldots u_i+v_i,\ldots,u_n) - L(u_1,u_2,\ldots,u_n)$$

$$= (-1)^{i-1}[\, b_{i-1}(u_{i-1},v_i) - \langle f_i,v_i\rangle] + (-1)^i b_i(v_i,u_{i+1})$$

$$= (-1)^{i-1}[\, b_{i-1}(u_{i-1},v_i) - b_i(v_i,u_{i+1}) - \langle f_i,v_i\rangle \,]$$

$$= 0$$

$(i = 2,3,\ldots,n-1)$

$$L(u_1,u_2,\ldots,u_n+v_n) - L(u_1,u_2,\ldots u_n)$$

$$= (-1)^{n-1}[\, b_{n-1}(u_{n-1},v_n) - \langle f_n,u_n\rangle \,]$$

$$= 0$$

Then $(u_1,u_2,\ldots,u_n)$ is also the solution of minimization problem (2.2).

Assume that $(u_1,u_2,\ldots,u_n)$ is the solution of problem (2.2). Due to the above discussion, it remains to prove the first equation of problem (1.4). Let ϵ be a positive number. The inequality

$$L(u_1+v_1,u_2,\ldots,u_n) >= L(u_1,u_2,\ldots,u_n) \quad \forall\, v_1 \in V_1$$

leads to

$$a(u_1,v_1) - b_1(v_1,u_2) - \langle f_1,v_1\rangle >= -\frac{\epsilon}{2}a(v_1,v_1) \qquad \forall\, \epsilon, v_1$$

Since ϵ may be arbitrarily small, it comes that

$$a(u_1,v_1) - b_1(v_1,u_2) - \langle f_1,v_1\rangle >= 0 \qquad \forall\, v_1 \in V_1$$

The above inequality is indeed an equality since the left hand side is linear to v_1. The proof is completed.

Example 2.1 (Hu-Washizu Principle)

We consider a linear elastic solid undergoing infinitesimal deformation. The governing equations and boundary conditions are well known as follows : Find $(\sigma, \epsilon, u) \in T \times E \times V$ such that

$$
(2.4) \qquad \begin{cases} \sigma_{ij} = a_{ijkl}\,\epsilon_{kl} & \text{in } \Omega \\ \epsilon_{ij} = u_{(i,j)} & \text{in } \Omega \\ u_i = \bar{u}_i & \text{at } S \\ \sigma_{ij,j} + \bar{f}_i = 0 & \text{in } \Omega \\ \sigma_{ij}\, n_j \triangleq t_i = \bar{t}_i & \text{at } S \end{cases}
$$

If the material is isotropic, the coefficients of elasticity a_{ijkl} have the properties of symmetry and ellipticity :

$$
(2.5a) \qquad a_{ijkl} = a_{ijlk} = a_{klij}
$$

$$
(2.5b) \qquad a_{ijkl}\,\epsilon_{ij}\,\epsilon_{kl} \geqslant \alpha\, \epsilon_{ij}\,\epsilon_{ij} \quad \forall\, \epsilon_{ij} \quad ,
$$

where the positive number α depends only on the property of the material.

By undertaking the concept of weak solution of a partial differential equation systerm, the weak form of (2.4) will lead to the following multi-field variational problem : Find $(\epsilon, \sigma, u) \in E \times T \times V$ such that

$$
(2.6) \qquad \begin{cases} a(\epsilon, e) - l(e, \sigma) = 0 & \forall\ e \in E \\ l(\epsilon, \tau) - p(\tau, u) = \langle g, \tau \rangle & \forall\ \tau \in T \\ \qquad\qquad p(\sigma, v) = \langle f, v \rangle & \forall\ v \in V \quad , \end{cases}
$$

where the bilinear forms and linear functionals are defined by

$$
\begin{aligned}
a(\epsilon, e) &= \int_\Omega a_{ijkl}\,\epsilon_{kl}\, e_{ij}\, d\Omega \\
l(\epsilon, \tau) &= \int_\Omega \epsilon_{ij}\,\tau_{ij}\, d\Omega \\
p(\tau, v) &= \int_\Omega \tau_{ij}\, v_{(i,j)}\, d\Omega - \int_{S_u} \tau_{ij}\, n_j\, u_i\, ds \\
\langle f, v \rangle &= \int_\Omega \bar{f}_i v_i\, d\Omega + \int_{S_t} \bar{t}_i v_i\, ds \\
\langle g, \ \rangle &= \int_{S_u} \sigma_{ij} n_j \bar{u}_i\, ds \quad .
\end{aligned}
$$

According to Theorem 2.1 the equivalent minimization problem(2.2) with n=3 will employ the following functional

$$
L(\epsilon, \sigma, u) = \frac{1}{2} a(\epsilon, \epsilon) - l(\epsilon, \sigma) + \langle g, \sigma \rangle + p(\sigma, u) - \langle f, u \rangle
$$

i.e.

$$
L(\epsilon, \sigma, u) = \int_\Omega \left[\frac{1}{2} a_{ijkl}\,\epsilon_{kl}\,\epsilon_{ij} + \sigma_{ij}(u_{(i,j)} - \epsilon_{ij}) - \bar{f}_i u_i \right] d\Omega + \int_{S_u} \sigma_{ij}\, n_j\, (\bar{u}_i - u_i)\, ds - \int_{S_t} \bar{t}_i u_i\, ds \quad .
$$

This functional is exactly the same as that employed in the well-known Hu-Washizu principle in linear elasticity.

3. EXISTENCE, UNIQUENESS, AND CONVERGENCE

Lemma 1 (Girault and Raviart [5])

The following three properties are equivalent :

(A) The bilinear form b(.,.) satisfies the LBB-condition, i.e.

$$\sup_{\forall v \in V} \frac{b(v,p)}{\|v\|_V} \geq \beta \|p\|_P \qquad \forall p \in P \tag{3.1}$$

(B) The operator B^* is an isomorphism from P onto $(\operatorname{Ker}(B))^{\circ}$, therefore

$$\|B^* p\|_{V^*} >= \beta \|p\|_P \qquad \forall \ p \in P \tag{3.2}$$

(C) The operator B is an isomorphism from $(\operatorname{Ker}(B))_{\perp}$ onto P^*, therefore

$$\|B v\|_{P^*} >= \beta \|v\|_V \qquad \forall \ v \in (\operatorname{Ker}(B))_{\perp} \tag{3.3}$$

where $(\operatorname{Ker}(B))_{\perp}$ and $(\operatorname{Ker}(B))^{\circ}$ are defined by

$$(\operatorname{Ker}(B))_{\perp} = \{ v \in V, \ (v, v_0) = 0 \quad \forall v_0 \in \operatorname{Ker}(B) \}$$

$$(\operatorname{Ker}(B))^{\circ} = \{ f \in V, \langle f, v_0 \rangle = 0 \quad \forall v_0 \in \operatorname{Ker}(B) \},$$

(u,v) denotes the inner product of u and v, and $\langle f,.\rangle$ denotes the continuous linear operator in V^*.

Theorem 2.1 (existence, uniqueness, and stability)

Suppose that the following LBB-conditions and ellipticity condition are satisfied

$$\begin{cases} \sup\limits_{\forall v_{k-1} \in V_{k-1}} \dfrac{b_{k-1}(v_{k-1}, v_k)}{\|v_{k-1}\|} \geq \beta \|v_k\| & \forall v_k \in \operatorname{Ker}^*(B_k); \ k=2,3,\ldots,n-1,n \\ a(v_1, v_1) >= \alpha \|v_1\|^2_{V_1} & \forall v_1 \in \operatorname{Ker}^*(B_1) \end{cases} \tag{3.4}$$

where

$$\operatorname{Ker}^*(B_k) = \{ v_k \in V_k, \ b_k(v_k, v_{k+1}) = 0 \quad \forall v_{k+1} \in \operatorname{Ker}^*(B_{k+1}) \}$$

$$(k = 1,2, \ldots ,n-1)$$

$$\operatorname{Ker}^*(B_n) \triangleq V_n$$

Then we have the following results :

(1) Multi-field variational problem (1.4) has a unique solution $(u_1, u_2, \ldots, u_n) \in V_1 \times V_2 \times \ldots V_n$.

(2) There exists a positive number C suth that

$$\|u\| <= C \|f\| \quad , \tag{3.5}$$

where

$$\|u\| \triangleq \sum_{i=1}^{n} \|u_i\|_{V_i} \quad ; \quad \|f\| \triangleq \sum_{i=1}^{n} \|f_i\|_{V_i^*} \; .$$

Therefore the mapping $\gamma : f \triangleq (f_1, f_2, \ldots, f_n)^T \text{ --> } u \triangleq (u_1, u_2, \ldots, u_n)^T$ is an isomorphism from $V_1^* \times V_2^* \times \ldots \times V_n^*$ onto $V_1 \times V_2 \times \ldots \times V_n$.

Proof.
In the case of n=0 , this is exactly the Lax-Milgram theorem [5]. Then we prove this theorem by deduction. Suppose that the results for nth problem is correct already. Let us investigate the (n+1)th problem. Notice that the last two equations become

$$B_{n-1} u_{n-1} - B_n^* u_{n+1} = f_n \qquad \text{in } V_n^*$$
$$B_n u_n = f_{n+1} \qquad \text{in } V_{n+1}^* \; .$$

(1)
According to LBB-condition of b (.,.) and Lemma 1 (C), there exists a unique $(u_n)_\perp \in \mathrm{Ker}(B_n)_\perp$ such that

$$B_n (u_n)_\perp = f_{n+1} \qquad \text{in } V_{n+1}$$

Let $(u_n)_o$ be arbitrary element in $\mathrm{Ker}(B_n)$ and denote $(u_n)_\perp + (u_n)_o = u_n \in V_n$. Then we have

$$B_n u_n = B_n (u_n)_\perp = f_{n+1} \qquad \text{in } V_{n+1}^*$$

It remains to find an appropriate $(u_n)_o \in \mathrm{Ker}(B_n)$, such that the first n equations are satisfied.

Notice that the unknown variable u_{n+1} is involved only in the nth equation. Due to the LBB-condition of $b_n(.,.)$ and Lemma 1 (B), the equation $B_n^* u_{n+1} = B_{n-1} u_{n-1} - f_n$ has a unique solution u_{n+1} if and only if $B_{n-1} u_{n-1} - f_n$ belongs to subspace $(\mathrm{Ker}(B_n))^o$,i.e.

$$B_{n-1} u_{n-1} = f_n \qquad \text{in } (\mathrm{Ker}(B\;))^* \; .$$

Then it suffices to solve the following problem : Find $(u_1, u_2, \ldots u_{n-1}, (u_n)_o) \in V_1 \times V_2 \times \ldots V_{n-1} \times \mathrm{Ker}(B_n)$, such that

$$\begin{bmatrix} A & -B_1^* & & & \\ B_1 & 0 & -B_2^* & & \\ & \cdots & \cdots & \cdots & \\ & & B_{n-2} & 0 & -B_{n-1}^* \\ & & & B_{n-1} & 0 \end{bmatrix} \begin{bmatrix} u_1 \\ u_2 \\ \cdot \\ u_{n-1} \\ u_n \end{bmatrix} = \begin{bmatrix} f_1 \\ f_2 \\ \cdot \\ f_{n-1}^* \\ f_n \end{bmatrix} \quad \text{in} \quad \begin{bmatrix} V_1 \\ V_2 \\ \cdot \\ V_{n-1} \\ V_n \end{bmatrix}^*$$

Where f_{n-1}^* defined by

$$f_{n-1}^* = f_{n-1} + B_{n-1}(u_n)_\perp$$

is a fixed element in V_{n-1}^* since $(u_n)_\perp$ is given by f_n. The above equation system is the case of nth problem. Then the existence and uniqueness of $u_1, u_2, \ldots, (u_n)_0$, therefore u_n, are proved.

(2)

Due to the assumption of deduction, one may easily get the astimate that

$$\sum_{i=1}^{n-1} \|u_i\|_{V_i} + \|(u_n)_0\|_{V_n} \leq c\Big[\sum_{i=1}^{n} \|f_i\|_{V_i^*} + \|B_{n-1}\| \, \|(u_n)_\perp\|_{V_n}\Big]$$

From Lemma 1 (3.3), we have

$$\|(u_n)_\perp\|_{V_n} <= \beta^{-1} \|B_n(u_n)_\perp\|_{V_{n+1}^*} = \beta^{-1} \|f_{n+1}\|_{V_{n+1}^*}$$

Then we have

$$\begin{aligned} \sum_{i=1}^{n} \|u_i\|_{V_i} &\leq \sum_{i=1}^{n-1} \|u_i\|_{V_i} + \|(u_n)_\perp\|_{V_n} + \|(u_n)_0\|_{V_n} \\ &\leq C\sum_{i=1}^{n} \|f_i\|_{V_i^*} + (c\|B_{n-1}\| + 1)\beta^{-1} \|f_{n+1}\|_{V_{n+1}^*} \\ &\leq C_1 \sum_{i=1}^{n+1} \|f_i\|_{V_i^*} . \end{aligned}$$

Moreover we know from Lemma 1 (3.2) that

$$\begin{aligned} \|u_{n+1}\|_{V_{n+1}} &\leq \beta^{-1} \|B_n^* u_{n+1}\|_{V_n^*} \\ &= \beta^{-1} \|B_{n-1} u_{n-1} - f_n\|_{V_n^*} \\ &\leq \beta^{-1}\big(\|B_{n-1}\| \, \|u_{n-1}\|_{V_{n-1}} + \|f_n\|_{V_n^*} \big) . \end{aligned}$$

The above two inequalities lead immediately to estimate

$$\sum_{i=1}^{n+1} \|u_i\|_{V_i} \leq C_2 \sum_{i=1}^{n+1} \|f_i\|_{V_i^*} .$$

The proof is completed.

As is well known, a finite element approach is to find the solution $(u_1,u_2,\dots,u_n)^h$ in a finite dimensional subspace $V_1^h \times V_2^h \times \dots V_n^h$ instead of the infinite dimensional Hilbert space $V_1 \times V_2 \times \dots V_n$. Thus the finite element approximation of variational problem (1.4) may be stated in the following problem : Find $(u_1,u_2,\dots,u_n)^h \in V_1^h \times V_2^h \times \dots V_n^h$, such that

$$\begin{bmatrix} A & -B_1^* & & & \\ B_1 & 0 & -B_2^* & & \\ & \cdots & \cdots & & \\ & & \cdots & \cdots & \\ & & & B_{n-1} & 0 \end{bmatrix} \begin{bmatrix} u_1 \\ u_2 \\ \cdot \\ \cdot \\ u_n \end{bmatrix}^h = \begin{bmatrix} f_1 \\ f_2 \\ \cdot \\ \cdot \\ f_n \end{bmatrix} \quad \text{in} \quad \begin{bmatrix} V_1 \\ V_2 \\ \cdot \\ \cdot \\ V_n \end{bmatrix}^{*h} , \tag{3.6}$$

where the h is the discretization parameter which depends only on the geometric discretization of the domain Ω .

Theorem 2.2 (convergence)

In addition to the assumption (3.4), we assume that

$$\text{(3.7)} \begin{cases} \sup\limits_{\forall v_i^h \in V_i^h} \dfrac{b_i(v_i^h, v_{i+1}^h)}{\|v_i^h\|_{V_i}} \geq \beta \|v_{i+1}^h\|_{V_{i+1}} & \forall\, v_{i+1}^h \in \operatorname{Ker}_h^*(B_i);\ i=1,2,\dots,n-1 \\ a(v_1^h, v_1^h) >= \alpha \|v_1^h\|_{V_1}^2 & \forall\, v_1^h \in \operatorname{Ker}_h^*(B_1) \end{cases}$$

Then we have the following results :

(1) Problem (1.4) and (3.6) have unique solutions $(u_1,u_2,\dots u_n)$ and $(u_1,u_2,\dots u_n)^h$ respectively.

(2) There exists a positive number C such that

$$\|u - u^h\| <= C \operatorname{dis}\{u, V_h\} \tag{3.8}$$

where

$$\|u - u^h\| = \sum_{i=1}^{n} \|u_i - u_i^h\|_{V_i} \tag{3.9a}$$

$$\operatorname{dis}\{u, V_h\} = \sum_{i=1}^{n} \inf_{\forall v_i \in V_i^h} \|u_i - v_i^h\|_{V_i} \tag{3.9b}$$

Proof.

Due to Theorem 2.1 the existence and uniqueness are already provided. It remains to prove the estimate of the error.

Denote that

$$(3.10) \qquad T = \begin{bmatrix} A & -B_1^* & & \\ B_1 & 0 & -B_2^* & \\ & \cdots\cdots\cdots & & \\ & \cdots\cdots\cdots & & \\ & & B_{n-1} & 0 \end{bmatrix}$$

Then the problems (1.4) and (3.6) become that

$$T u = f \qquad \text{in } V^*$$

$$T u^h = f^i \qquad \text{in } V_h^*$$

It may be easily seen that for an arbitrary, but fixed $(v_1, v_2, \ldots v_n)^h \in V_1^h \times V_2^h \times \ldots V_n^h$, one has

$$T(u^h - v^h) = T(u - v^h) \qquad \text{in } V_h^*$$

According to estimate (3.5) and definition (3.10) we have

$$\|u^h - v^h\| <= C \;\|T(u - v^h)\|$$

$$<= C \{ (\|A\| + \|B_1\|) \|u_1 - v_1^h\|_{V_1}$$

$$+ \sum_{i=1}^{n-2} (\|B_i^*\| + \|B_{i+1}\|) \;\|u_i - v_i^h\|_{V_i} + \|B_{n-1}\| \|u_{n-1} - v_{n-1}^h\|_{V_{n-1}}$$

$$<= C_1 \sum_{i=1}^{n} \|u_i - v_i^h\|_{V_i}$$

Therefore we get that

$$\|u - u^h\| \leq \|u - v^h\| + \|u^h - v^h\| \leq (C_1 + 1) \sum_{i=1}^{n} \|u_i - v_i^h\|_{V_i}$$

Since $(v_1, v_2, \ldots, v_n)^h$ is arbitrary, the above inequality leads to estimate (3.8) directly. The proof is completed.

Example (simplified hybrid displacement method)

The abstract variational statement of P.Tong's simplified hybrid displacement method [6] is derived as follows :

$$(3.11) \quad \begin{cases} \text{Find } (u, \tilde{u}_\rho) \in V_0 \times \tilde{V}(\rho), \text{ s.t.} \\ a(u,v) + b(v, \tilde{u}_\rho) = \langle f, v\rangle \qquad \forall \quad v \in V_0 \\ \qquad\qquad b(u, \tilde{v}_\rho) = 0 \qquad\qquad \forall \quad \tilde{v}_\rho \in \tilde{V}(\rho) \end{cases}$$

where

$$a(u,v) = \sum_m \left\{ \int_{\Omega_m} \sigma_{ij}(u)\, \epsilon_{ij}(v)\, d\Omega - \int_{\rho_m} (u_i \sigma_{ij}(v) n_j + v_i \sigma_{ij}(u) n_j)\, ds \right\}$$

$$b(u, \tilde{v}_\rho) = \sum_m \int_{\rho_m} \sigma_{ij}(u) n_j \tilde{v}_{i\rho}\, ds \quad .$$

The associated functional of problem (3.11) is given by

$$(3.12)\qquad L(u,\tilde{u}_\rho) = \frac{1}{2} a(u,u) - \langle f,u\rangle + b(u,\tilde{u}_\rho)$$
$$= \sum_m \{ \int_{\Omega_m} (\frac{1}{2} \sigma_{ij}(u)\, \epsilon_{ij}(u) - \bar{f}_i u_i)\, d\Omega - \int_{S_{tm}} \bar{t}_i u_i \, ds - \int_{\rho_m} \sigma_{ij}(u) n_j (u_i - \tilde{u}_{i\rho})\, ds \} .$$

This is the same functional employed P.Tong in the "simplified" hybrid displacement method. Now this method can be seen to be unstable, since the bilinear form a(u,v) is not elliptic any more. As the ellipticity is violated, the functional (3.12) has even no minimization property. To see that let us assume u and $\tilde{u}_\rho$ are the solution of elasticity problem. Then one may derive that

$$L(u+v,\tilde{u}_\rho) - L(u,\tilde{u}_\rho)$$
$$= \sum_m \{ \int_{\Omega_m} \frac{1}{2} \sigma_{ij}(v)\, \epsilon_{ij}(v)\, d\Omega - \int_{\rho_m} \sigma_{ij}(v) n_j v_i \, ds \} .$$

This term is not necessarily always positive.

Acknowlegement --- The results presented herein were obtained under the suggestions and advicements from Prof. S.N. Atluri. The author is heartily grateful to his advisor Prof. Atluri.

REFERENCES

[1] F. Brezzi, On the existence, uniquemess, and approximation of saddle-point problem arising from Lagrange multipliers. R.A.I.R.O. 8-R2, pp,129-151 (1974)

[2] L-A. Ying and S.N. Atluri, A hybrid finite element method for Stokes flow : Part II --- stability and convergence studies. Comp. Meth. Appl. Engng. 36, pp. 39-60 (1983)

[3] W.M. Xue, L.A. Karlovitz,and S.N.Atluri, On the existence and stability conditions for mixed-hybrid finite element solutions based on Reissner's variational principle. Int. J. Solids Structures Vol.21, NO. 1, pp. 97-116 (1985)

[4] W.M. Xue, Existence, uniqueness, and stability conditions for generalmixed-hybrid finite element methods in linear elasticity, Ph.D. thesis, Georgia Institute of Technology, U.S.A. (1984)

[5] V.Girault and P.A.Raviart, Finite Element Approximations of the Navier-Stokes Equations. Lecture Notes in Mathematics, Vol. 479, Springer-Verlag (1979)

[6] P. Tong, New displacement hybrid finite element models for solid continua, Int. J. Num. Meth. in Engng. Vol.2, pp.73-83 (1970)

SOLUTIONS WITH DETONATION AND DEFLAGRATION WAVES*

Ying Lung-an

Peking University

§1 Introduction

Friedrichs has studied the following system of ordinary differential equations[1]

$$\rho u = M,$$
$$\rho u^2 + p - \mu \frac{\partial u}{\partial x} = P,$$
$$\rho u\left(\frac{1}{2}u^2 + I\right) - \mu u \frac{\partial u}{\partial x} - \lambda \frac{\partial T}{\partial x} = E,$$
$$-u\frac{\partial Z}{\partial x} = K,$$

which is a mathematical model for combustion problems with two species, where u is velocity, ρ is density, p is pressure, T is temperature, Z is the factor of unburnt gas. I is enthalpy, and constants M, P, E, K denote the flow of mass, pressure, energy, the rate of chemical reaction respectively, and constants μ and λ are viscosity and heat conductivity. This system is the formulation of one dimensional travelling waves, that is, all physical quantities are invariable with respect to a moving coordinate system.

To discuss general process, Chorin, Teng, Liu[2,3] considered the one dimensional compressible gas dynamic system

$$\frac{\partial \rho}{\partial t} + \frac{\partial}{\partial x}(\rho u) = 0,$$
$$\frac{\partial}{\partial t}(\rho u) + \frac{\partial}{\partial x}(\rho u^2 + p) = 0,$$
$$\frac{\partial e}{\partial t} + \frac{\partial}{\partial x}(eu + pu) = 0,$$

* This work was suppored by the Science Foundation of Academia Sinica, No. (84)-176

$$e = \rho\varepsilon + \frac{1}{2}\rho u^2, \quad \varepsilon = \frac{1}{\gamma - 1}\frac{p}{\rho},$$

and replaced the expression of ε by

$$\varepsilon_1 = \frac{1}{\gamma_1 - 1}\frac{p_1}{\rho_1} + q_1, \quad \varepsilon_0 = \frac{1}{\gamma_0 - 1}\frac{p_0}{\rho_0} + q_0,$$

where q_i is binding energy, ε_i is internal energy, i=0,1, i=0 represents unburnt gas, and i=1 represents burnt gas. They solved this system by means of the Glimm's scheme, to this end, they gave the structure of the solutions of Riemann problem, but did not consider admissibility. Therefore, the solutions are not unique.

The original system is very complicated, so only little about it has been known. Therefore it is interesting to consider model system. Majda gave a model system [4]

$$\frac{\partial(u + qZ)}{\partial t} + \frac{\partial f(u)}{\partial x} = \nu\frac{\partial^2 u}{\partial x^2}, \tag{1}$$

$$\frac{\partial Z}{\partial t} = -K\phi(u)Z, \tag{2}$$

where u is a lumped variable representing density, velocity, temperature, etc.,

$$\phi(u) = \begin{cases} 0, & u < 0, \\ 1, & u \geqslant 0, \end{cases}$$

and $f' > 0$, $f'' > 0$. He indicated that the above system admits strong and weak detonation travelling waves, and discussed the conditions for generating these waves.

We relaxed the assumption $f' > 0$, then (1) (2) admits some deflagration waves. The role which (1) (2) plays for combustion is the same as Burger's equation does for compressible fluids.

We are extremely interested in the case when $\nu = +0$ and $K = +\infty$, because shock waves, contact discontinuities, detonation waves and deflagration waves are mathematical discontinuous curves for this case. Certainly, the meaning of solutions should be clarified.

In this paper we consider two classic medels of (1) (2), the c-J model and the Z-N-D model, at first, then give some results for the initial value problem of system (1) (2). We will always assume that $f''(u) > 0$..

§2 The Chapman-Jouguet model

We consider equation

$$\frac{\partial(u+qZ)}{\partial t}+\frac{\partial f(u)}{\partial x}=0 \tag{3}$$

instead of (1), and let $K=+\infty$ formally in (2) and obtain

$$\frac{\partial Z(x,t)}{\partial t}\begin{cases}=0, & u(x,t)<0,\\ \leq 0, & u(x,t)\geq 0,\end{cases} \tag{4}$$

$$\phi(u(x,t))Z(x,t)=0. \tag{5}$$

Then we consider travelling waves

$$(u,Z)=\begin{cases}(u_0\ Z_0), & x-st<0,\\ (u_1,Z_1), & x-st>0,\end{cases} \tag{6}$$

where s, u_0, Z_0, u_1, Z_1 are all constants, solutions (6) satisfy (3)-(5) in the sense of distributions. By substituting (6) into equation (3) we get

$$s(u_0+qZ_0)-f(u_0)=s(u_1+qZ_1)-f(u_1). \tag{7}$$

There are three cases:

a. $s = 0$

By (7),

$$f(u_0) = f(u_1) \ .$$

By (5), $Z_i = 0$ as $u_i \geqslant 0$. If $u_0 = u_1$, $Z_0 \neq Z_1$, then line x=0 is defined as a contact discontinuity, if $u_0 > u_1$, $Z_0 = Z_1$, then it is defined as a shock wave, if $u_0 < u_1$, $Z_0 = Z_1$, then it is defined as a rarefaction shock wave. If u and Z are discontinuous at x=0, then line x=0 is the coincidence of two discontinuities.

b. $s > 0$

By (4) (5), either $Z_0 = Z_1 \neq 0$, or $Z_0 = 0$. In the former case

$$s = \frac{f(u_0) - f(u_1)}{u_0 - u_1} ,$$

line x=st is either a shock wave or a rarefaction shock wave, in the latter case

$$s = \frac{f(u_0) - f(u_1)}{u_0 - (u_1 + q Z_1)} ,$$

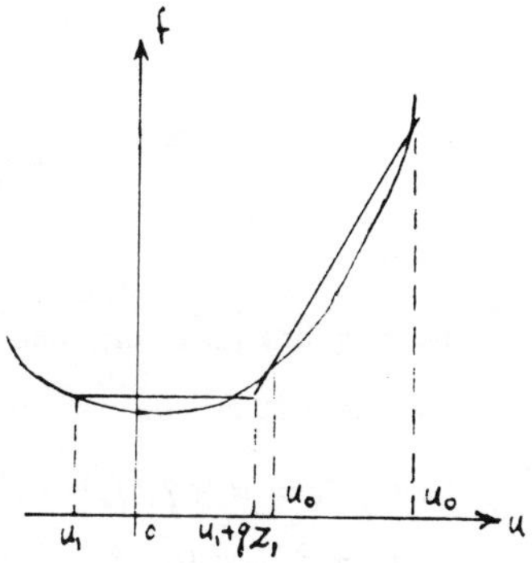

Figure 1

if the denominator and numerator vanish together, then s is indefinite. It is easy to see that $s > f'(u_1)$. Because $f'(u_1)$ is the eigenvalue at the positive side of discontinuity, this property corresponds to that the relative speed of the flow in front of the wave is supersonic. Such kind of discontinuity is defined as a detonation wave. The two possible positions of u_0 are shown at Figure 1, which correspond to $f'(u_0) > s$ and $f'(u_0) < s$, and are defined as strong and weak detonation waves. For the critical case, $f'(u_0)=s$, it is defined as a Chapman-Jouguet detonation wave.

c. $s<0$

By (4) (5), either $Z_0 = Z_1 \neq 0$, or $Z_1 = 0$. In the former case line $x=st$ is also a shock wave or a rarefaction shock wave, in the latter case

$$s = \frac{f(u_1) - f(u_0)}{u_1 - (u_0 + q Z_0)} \quad ,$$

then $f'(u_1) > s$, it is defined as a deflagration wave. If $f'(u_0) < s$, it is defined as a strong deflagration wave, just the same, we have weak deflagration wave for case $f'(u_0) > s$ and Chapman-Jouguet deflagration wave for critical case $f'(u_0)=s$.

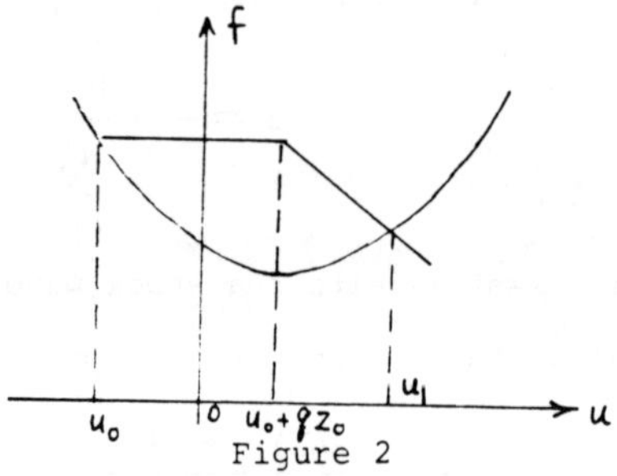

Figure 2

It is known that the weak solutions of the initial value problem of single conservation law

$$\frac{\partial u}{\partial t} + \frac{\partial f(u)}{\partial x} = 0 \qquad (8)$$

are not unique, and the rarefaction shock wave is not admissible. Now the situation is more complicated. We have noticed that if $u_0 + qZ_0 = u_1 + qZ_1$, $f(u_0)=f(u_1)$ in (7), then s is arbitrary, and if $u_0 + qZ_0 = u_1 + qZ_1$, but $f(u_0) \neq f(u_1)$, then (7) has no solutions. On the contrary, we will see in the following that Riemann problem always admits unique admissible solution.

§3 The Zeldovich-von Neumann-Doring model

We consider equations (2) (3), where K is finite. To simplify our discussion, we replace the original function ϕ by a smooth function, still denoted by ϕ, we take a constant $\beta<0$ and let

$$\phi(u)=\begin{cases} 0, & u\leq\beta, \\ \text{monotonic and smooth}, & \beta\leq u\leq 0, \\ 1, & u\geq 0. \end{cases}$$

Let u=u(x-st), z=z(x-st) be travelling waves, where s is constant, and we assume that the entropy condition

$$u(\xi-0)\geq u(\xi+0), \qquad \forall\,\xi\in(-\infty,+\infty) \tag{9}$$

is satisfied. Thus only shock waves and contact discontinuities are admissible. Equations (2) (3) then become

$$(f(u)-su)'=sq\,z' \tag{10}$$

$$sz'=K\phi(u)z\,. \tag{11}$$

If $s\neq 0$, we may integrate (11) once and obtain

$$z=C\exp\left\{\int K\phi(u(\xi))\,d\xi/s\right\}, \tag{12}$$

where C is a constant. We consider boundary value

$$(u,z)=\begin{cases}(u_0,z_0), & z=-\infty, \\ (u_1,z_1), & z=+\infty\end{cases} \tag{13}$$

for equations (10) (11). It is proved that [5]

a. If $u_0 \geq 0 > u_1$, $Z_0 = 0$, $Z_1 \neq 0$, $f(u_0) = f(u_1)$, then problem (10) (11) (13) admits solutions

s = 0,

$$u(\xi) = \begin{cases} u_0, & \xi < \xi_0, \\ u_1, & \xi > \xi_0, \end{cases}$$

$$Z(\xi) = \begin{cases} 0, & \xi < \xi_0, \\ \text{any, but } Z \to Z_1 \text{ as } \xi \to +\infty, & \xi > \xi_0, \end{cases}$$

where ξ_0 is a constant, provided $\beta > u_1$. And all solutions are in the above form.

b. If $u_0 \geq 0 > u_1$, $Z_0 = 0$, $Z_1 \neq 0$, $f(u_0) \neq f(u_1)$, then problem (10)(11)(13) admits solutions iff

$$f'(u_0) \geq \frac{f(u_0) - f(u_1)}{u_0 - (u_1 + \rho Z_1)} > 0 ,$$

provided $\beta > u_1$, s satisfies (7). We take $u_2 > u_1$ such that

$$s = \frac{f(u_2) - f(u_1)}{u_2 - u_1} ,$$

and let F(u) = f(u)-su, then

$$Z(\xi) = \begin{cases} Z_1 \exp\{K(\xi - \xi_0)/s\}, & \xi < \xi_0, \\ Z_1, & \xi \geq \xi_0, \end{cases}$$

$$u(\xi) = \begin{cases} F^{-1}(F(u_2) + s\rho Z_1 \exp\{K(\xi - \xi_0)/s\} - s\rho Z_1), & \xi < \xi_0, \\ u_1, & \xi \geq \xi_0. \end{cases}$$

c. If $u_1 \geq 0 > u_0$, then problem (10)(11)(13) admits no solution.

All the above expressions are independent of β, we may let $\beta \to -0$, which implies that the solutions are the same even if ϕ is the discontinuous function given in §1. We define admissible solutions for $K = +\infty$ by passing to limit as $K \to +\infty$. It can be proved that only strong and Chapman-Jouguet detonation waves are admissible among detonation and deflagration travelling waves. Besides the above three cases, all other cases are trivial.

If we consider the initial value problem for (2)(3) and only consider the admissible solution in the sense of Lax and Vol'pert[6,7], then it is proved that[5]

Theorem 1 The admissible solution of equations (2)(3) and initial value

$$u(x,0) = u_0(x), \quad z(x,0) = z_0(x) \tag{14}$$

exists and is unique on the half plane $t \geq 0$, where u_0, z_0 are functions with bounded variation and $0 \leq z_0(x) \leq 1$.

§4 The Riemann Problem

We consider a special case of initial value (14)

$$(u(x,0), z(x,0)) = \begin{cases} (u_0, z_0), & x < 0, \\ (u_1, z_1), & x > 0. \end{cases} \tag{15}$$

By Theorem 1, there is a unique solution. We define admissible solutions for (1)(2)(15) as $\nu = +0$, $K = +\infty$ by passing to limit as $\beta \to -0$ $K \to +\infty$. If $f''(u) \geq \alpha > 0$ for positive u, where α is a positive constant then the admissible solution of the above Riemann problem exists and is unique. We illustrate the solution in the following, the detail is in [5].

1. $u_0 \geqslant 0 > u_1$, $z_1 > 0$, $f(u_0) > f(u_1)$

a. $u_0 > u_1 + q z_1$, $f'(u_0) > \dfrac{f(u_0) - f(u_1)}{u_0 - (u_1 + q z_1)}$

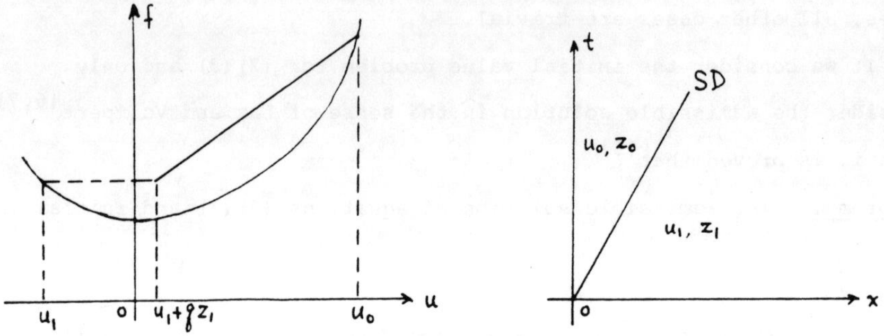

Figure 3

b. $u_0 \leqslant u_1 + q z_1$, or $f'(u_0) \leqslant \dfrac{f(u_0) - f(u_1)}{u_0 - (u_1 + q z_1)}$

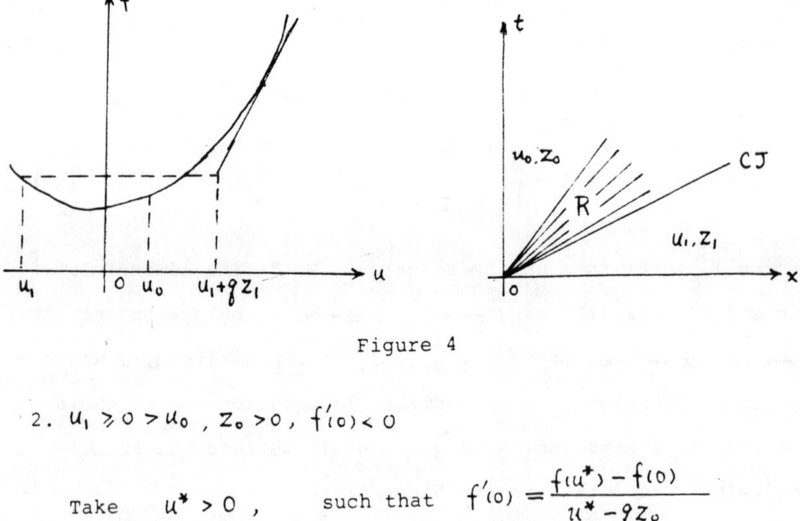

Figure 4

2. $u_1 \geqslant 0 > u_0$, $z_0 > 0$, $f'(0) < 0$

Take $u^* > 0$, such that $f'(0) = \dfrac{f(u^*) - f(0)}{u^* - q z_0}$

a. $u^* > u_1$

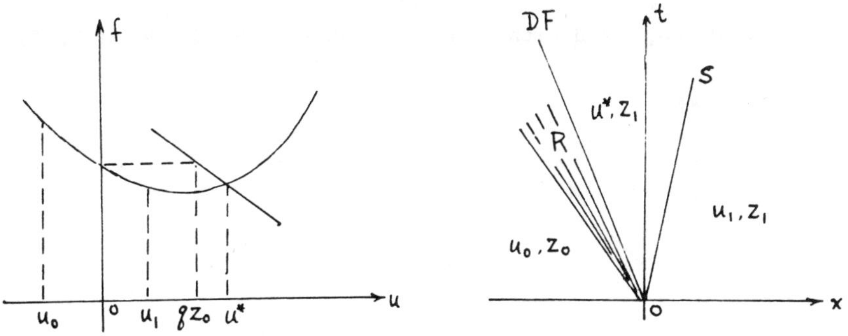

Figure 5

b. $u^* \leqslant u_1$

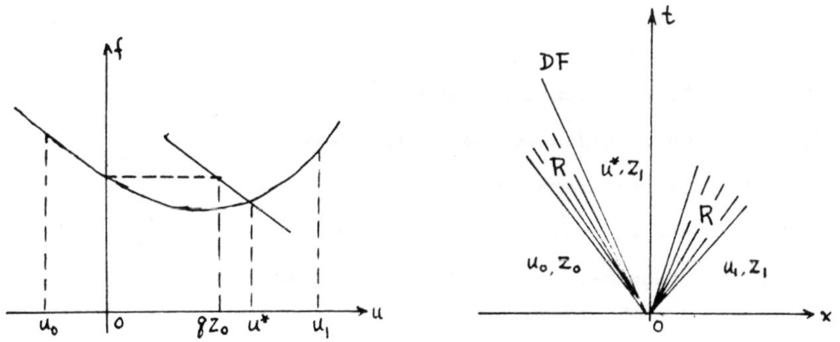

Figure 6

3. The other cases

a. $u_1 \geqslant u_0$ and either $f'(0) \geqslant 0$ or $f'(0) < 0,\ 0 \bar{\in} (u_0, u_1]$

or $Z_0 = Z_1 = 0$

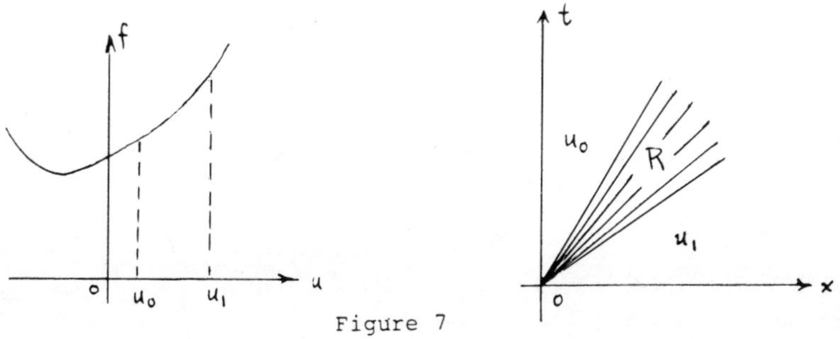

Figure 7

b. $u_0 > u_1$ and either $f(u_0) \leq f(u_1)$, or $0 \bar{\in} (u_1, u_0]$, or $Z_0 = Z_1 = 0$

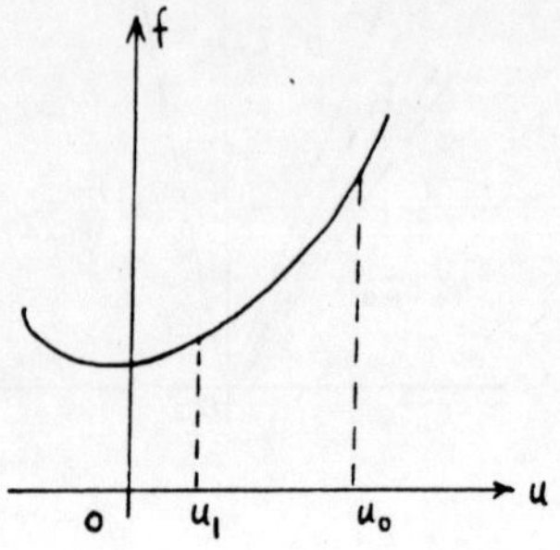

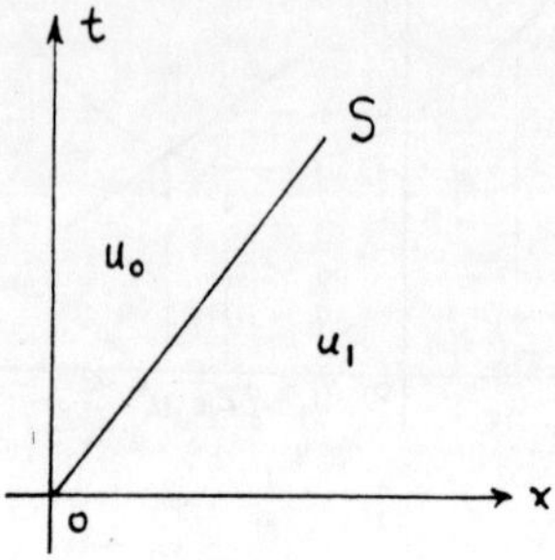

Figure 8

Abbreviations:

SD: Stong detonation wave,

CJ: Chapman-Jouguet detonation wave,

R: central rarefaction wave,

S: shock wave,

DF: Chapman-Jouguet deflagration wave.

§5. The general initial value problem

We consider problem (1)(2)(14) at the case $\nu = +0$, $K = +\infty$. Let u_0, Z_0 be bounded measurable functions, and for almost every x, if $u_0(x) \geq 0$, then $Z_0(x) = 0$. We define the solution of this problem as follows:

Problem P To find bounded measurable functions $u(x,t)$, $z(x,t)$, defined on $t \geq 0$, such that $t \mapsto u(.,t)$ is continuous from $[0,\infty)$ to $L^1_{loc}(-\infty,+\infty)$, and

$$\lim_{h \to +0} \frac{1}{h} \int_{x-h}^{x} u(\xi,t)\,d\xi = u^-(x,t), \qquad \forall\, x \in (-\infty,+\infty),\ t \geq 0,$$

exists, and for any smooth function $\varphi(x,t)$ with compact support on $t \geqslant 0$

$$\iint_{t\geq 0} \left\{ \frac{\partial \varphi}{\partial t}(u+\rho z) + \frac{\partial \varphi}{\partial x} f(u) \right\} dx\,dt + \int_{-\infty}^{+\infty} \{ u_0(x) + \rho z_0(x) \} \varphi(x,0)\, dx = 0,$$

moreover, for any non-negative smooth function $\varphi(x,t)$ with compact support on $t \geqslant 0$,

$$\iint_{t\geq 0} \frac{\partial \varphi}{\partial t} z\, dx\,dt + \int_{-\infty}^{+\infty} z_0(x)\varphi(x,0)\, dx \geqslant 0,$$

and finally with

$$v(x,t) = \sup_{0 \leqslant \tau \leqslant t} u^-(x,\tau)$$

we have

$$z(x,t) = \begin{cases} 0, & \text{if } v(x,t) > 0, \\ z_0(x), & \text{if } v(x,t) < 0. \end{cases}$$

<u>Problem Q</u> To find solutions $u(x,t)$, $z(x,t)$ of Problem P which satisfy, in addition, $z(x,t) = 0$ if $v(x,t) = 0$

We define an auxiliary function ψ, if $f'(0) \geqslant 0$, then let

$$\psi(u) = \begin{cases} K_0 u & u \leqslant 0, \\ u & u > 0, \end{cases}$$

otherwise

$$\psi(u) = \begin{cases} K_0 u, & u \leqslant 0, \\ u - \dfrac{f(u) - f(0)}{f'(0)} & u > 0, \end{cases}$$

where K_0 is a suitable constant. By means of the solutions in §4 and using approximate method, we can prove[8,9].

Theorem 2 If $\psi(u_0(x))$ and $Z_0(x)$ are functions with bounded variation, then the solution of Problem P exists, and $\psi \circ u, Z \in BV$.

To prove the existence of the solution of Problem Q, we need one more definition and hypothesis.

Definition It is said that $u_0(x)$ assumes the value $\tilde{u}$ at point x, if one of the following holds: either $u_0(x-0) = \tilde{u}$, or $u_0(x+0) = \tilde{u}$, or $u_0(x+0) > u_0(x-0)$ and $\tilde{u} \in (u_0(x-0), u_0(x+0))$.

Hypothesis A If $u_0(x)$ assumes the value 0 at a point x, then there is an interval $[b,c] \ni x$ such that $u_0(x) \geqslant 0$ on (b,c).

Theorem 3 If $\psi(u_0(x))$ and $Z_0(x)$ are functions with bounded variation and Hypothesis A holds, then the solution of Problem Q exists, and $\psi \circ u, Z \in BV$.

We do not know if Hypothesis A is necessary, perhaps we can get stronger results by improving our technique.

§6 Some open problems

a. If the solutions in Theorem 1 tend to a limit when $\beta \to -0$, $K \to +\infty$, then this is a natural way to define the solution of problem (1)(2)(4) at the case $\nu = +0$, $K = +\infty$. Unfortunately we still do not know if it is true.

b. We do not know how to give an "entropy condition" for Problem Q, so uniqueness is unknown.

c. It is known that the limit depends on the relation of K and ν when $K \to +\infty$ and $\nu \to +0$. Therefore it is interesting that if $K\nu$ is a constant, the limit exists or not, and how the limit varies along with constant $K\nu$.

References

[1] Courant R., Friedrichs K.O., Supersonic Flow and Shock Waves, Interscience Publishers Inc., New York, 1948.

[2] Chorin A., Random choice methods with applications to reacting gas flow, J. Comp. Phys., 25 (1977), 253-272.

[3] Teng Z.H., Chorin A., Liu T.P., Riemann problems for reacting gas, with applications to transition, SIAM J. Appl. Math., 42 (1982), 946-981.

[4] Majda A., A qualitative model for dynamic combustion, SIAM J. Appl. Math., 41 (1981), 70-93.

[5] Ying L.A., Teng Z.H., Riemann problem for a reacting and convection hyperbolic system, J. Appr. Theory Appl., 1,1 (1984), 95-122.

[6] Lax P., Shock waves and entropy, in contributions to Nonlinear Functional Analysis (E.Zarantonello ed.) Academic Press, New York, 1971, 603-634.

[7] Vol'pert A.I., The space BV and quasilinear equations, Math. USSR Sb. 2 (1967), 257-302.

[8] Ying L.A., Teng Z.H., A hyperbolic model of combustion, in Nonlinear Partial Differential Equations in Applied Science, Kinokyniya, North Holland, 1983, 409-434.

[9] Ying L.A., Teng Z.H., Global existence theorems for solutions with detonations and deflagrations (to appear).

SOME RESULTS FOR CONSERVATION LAWS IN TWO SPACE DIMENSIONS

Chang Tung

Institute of Mathematics, Academia Sinica
Beijing, China

1. Chang Tung and Chen Gui-Qiang, Some Fundamental Concepts for Conservation Laws in Two Space Dimensions

Lax[1] and others[2-5] had established some fundamental concepts for system of conservation laws in one space dimension, such as strictly hyperbolicity, genuine nonlinearity (convexity), linear degenerate, entropy condition (stability condition for convex case, stability condition (E) for non-convex case) etc. We have extended all of these concepts to system of conservation laws in two space dimensions as follows:

$$u_t + f(u)_x + g(u)_y = 0 \tag{1}$$

where $u = (u_1, u_2, \ldots, u_n)$, $f = (f_1, f_2, \ldots, f_n)$, $g = (g_1, g_2, \ldots, g_n)$.

This is a necessary preliminary for advanced study.

Stability condition (E), the entropy condition for system without convexity, has been obtained by Gel'fand[2], Oleinik and Kalasnikov[3] for scalar one spatial dimension conservation law, by Chang and Hsiao[4], and Liu[5] for system of one spatial dimension case, and by Kruzkov[6] for scalar multi-dimension case. We have extended them to system of two spatial dimension case.

2. Chang Tung and Yu-Xi Zheng, Two-Dimensional Riemann Problem for a Single Conservation Law

The Riemann problem has been a key to the development of the theory of one-dimensional conservation laws. We consider the Riemann problem for a single conservation law (1) $(n=2)$, i.e. (1) with initial data constant in each quadrant of the (x, y) plane. This problem has been solved by Wagner[9] under the assumptions $f(u) \equiv g(u)$, $f''(u) \neq 0$ which state that the problem can be transformed to be of one space dimension through the transformation $x' = \frac{x+y}{2}$, $y' = \frac{x-y}{2}$. Instead of the assumption $f \equiv g$, we consider the simplest genuine two-dimensional case, i.e. $(f''(u)/g''(u))' \neq 0$ which means (1) has at most one inflexion point in each direction (μ, ν). Under this assumption we solved the Riemann problem completely. We constructed the solution by characteristic method.

3. Chang Tung and Chen Gui-Qiang, Diffraction of Planar Shock along a Compressive Corner.

We consider the adiabatic flow of ideal gas. When a planar shock strike on a wedge, four different types of construction of solution have been observed in the laboratory. They are regular, single-Mach, complex-Mach, double-Mach reflexions. The transition criteria between different types were first studied by von Neumann[7].

We have obtained the analytical representation of the necessary and sufficient condition for the existence and uniqueness of stability solution of regular reflexion in the neighbourhood of reflexion point on the rigid wall. It is

$$\tan^2\theta \geq \frac{1}{\eta}\left(\sqrt[3]{-\frac{q}{2} + \sqrt{\frac{q^2}{4} + \frac{p^3}{27}}} + \sqrt[3]{-\frac{q}{2} - \sqrt{\frac{q^2}{4} + \frac{p^3}{27}}} - 1\right) > 0$$

where

θ - half of the angle of the wedge,

$\eta > 1$ - the ratio of densities on both sides of the incident shock,

$$p = \frac{3ac - b^2}{3a^2} \ (< 0) \quad ,$$

$$q = \frac{2b^3 - 9abc + 27a^2 d}{27a^3} \ (< 0) \quad ,$$

$$a = \eta^2 (> 0) \quad ,$$

$$b = -\eta((3\eta - 2) + (\gamma - 1)(\eta - 1)((\gamma + 1)(\eta - 1) + 2)) \ (< 0) \quad ,$$

$$c = -(\eta - 1)(2(\gamma + 1)\eta(\eta - 1) + (\eta + 1)) \ (< 0) \quad ,$$

$$d = -(\eta - 1)^2 \ (< 0) \quad .$$

This is a refinement of von Neumann's detachment criterion for transition between regular reflexion and Mach reflexion.

We have given a new transition criterion between complex-Mach reflexion and double-Mach reflexion. We think that the concept of "kink point" in the old transition criterion appearing in Ref. 8 is not clear enough.

References

1. Lax, P.D., Commun. Pure Appl. Math. 10 (1957) 537-566.
2. Gel'fand, I., Uspehi Mat. Nauk 14 (1959) 87-158.
3. Oleinik, O. and Kalasnikov, A., Proc. Diff. Eqs. (1960), 133-137, Yerevan.
4. Chang Tung and Hsiao Ling, Acta Math. Sinica 20 (1977) 229-231 (in Chinese).
5. Liu, T.P., Mem. Amer. Soc., No. 240, Providence, Amer. Math. Soc. (1981).
6. Kruzkov, N., Math. Sbornik 83 (123) (1970).
7. Neumann, J.von, Collected Works, Vol. 6, Pergamon Press (1963).
8. Ben-Dor, G. and Glass, I., J. Fluid Mech. 92 (1979) 459-496.
9. Wagner, D., J. Math. Anal., Vol. 14. No. 3, 534-559, SIAM (1983).

EQUILIBRIUM SOLUTIONS OF A KIND OF THE RESTRCTED PROBLEM OF 3+1 BODIES

Zhang Xiang-Ling (張翔齡)

(Department of Mathematics. Leshan Normal College
Sichaun, China.)

ABSTRACT

This paper discusses the motion of the fourth body P when three primaries P_1, P_2, P_3 satisfy the Lagrange equilateral solution. Some equilibrium solutions of P are found and discussed. As application, the topological structure of the region of the possible motion of P is given, when $m_1 = m_2 = m_3$ where m_i is the mass of Pi (i = 1, 2, 3).

I Introduction

For the research of the planar n-body problem, S. Smale pointed out we know little about the number of the equilibrium solutions and their properties[1]. But they are the most important. Recently, V. Szebehely and L. Whipple have turned the classical restricted problem of 3-body to be the restricted problem of n+ν bodies. And L. Whipple studied the equilibrium solutions of restricted problem of 2+2 bodies[2][3].

Here in this paper are studied the equilibrium solutions and their properties of a kind of restricted problem of 3+1 bodies. As application, the topological structure of the possible region of the fourth body is given when the masses of three primaries are equal each other.

We use P_1, P_2 and P_3 to denote the three primaries, their masses are m_1, m_2, and m_3, respectively. While P represents the fourth body. Its mass is m and $m \ll m_i$ (i = 1, 2, 3). Neglect the influence of P to P_i (i = 1, 2, 3).

This paper studies the equilibrium solutions of P when $m_2 = m_3$

and P_1, P_2 and P_3 satisfied the Lagrarge equilateral solution.

II The Equation of Motion

When P_1, P_2, P_3 satisfy Lagrange equilateral solution, they move on circular orbits around their center of mass. Their angular velocity is $\omega = (GM/r^3)^{\frac{1}{2}}$, where $M = m_1 + m_2 + m_3$, r is the side of the equilateral triangle, G is the constant of gravitation.[4] Let $M = m_1 + m_2 + m_3 = 1$, $r = 1$. We may select the unit of time to make $G = 1$ and $\omega = 1$. Let $\mu = m_1/m_1 + m_2 + m_3 = m_1/m_1 + 2m_1$, then $m_2/m_1 + m_2 + m_3 = m_3/m_1 + m_2 + m_3 = \frac{1}{2}(1 - \mu)$.

Now we introduce the rotating coordinate system O-xy. Let the center of mass of P_i (i = 1, 2, 3) be the origin. The line that links P_1 with the mid-point of side P_2P_3 is the x-axis of the rotating coordinate system. The angular velocity of the coordinate system around O is equal to 1. Then the coordinates of P_1, P_2, P_3 and P are $(\frac{\sqrt{3}}{2}(1-\mu), 0)$, $(-\frac{\sqrt{3}}{2}\mu, \frac{1}{2})$, $(-\frac{\sqrt{3}}{2}\mu, -\frac{1}{2})$ and (x, y) respectively in this coordinate system (Fig. 1).

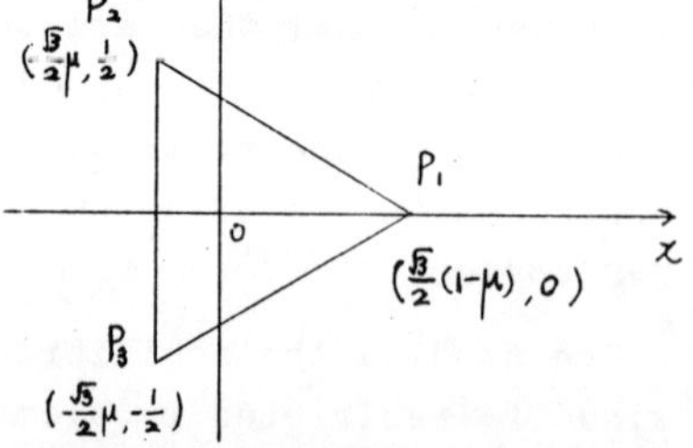

(Fig. 1)

The equation of motion of P is

$$\begin{cases} \ddot{x} - 2\dot{y} = \dfrac{\partial \Omega}{\partial x} \\ \ddot{y} + 2\dot{x} = \dfrac{\partial \Omega}{\partial y} \end{cases} \tag{2.1}$$

Where

$$\Omega(x, y) = \frac{1}{2}(x^2 + y^2) + \frac{\mu}{r_1} + \frac{\frac{1}{2}(1-\mu)}{r_2} + \frac{\frac{1}{2}(1-\mu)}{r_3}$$

$$r_1 = \sqrt{(x - \frac{\sqrt{3}}{2}(1-\mu))^2 + y^2} \quad ;$$

$$r_2 = \sqrt{(x + \frac{\sqrt{3}}{2}\mu)^2 + (y - \frac{1}{2})^2} \quad ;$$

$$r_3 = \sqrt{(x + \frac{\sqrt{3}}{2}\mu)^2 + (y + \frac{1}{2})^2} \quad ;$$

$$0 \leqslant \mu \leqslant 1$$

By Eq. (2.1), we can obtain an integral of the equation of motion

$$v^2 = \dot{x}^2 + y^2 = 2\Omega(x, y) - c$$

C represents the integral constant, V is the velocity of P about coordinate system $0 - xy$.

III The Equilibrium Solutions

The equilibrium solutions of Eq. (2. 1) satisfy

$$\frac{\partial \Omega}{\partial x} = 0 , \qquad \frac{\partial \Omega}{\partial y} = 0.$$

It is also written as

$$x - \frac{x - \frac{\sqrt{3}}{2}(1-\mu)}{r_1^3} - \frac{\frac{1}{2}(1-\mu)(x + \frac{\sqrt{3}}{2}\mu)}{r_2^3} - \frac{\frac{1}{2}(1-\mu)(x + \frac{\sqrt{3}}{2}\mu)}{r_3^3} = 0 \qquad (3.1)$$

$$y - \frac{\mu y}{r_1^3} - \frac{\frac{1}{2}(1-\mu)(y - \frac{1}{2})}{r_2^3} - \frac{\frac{1}{2}(1-\mu)(y + \frac{1}{2})}{r_3^3} = 0$$

It is easy to show that the points located on $y = 0$ satisfy the second equation of (3.1). In this paper, we study only the equilibrium solutions located on $y = 0$.

Thus $r_2 = r_3 = \sqrt{(x + \frac{\sqrt{3}}{2}\mu)^2 + \frac{1}{4}}$, $r_1 = \sqrt{\left[x - \frac{\sqrt{3}}{2}(1-\mu)\right]^2}$ and

$$\Omega_x(x,0) = x - \frac{x - \frac{\sqrt{3}}{2}(1-\mu)}{r_1^3} - \frac{(1-\mu)(x + \frac{\sqrt{3}}{2}\mu)}{r_2^3} = 0 \qquad (3.2)$$

Following properties are the principal results of this paper. But all proofs of these properties are omitted.

Property 1

(i) In $(-\infty, -\frac{\sqrt{3}}{2}\mu)$, $\forall\mu$, there is a unique equilibrium solution of Eq. (2.1). Denote it by $L_1: (x_{L_1}, 0)$.

(ii) In $(-\frac{\sqrt{3}}{2}\mu, \frac{\sqrt{3}}{2}(1-\mu))$, $\exists\,\mu_0$ which satisfies $0 < \mu_0 < 1$.

When $\mu < \mu_0$ there are two equilibrium solutions of Eq. (2.1). Denote them by $L_2: (x_{L_2}, 0)$, $L_3: (x_{L_3}, 0)$.

When $\mu = \mu_0$ there is an equilibrium solution of Eq. (2.1). And $L_2 = L_3$.

When $\mu > \mu_0$ there is not any equilibrium solution of Eq. (2.1).

And $\mu_0 = 0.423461$.

(iii) In $(\frac{\sqrt{3}}{2}(1-\mu), +\infty)$, $\forall\mu$, there is unique equilibrium solution of Eq. (2.1). Denote it by $L_4: (x_{L_4}, 0)$.

Property 2

(i) $\forall\mu \in [0,1]$, the minima of the function $\Omega(x, y)$ occurs at L_1. L_4 is saddle point of $\Omega(x, y)$.

(ii) When $0 < \mu < \mu_0$, L_2 is saddle point of $\Omega(x, y)$, and the minima of the function $\Omega(x, y)$ occurs at L_3.

(iii) When $\mu = \mu_0$, L_2, L_3 ($L_2 = L_3$) are saddle points of $\Omega(x, y)$.

Property 3

(i) $\Omega(x_{L_4}, 0) \geqslant \Omega(x_{L_1}, 0)$.

(ii) When $0 \leq \mu \leq \mu_0$,

$$\Omega(x_{L_2}, 0) \geq \Omega(x_{L_3}, 0) \geq \Omega(x_{L_4}, 0) \geq \Omega(x_{L_1}, 0).$$

The first equality holds, if $\mu = \mu_0$. The other equalities hold, if $\mu = 0$.

IV Application

Using the results mentioned above, we can obtain the structure of the possible region of P in term of the energy of P, when $m_1 = m_2 = m_3$. Notice the following properties, first.

Property 1

When $m_1 = m_2 = m_3$, there are ten equilibrium solutions of P. Divide them into four classes (Fig. 2): $\{L_1, L_1', L_1''\}$ $\{L_2, L_2', L_2''\}$, $\{L_3\}$, $\{L_4, L_4', L_4''\}$.

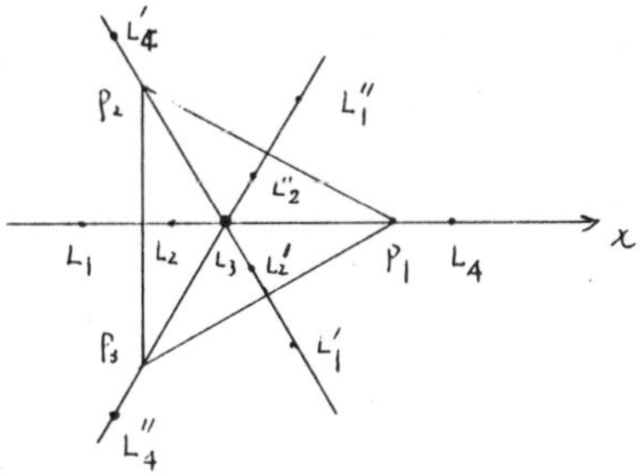

(Fig. 2)

Proof: It is proved easily by the results of (III), symmetry of the system and continuation of the functions $\Omega_x(x, y)$, $\Omega_y(x,y)$.

Property 2

$$\Omega(L_2)=\Omega(L_2')=\Omega(L_2'')>\Omega(L_3)>\Omega(L_4)=\Omega(L_4')=\Omega(L_4'')>\Omega(L_1) =\Omega(L_1')=\Omega(L_1''),$$

if $0 \leq \mu \leq \mu_0$

Proof: It is a corollary of the property 3 of (III).

The region of possible motion of the fourth body P is the point set

$$R = \left\{ (x, y) \;\middle|\; 2\Omega(x, y) > C \right\}$$

in xy - plane, where c is determined by the energy of P. The curve $2\Omega(x, y) = 0$ is called the curve of zero velocity.

Property 3

(i) When $C > 2\Omega(L_2) = 2\Omega(L_2') = 2\Omega(L_2'') = C_2$, the region of possible motion of P has four connected components. They are shown schematically on Fig. 3 by shading.

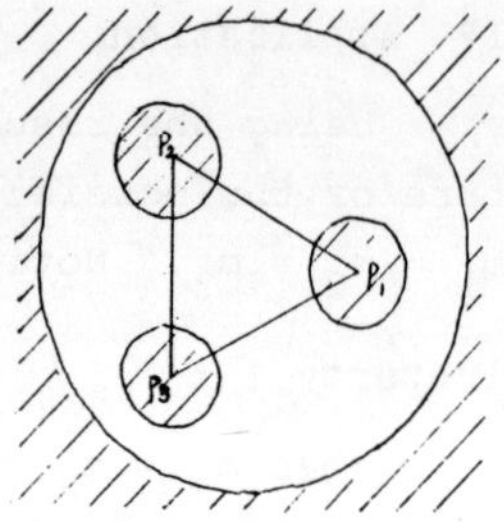

Fig. 3

(ii) When $C_2 \geqslant C \geqslant 2\Omega(L_3) = C_3$ the region of possible motion of P has two connected components They are shown schematically on Fig. 4 by shading

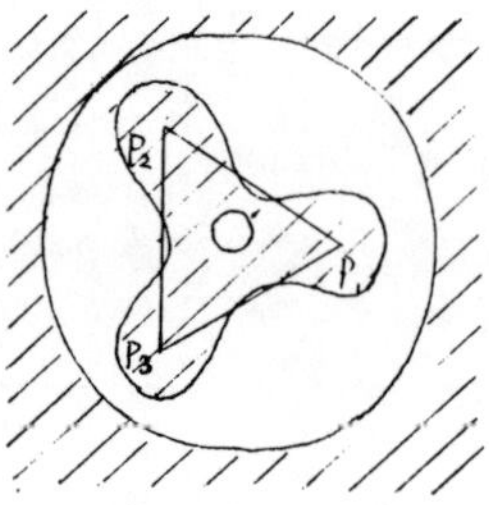

Fig. 4

(iii) When

$$C_3 \geqslant C \geqslant 2\Omega(L_4) = 2\Omega(L_4') = 2\Omega(L_4'') = C_4,$$

the region of possible motion of P has two connected components. They are shown schematically on Fig. 5 by shading.

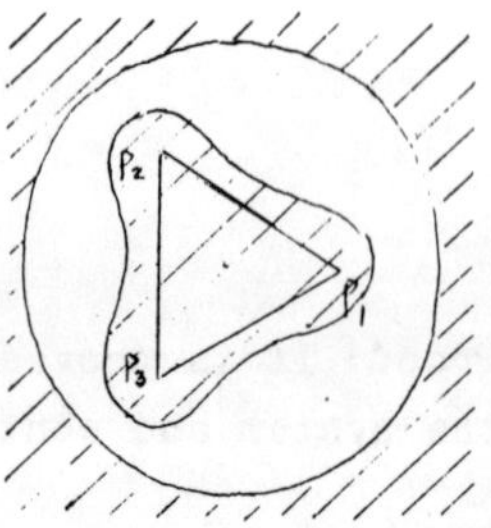

Fig. 5

(IV) When

$$C_4 \geqslant C \geqslant 2\Omega(L_1) = 2\Omega(L_1') = 2\Omega(L_1'') = C_1,$$

the region of possible motion of P has one connected component. It is shown schematically on Fig. 6 by Shading.

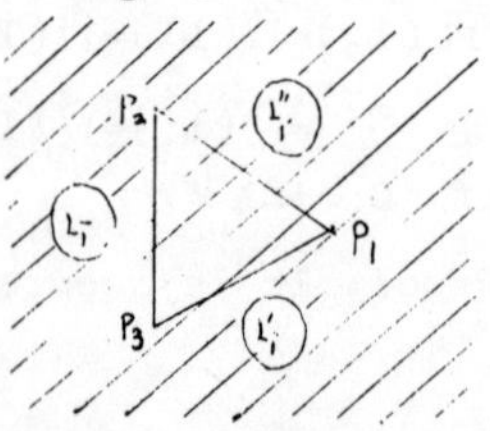

Fig. 6

(V) When $C \leqslant 2\Omega(L_1) = 2\Omega(L_1') = 2\Omega(L_1'') = C_1$, the region of possible motion of P is xy - plane.

Thus, we complete the discussion.

References

1 Smale, S., Inventions Math, 10 (1970), 305.

2 Whipple, A, L. and Szebehely, V., Celest. Mech 32 (1984), 137.

3 Whipple, A. L., Celest. Mech 33 (1984), 271.

4 Siegel C. L. and Moser, J. K., Lectures on Celestial Mechanics (1971).

Robustness of Optimal Control for Linear-quadratic Regulators*

DA-ZHONG ZHENG

Department of Automation
Tsinghua University
Beijing, China

Abstract - The purpose of this paper is to investigate the problem of optimization of linear-quadratic regulator systems in the presence of parameter perturbations. The permitted region is obtained for allowable perturbation parameters such that the resulting closed-loop system remains optimal. An algorithm is also given for its construction. The algorithm has recursive form and is therefore suitable for calculation by computer. The size of the permitted region is taken as measures of the robustness of optimal control for linear-quadratic regulator systems. It will further be shown by examples that the size of the permitted region can be expanded by increasing the degree of the positive-definite of the weighting matrices in the quadratic performance index or using the performance index with prescribed degree of stability. Thereby it gives the possibility to help a designer to select appropriate weighting matrices in the quadratic performance index or appropriate degree of stability to attain a robust design.

I. INTRODUCTION

It is well known that the optimal linear-quadratic regulator has many important and interesting properties. However, the usefulness of these properties of a linear-quadratic regulator design is very much dependent on the availability of an accurate mathmatical model of the system. In the real world, the situation of parameter perturbations in the system is not avoidable because of modelling errors and parameter variations. The parameter perturbations are often very large such that the performance index and other properties of a practical regulator are very different from those in design. Therefore, it is necessary to investigate the robustness of the optimal linear-quadratic regulator systems in the presence of system parameter perturbations.

During the past decade, many papers have been published in which the robustness of the optimal regulator has been discussed and studied. However, these papers dealt with only the stability and

*Projects supported by the Sciences Fund of the Chinses Academy of Sciences

suboptimization for the linear-quadratic regulator systems in the presence of perturbations. Examples are the works of Patel et al[1], Šiljak [2,3], Krtolica and Šiljak[4], and Sezer and Šiljak[5].

This paper addresses the problem of the robustness of optimal control for the linear-quadratic regulators in terms of the permitted region for allowable perturbation parameters such that the resulting closed-loop system remains optimal. The permitted region can be easily determined by using the algorithm given in this paper. It is obivous that the larger the size of the permitted region, the larger the possibility of the closed-loop system remaining optimal in the presence of perturbations. Therefore, the size of the permitted region can be used as a measure for the system robustness to the plant parameter perturbations. Relations between size of the permitted region and the weighting matrices in the quadratic performance index or the degree of stability prescribed in performance index are discussed, thereby helping a designer to select appropriate weighting matrices or appropriate degree of stability in the quadratic performance index to attain a robust design.

This paper is organized as follows. In Section II, the concept on the permitted region,for allowable perturbation parameters such that the resulting closed-loop system remains optimal, is introduced. Section III gives the conditions under which the quadratic regulator can remain optimal in the presence of parameter perturbations. In Section IV, an algorithm is given for determining the permitted region. In Section V, we introduce two ways to expand the size of the permitted region and give the examples for illustration. Finally, we conclude in Section VI.

II. THE PERMITTED REGION OF PERTURBATION PARAMETERS

Let us consider the standard linear-quadratic regulator problem. The system and the performance index to be minimized are expressed, respectively, by differential equations of the form

$$\delta_o: \quad \dot{x}=A_o x+Bu, \qquad x(0)=x_o \tag{1}$$

and

$$J=\int_o^{\infty}(x^T Q_o x+u^T Ru)dt \tag{2}$$

where $x(t)$ is an n-dimentional state and u is an r-dimentional input of δ_o; A_o and B are the constant marices and denote the system model parameters; R is symmetric positive definite matrix; Q is symmetric positive definite mateix or is a symmetric positive semidefinite matrix with the pair (A, $Q^{1/2}$) being observable; the pair (A,B) is controllable.

As is well known [6], the optimal control law and the optimal cost are given, respectively, by

$$u^*=-K_o x, \qquad K_o=R^{-1}B^T\bar{P} \tag{3}$$

and

$$J^*=x_o^T\bar{P}x_o, \qquad \forall x_o \neq 0 \tag{4}$$

where $\bar{P}$ is the unique symmetric positive definite solution of the following algebraic Riccati equation

$$\bar{P}A_o+A_o^T\bar{P} +Q_o -\bar{P}BR^{-1}B^T\bar{P} = 0 \tag{5}$$

The optimal closed-loop system is

$$\delta_{oc}: \quad \dot{x}=(A_o-BK_o)x, \qquad x(0)=x_o \tag{6}$$

It has been also proven that the optimal closed-loop system δ_{oc} has the following important and useful properties:

Theorem 1[7]: If $R=\mathrm{diag}(\rho_1, \ldots, \rho_r)$, then each feedback loop of δ_{oc} has

1) infinite gain margin,
2) at least $\pm 60^\circ$ phase margin, and
3) at least 50 percent gain reduction tolerance.

Theorem 2[6]: Suppose a vector valued nonlinearity Φ is inserted in the feedback path of δ_{oc} to yield a system $\hat{\delta}_{oc}$:

$$\hat{\delta}_{oc}: \quad \dot{x}=A_o x+B\Phi(\sigma), \; \sigma=-K_o x \tag{7}$$

Then the equilibrium x=0 of the system $\hat{\delta}_{oc}$ is asymptotically stable in the large for all $\Phi(\sigma)$ satisfying the sector condition

$$k_1\sigma^T R\sigma \le \sigma^T R\,\Phi(\sigma) \le k_2\sigma^T\sigma, \qquad \text{for all } \sigma \tag{8}$$

where $k_1 > 1/2$ and $k_2 < \infty$.

Now, let us consider a perturbation of δ_o

$$\delta_p: \quad \dot{x}=(A_o+A_p)x+Bu, \qquad x(0)=x_o \tag{9}$$

where A_p is a constant matrix and is expressed by

$$A_p=\begin{bmatrix} \varepsilon_{11} & \cdots & \varepsilon_{1n} \\ \vdots & & \vdots \\ \varepsilon_{n1} & \cdots & \varepsilon_{nn} \end{bmatrix} \tag{10}$$

In the engineering problem, usually only some elements in system matrix are pertuebed. Therefore, we assume that

$$\varepsilon_{ij}\begin{cases} \neq 0, & i=\alpha, \beta, \ldots, \gamma; \quad j=\xi, \eta, \ldots \varphi, \\ = 0 & \text{on the other locations} \end{cases} \tag{11}$$

where $\alpha, \beta, \ldots, \gamma \in [1,2,\ldots,n]$ and $\xi, \eta, \ldots, \varphi \in [1,2,\ldots,n]$.
If we apply the same control (3) to δ_p, then the closed-loop perturbed system is given by

$$\delta_{pc}: \quad \dot{x}=(A_o+A_p-BK_o)x, \qquad x(0)=x_o \tag{12}$$

In general, the cost and other properties of δ_{pc} are very different from the optimal cost (4) and properties of δ_{oc}. This leads to the concept of the permitted region for allowable perturbation parameters such that the closed-loop perturbed system δ_{pc} remains optimal.

Definition: Let $a_{ij}= a^o_{ij}+ \varepsilon_{ij}$ denote the allowable perturbation parameters in system matrix, where a^o_{ij} are the model parameters in A_o, ε_{ij} are the elements in A_p, and $i= \alpha, \beta, \ldots \gamma$, $j= \xi, \eta, \ldots, \varphi$.

Let $\mathcal{A}$ denote a parameter space with a_{ij} being its axes. Let $\mathcal{Q}^o$ denote the model parameter point in $\mathcal{A}$ with $a_{ij}=a_{ij}^o$. Then the permitted region of perturbation parameter of δ_{pc} is defined to be a largest subspace Ω around the model parameter point $\mathcal{Q}^o$ in $\mathcal{A}$. For all parameter points in Ω the closed-loop perturbed system δ_{pc} remains optimal in sense of that δ_{pc} has the same cost in (4) and the same properties given in Theorems 1 and 2 as δ_{oc}.

It is clear from the Definition that the permitted region Ω gives the allowable perturbation bounds for a_{ij} so that a designer can estimate how large of modeling errors can be accepted.

III. THE CONDITIONS UNDER WHICH δ_{pc} REMAINS OPTIMAL

In this section, we shall give the conditions under which the optimality of the closed-loop perturbed system δ_{pc} is guaranteed. The permitted region Ω can be constructed by using the conditions.

Theorem 3: The closed-loop perturbed system δ_{pc} remains optimal in the sense of that δ_{pc} has the same cost and the same properties as δ_{oc}, if the following condition holds:

1) the symmetric matrix $Q=Q_o-\sum_{i=\alpha}^{\gamma}\sum_{j=\xi}^{\varphi}\varepsilon_{ij}(\bar{P}E_{ij}+E_{ji}\bar{P})$ is positive definite; or

2) the symmetric matrix $Q=Q_o-\sum_{i=\alpha}^{\gamma}\sum_{j=\xi}^{\varphi}\varepsilon_{ij}(\bar{P}E_{ij}+E_{ji}\bar{P})$ is positive semidefinite and $(A_o, Q^{1/2})$ is observable.

Where Q_o is the weighting matrix in (2), $\bar{P}$ is the solution of (5), and E_{ij} is an nxn matrix with the (i,j)th element being 1 and the others being equal to zero.

Proof: Consider the linear-quadratic regulator problem in the presence of system parameter perturbations

$$\delta: \quad \dot{x}=(A_o+A_p)x + Bu, \qquad x(0)=x_o \tag{13}$$

$$J= \int_o^{\infty}(x^TQx + u^TRu)dt \tag{14}$$

where R is the same as in (2) and Q is given in Theorem 3. We assume that the perturbed system (13) remains to be controllable. Then the optimal control law u^o and the optimal cost J^o are given, respectively, by

$$u^o=-Kx, \quad K=R^{-1}B^TP \tag{15}$$

and

$$J^o=x_o^TPx_o, \qquad \forall x_o \neq 0 \tag{16}$$

where P is the unique symmetric positive definite solution of the following algebraic Riccati equation

$$P(A_o+A_p) + (A_o+A_p)^T P + Q - PBR^{-1}B^T P = 0 \tag{17}$$

We note that

$$PA_p + A_p^T P = \sum_{i=\alpha}^{\gamma} \sum_{j=\xi}^{\varphi} \mathcal{E}_{ij}(PE_{ij} + E_{ji}P) \tag{18}$$

and

$$Q + PA_p + A_p^T = Q_o + \sum_{i=\alpha}^{\gamma} \sum_{j=\xi}^{\varphi} \mathcal{E}_{ij}[(P-\bar{P})E_{ij} + E_{ji}(P-\bar{P})] \tag{19}$$

Therefore, substituting (19) into (17), we obtain

$$\left\{PA_o+A_o^T P+Q_o-PBR^{-1}B^T P\right\}+\left\{\sum_{i=\alpha}^{\gamma} \sum_{j=\xi}^{\varphi} \mathcal{E}_{ij}[(P-\bar{P})E_{ij}+E_{ji}(P-\bar{P})]\right\}= 0 \tag{20}$$

Considering (5) and (20), we know immediately that $P=\bar{P}$ is the unique solution of (17). It implies that

$$u^o=u^*, \quad K=K_o, \quad J^o=J^* \tag{21}$$

Thus, we can conclude that δ_{pc} remains optimal if the condition given in Theorem holds. This completes the proof.

Theorem 4: Let $\Omega_\mathcal{E}$ denote the set of the points, which satisfy the condition given in Theorem 3, for $\mathcal{E}_{ij}$ and be around the point $\mathcal{E}_{ij}=0$, where $i=\alpha, \beta, \ldots, \gamma$ and $j=\xi, \eta, \ldots, \varphi$. Then the permitted region Ω, for allowable perturbation parameters a_{ij} such that the closed-loop perturbed system δ_{pc} remains optimal, is given by

$$\Omega = \mathcal{Q}^o + \Omega_\mathcal{E} \tag{22}$$

proof: This follows directly from Theorem 3 and Definition.

IV. AN ALGORITHM FOR DETERMINING Ω

In this section, assuming that the matrices Q_o and $\bar{P}$ are known, we give an algorithm for determining the permitted region Ω. For simplification of writting, in algorithm let $\mathcal{E}_1, \mathcal{E}_2, \ldots, \mathcal{E}_N$ denote $\mathcal{E}_{ij}$ ($i=\alpha, \beta, \ldots, \gamma$; $j=\xi, \eta, \ldots, \varphi$) and let $E_1, E_2, \ldots, E_N$ denote E_{ij} ($i=\alpha, \beta, \ldots, \gamma$; $j=\xi, \eta, \ldots, \varphi$), where N denote the sum total of the perturbation parameters in system matrix.

The algorithm for determining the permitted region Ω is as follows.

Step 1: Calculate the matrices $(\bar{P}E_i+E_i^T\bar{P})$, where $i=1,2, \ldots, N$.
Step 2: Denote $\mathcal{E}_i$ to be

$$\mathcal{E}_i=(-1)^q h_i \mu_i, \quad h_i=1,2, \ldots, d_i, \quad i=1,2, \ldots, N-1 \tag{23}$$

where μ_i is a positive constant and q and d_i are positive and whole numbers. Usually, the values of μ_i and d_i should be taken according to the concrete situation of the given problem.

Step 3: Set $h_1=h_2=\ldots=h_{N-1}=0$ and $q=0$.

Step 4: Determine the maximum and minimum values of $\mathcal{E}_N$, denoted by $\mathcal{E}_{NM}(h_1, h_2, \ldots, h_{N-1})$ and $\mathcal{E}_{Nm}(h_1, h_2, \ldots, h_{N-1})$, such that for any $\mathcal{E}_N \in (\mathcal{E}_{Nm}, \mathcal{E}_{NM})$ the following matrix

$$\left[Q_o - \sum_{i=1}^{N-1} \mathcal{E}_i(\bar{P}E_i + E_i^T\bar{P})\right] - \mathcal{E}_N(\bar{P}E_N + E_N^T\bar{P}) \tag{24}$$

is symmetric positive definite.

Step 5: $h_1 + 1 \rightarrow h_1$

Step 6: If $h_1=d_1+1$, let $h_1=0$ and go to the next step. If $h_1 < d_1+1$, go to Step 4.

Step 7: $h_2 + 1 \rightarrow h_2$

Step 8: If $h_2=d_2+1$, let $h_2=0$ and go to the next step. If $h_2 < d_2+1$, go to Step 4.

....

Step 2N+1: $h_{N-1} + 1 \rightarrow h_{N-1}$

Step 2N+2: If $h_{N-1}=d_{N-1}+1$, let $h_{N-1}=0$ and go to the next step. If $h_{N-1} < d_{N-1}+1$, go to Step 4.

Step 2N+3: $q+1 \rightarrow q$.

Step 2N+4: If $q=2$, go to the next step. If $q < 2$, go to Step 4.

Step 2N+5: Calculate $a_i=a_i^o + \mathcal{E}_i$, where $\mathcal{E}_i=(-1)^q h_i \mu_i$ and $i=1,2,\ldots, N-1$. Calculate

$$a_{Nm}(a_1, a_2, \ldots, a_{N-1}) = a_N^o + \mathcal{E}_{Nm}(h_1, h_2, \ldots, h_{N-1})$$

$$a_{NM}(a_1, a_2, \ldots, a_{N-1}) = a_N^o + \mathcal{E}_{NM}(h_1, h_2, \ldots, h_{N-1})$$

The permitted region of perturbation parameters , denoted by Ω , can be determined from the data obtained above.

Step 2N+6: Stop.

It is easily seen that the algorithm has recursive form which facilitates computation by computer. The key step in this algorithm is to determine whether the symmetric matrix (24) is positive definite. There are some numerically stable approaches for solving this problem. Except this step, the computations in the algorithm are very simple. It seems that the algorithm offers a simple way to determine the permitted region for allowable perturbation parameters such that the closed-loop perturbed system δ_{pc} remains optimal. It should be noted that the algorithm given above is derived from the condition 1) in Theorem 3. The similar algorithm can be derived from the condition 2) in Theorem 3 and is omitted here.

V. THE WAYS FOR EXPANDING THE SIZE OF PERMITTED REGION

It is clear that the larger the size of permitted region obtained for allowable perturbation parameter such that the closed-loop perturbed system remains optimal, the better the robustness of the optimal linear-quadratic regulator system, It is expected that the designing optimal regulator system, except satisfying other prescribed performance indices, should have robustness as good as possible. Therefore, the problem of expanding the size of permitted region will certainly interest the control engineers. In this section, we shall introduce two ways for expanding the size of the permitted region.

Approach 1: Increase the degree of the positive-definite of the weighting matrix Q_o in (2).

For the positive definite matrices Q_1 and Q_o, the degree of positive-definite of Q_1 is called to be greater than Q_o, denoted by $Q_1 > Q_o$, if the matrix $Q_1 - Q_o$ is also a positive definite matrix. It is well known that if $Q_1 > Q_o$, then $P_1 > \bar{P}$, where P_1 is the solution of (5) for the weighting matrix being Q_1. Let

$$\Delta Q = Q_1 - Q_o, \qquad \Delta P = P_1 - \bar{P} \tag{25}$$

Then, if Q_o is replaced by Q_1, the symmetric matrix given in Theorem 3 is correspondingly replaced by the following symmetric matrix

$$\left[Q_o - \sum_{i=\alpha}^{\gamma} \sum_{j=\xi}^{\varphi} \mathcal{E}_{ij}(\bar{P}E_{ij}+E_{ji}\bar{P})\right] + \left[\Delta Q - \sum_{i=\alpha}^{\gamma} \sum_{j=\xi}^{\varphi} \mathcal{E}_{ij}(\Delta P E_{ij}+E_{ji}\Delta P)\right] \tag{26}$$

It is shown by many examples that the second part of the matrix given in (26) is often conducive to improve the degree of positive-definite of the matrix to meet the condition in Theorem 3. Thus, we can conclude that in many cases, an increase in degree of posituve definite of the weighting matrix Q_o is useful for expanding the size of the permitted region. Now we give an exmple for illustration.

Exampe 1: Consider a linear-quadratic regulator problem described by

$$\dot{x} = A_o x + Bu, \tag{27}$$

$$J = \int_o^{\infty} (x^T Q_o x + u^T R u)dt \tag{28}$$

where

$$A_o = \begin{bmatrix} 1 & 0 & 3 & 2 \\ 4 & 1 & 1 & 3 \\ 2 & 3 & 0 & 4 \\ 0 & 1 & 4 & 1 \end{bmatrix}, \quad B = \begin{bmatrix} 1 & 0 \\ 0 & 1 \\ 1 & 0 \\ 0 & 1 \end{bmatrix}, \quad R = \begin{bmatrix} 5 & -2 \\ -2 & 1 \end{bmatrix}$$

It is obvious that the pair (A_o, B) is controllable. Now we assume that the elements a_{13} and a_{42}, whose model values are $a_{13}^o = 3$ and $a_{42}^o = 1$, in system matrix are perturbation parameters. The weighting matrix Q_o in (28) is chosen to be 9 schemes as follows:

#1 $$Q_{o1} = \begin{bmatrix} 4 & 1 & 0 & 1 \\ 1 & 2 & 1 & 0 \\ 0 & 1 & 3 & 1 \\ 1 & 0 & 1 & 4 \end{bmatrix}$$

#2 $$Q_{o2} = \begin{bmatrix} 6 & 1 & 1 & 1 \\ 1 & 3 & 1 & 1 \\ 1 & 1 & 4 & 1 \\ 1 & 1 & 1 & 7 \end{bmatrix}$$

#3 $Q_{o3}=\begin{bmatrix}10&1&3&1\\1&5&1&3\\3&1&6&1\\1&3&1&13\end{bmatrix}$ #4 $Q_{o4}=\begin{bmatrix}12&1&4&1\\1&6&1&4\\4&1&7&1\\1&4&1&16\end{bmatrix}$

#5 $Q_{o5}=\begin{bmatrix}14&1&5&1\\1&7&1&5\\5&1&8&1\\1&5&1&19\end{bmatrix}$ #6 $Q_{o6}=\begin{bmatrix}16&1&6&1\\1&8&1&6\\6&1&9&1\\1&6&1&22\end{bmatrix}$

#7 $Q_{o7}=\begin{bmatrix}18&1&7&1\\1&9&1&7\\7&1&10&1\\1&7&1&25\end{bmatrix}$ #8 $Q_{o8}=\begin{bmatrix}20&1&8&1\\1&10&1&8\\8&1&11&1\\1&8&1&28\end{bmatrix}$

#9 $Q_{o9}=\begin{bmatrix}22&1&9&1\\1&11&1&9\\9&1&12&1\\1&9&1&31\end{bmatrix}$

where Q_{oi} are positive definite and $Q_{o,i+1} > Q_{oi}$ hold for $i=1,2, \dots ,8$.

By using the algorithm given in Section IV, the permitted region can be easily obtained for the perturbation parameters a_{13} and a_{42} such that the resulting closed-loop perturbed system remains optimal. these regions are shown in Fig. 1 for 9 schemes of Q_o listed above. It can be seen from Fig. 1 that generally speaking, the greater the degree of positive-definite of weighting matrix Q_o in performance index (28), the larger the size of the permitted region of perturbation parameters a_{13} and a_{42}. Therefore, the effect of increasing the degree of positive-definite of Q_o on the expansion of size of the permitted region is obvious. However, it should be noted that on the other hand, this leads to an increase of the optimal cost.

Approach 2: Use the optimal control law for linear-quadratic regulator problem with prescribed degree of stability instead of the control law given in (3).

The linear-quadratic regulator problem with prescribed degree of stability is described by

$$\delta_o: \quad \dot{x}=A_o x + Bu, \qquad x(0)=x_o \tag{29}$$

$$J=\int_0^{\infty} e^{2\bar{\alpha}t}(x^T Q_o x + u^T Ru)dt \tag{30}$$

where the real number $\bar{\alpha} \gg 0$ and is used to prescribe the degree of stability of the designing optimal regulator system. It is well known that the optimization of the performance index (30) with the system model (29) yields the optimal control law

$$\bar{u}^* = -K^* x, \qquad K^* = R^{-1}B^T P \tag{31}$$

where $P > 0$ is the solution of the following algebraic Riccati equation:

$$P(A_o + \bar{\alpha}I) + (A_o + \bar{\alpha}I)^T P + Q_o - PBR^{-1}B^T P = 0 \tag{32}$$

If we apply the control law (31) to perturbed system δ_p, then the matrix in the condition of Theorem 3 should read as follows

$$Q = Q_o + 2\bar{\alpha} P - \sum_{i=\alpha}^{\gamma} \sum_{j=\xi}^{\varphi} \mathcal{E}_{ij}(PE_{ij}+E_{ji}P) \tag{33}$$

If the condition on Q given in Theorem 3 is satisfied, the optimality is guaranteed for the cloesd-loop perturbed system. However, the degree of stability prescribed by $\bar{\alpha}$ can not be guaranteed in this case. It is also shown by many exmples that in many cases, the size of the permitted region can be expanded by using this approach. Now we give the another exmple for illustrating this approach.

Example 2: Consider a linear-quadratic regulator problem described by

$$\dot{x} = A_o x + Bu \tag{34}$$

$$J = \int_o^\infty e^{2\bar{\alpha}t}(x^T Q_o x + u^T Ru)dt \tag{35}$$

where

$$A_o=\begin{bmatrix}1 & 0 & 3 & 2\\ 4 & 1 & 1 & 3\\ 2 & 3 & 0 & 4\\ 0 & 1 & 4 & 1\end{bmatrix}, \quad B=\begin{bmatrix}1 & 0\\ 0 & 1\\ 1 & 0\\ 0 & 1\end{bmatrix}, \quad Q_o=\begin{bmatrix}4 & 1 & 0 & 1\\ 1 & 2 & 1 & 0\\ 0 & 1 & 3 & 1\\ 1 & 0 & 1 & 4\end{bmatrix}, \quad R=\begin{bmatrix}5 & -2\\ -2 & 1\end{bmatrix}$$

It is clear that the pair (A_o, B) is controllable and Q_o and R are positive definite. Now we assume that the elements a_{13} and a_{42} , whose model values are $a_{13}^o=3$ and $a_{42}^o=1$, are the perturbation parameters in system matrix. The degree of stability is chosen to be $\bar{\alpha}=1,2, \ldots,9$. The permitted regiogs can be similarly determined and are shown in Fig. 2 for the cases of $\bar{\alpha}=1,2, \ldots ,9$. In order to make a comparison between the sizes of the permitted regions obtained in the cases $\bar{\alpha}=0$ and $\bar{\alpha}\neq o$, the permitted region associated with the case of $\bar{\alpha}=o$ is also given in Fig. 2. It is obvious that the size of the permitted region is expanded rapidly, as the value of $\bar{\alpha}$ increases. However, the optimal cost will rapidly increase simultaneously. Although this leads to an increase of the optimal cost, this approach is still a way to expand the size of the permitted region for perturbation parameters.

VI. CONCLUSIONS

By making use of the optimization property of the linear-quadratic regulator system, a condition has been obtained for the optimality of an optimal closed-loop system in the presence of parameter perturbations. An algorithm based on this condition has been also given to determine the permitted region for allowable perturbation parameters, for each parameter perturbation in which the closed-loop perturbed system remains optimal. It is natural that the size of the permitted region can be taken as a measure of robustness of the optimal linear-quadratic regulator systems.

From the point of view of engineering, satisfying the prescribed performance index, an optimal linear-quadratic regulator system should be designed to make its permitted region of perturbation parameters to be as large as possible. Two approaches have been introduced for expanding the size of the permitted region. Their effectiveness has been shown by exmples.

An important implication of the results given in this paper is that it gives the possibility to help a designer to select an appropriate weighting matrix Q in the quadratic performance index or an appropriate degree $\bar{\alpha}$ of stability to attain a robust design. This fact increases considerably the confidence in applicability of optimal linear-quadratic theory to the practical systems in which the parameter perturbations, due to both the modeling errors and parameter variations, are not avoidable.

REFERENCES

[1] R. V. Patel, M. Toda, and B. Sridhar, "Robustness of linear quadratic state feedback designs in the presence of system uncertainty," IEEE Trans. Automatic Control, vol. AC-22, pp. 945-949, 1977.

[2] D. D. Šiljak, Large-Scale Dynamic Systems: Stability and Structure, New York: North-Holland, 1978.

[3] D. D. Šiljak, "Overlapping decentralized control," in Large Scale Systems Engineering Applications, M. Singh and A. Titli, Eds. Amsterdam: North-Holland, pp. 145-166,1979.

[4] R. Krtolica and D. D. Šiljak, "Suboptimality of decentralized control and estimation," IEEE Trans. Automatic Control, vol. AC-25, pp. 76-83, 1980.

[5] M. E. Sezer and D. D. Šiljak, "Robustness of suboptimal control: gain and phase margins," IEEE Trans. Automatic Control, vol. AC-26, pp. 907-911, 1981.

[6] B. D. O. Anderson and J. B. Moore, Linear Optimal Control, Englewood Cliffs, NJ: Prentice-Hall, 1971.

[7] M. G. Safanov and M. Athans, "Gain and phase margins for multiloop LQG regulators," IEEE Trans. Automatic Control, vol. AC-22, pp. 173-179, 1977.

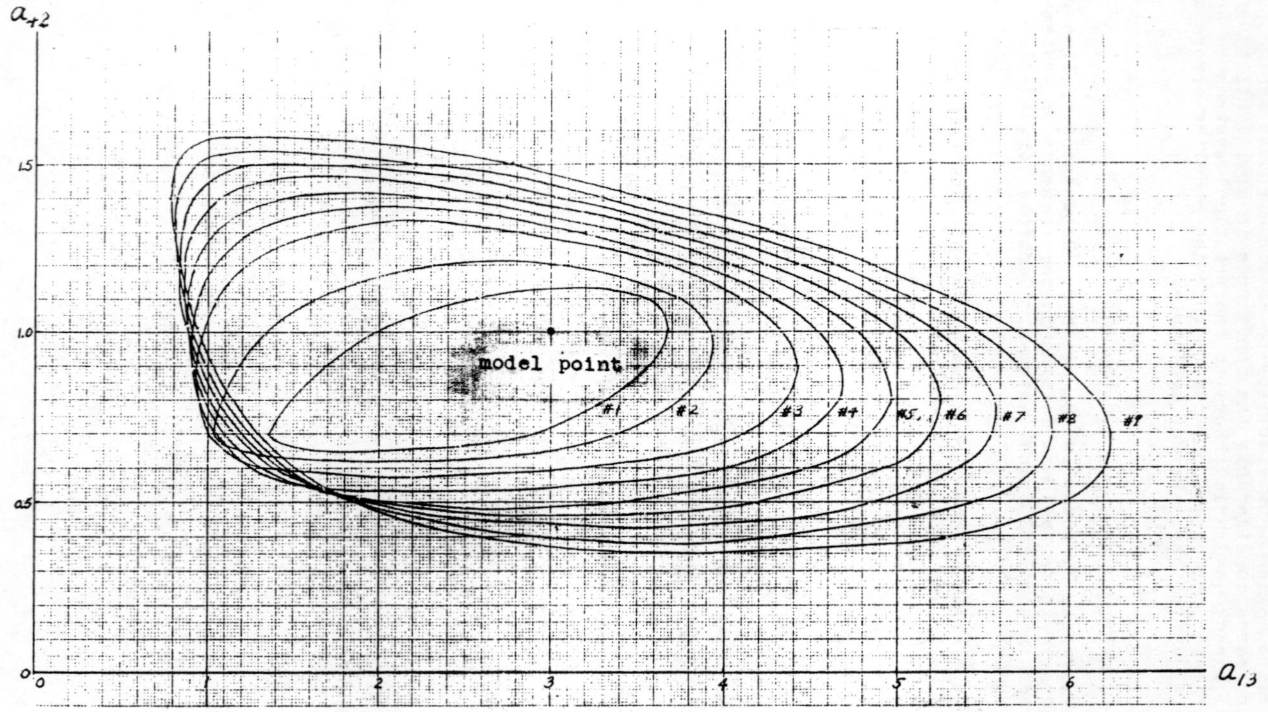

Fig. 1 Permitted regions of the perturbation parameters a_{13} and a_{42} for matrices Q_{01} with different degree of positive-definite

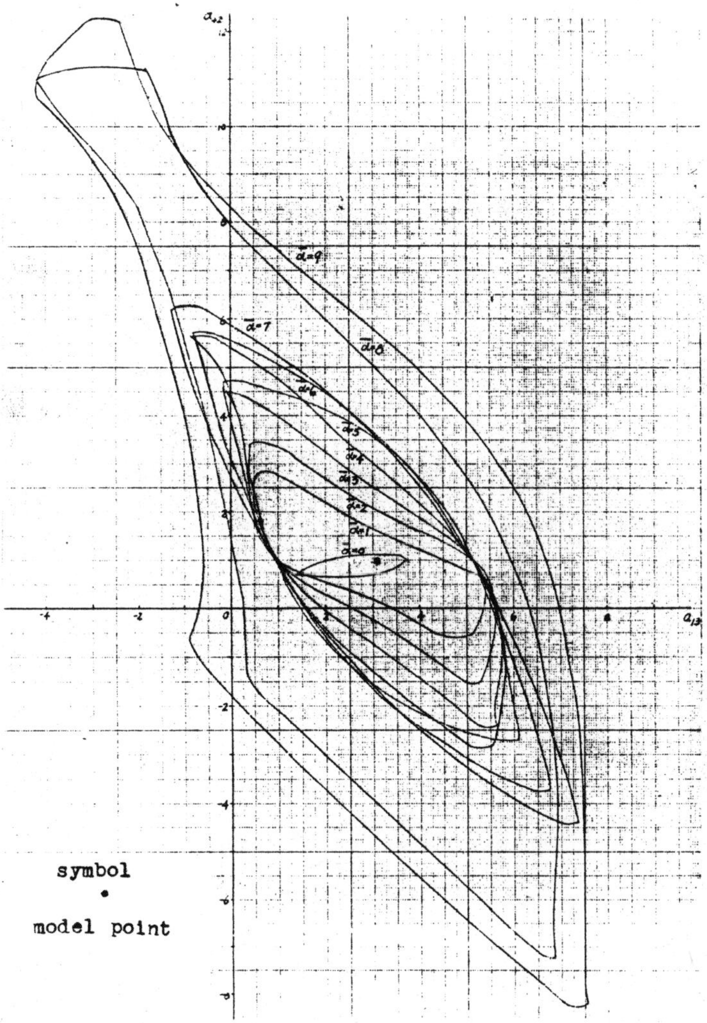

Fig. 2 Permitted regions of the perturbation parameters a_{13} and a_{42} for different values of $\bar{\alpha}$

List of Participants

	Name	Place and/or Address
1.	Albrecht, J.	Institut für Mathematik der TU Clausthal Erzstraße 1, D3392 Clausthal-Zellerfeld, FRG
2.	An Lianjun	Dept. of Appl. Math., Tsinghua Univ., Beijing, China
3.	Cao Cewen	Dept. of Math., Zhengzhou Univ., Zhengzhou, Henan, China
4.	Chavent, G.	Université de Paris IX-Dauphine-75775 Paris Cédex, France and INRIA B.P. 105 78150 Le Chesnay Cédex, France
5.	Chen Yazhe	Dept. of Math., Peking Univ., Beijing, China
6.	Chen Yunmei	Dept. of Appl. Math., Tongji Univ., Shanghai, China
7.	Collatz, L.	Institute of Math., University of Hamburg, 2 Hamburg 67, Eulenkrugstr. 84, FRG
8.	Dai Jiazun	Dept. of Math., Nanjing Institute of Aeronautics and Astronautics, Nanjing, Jiangsu, China
9.	Ding Zhengzhong	Dept. of Math., Hangzhou Institute of Commerce, Hangzhou, Zhejiang, China

10.	Dong Guangchang	Dept. of Math., Zhejiang Univ., Hangzhou, Zhejiang, China
11.	Dong Jinzhu (Tung Chin-Chu)	Graduate School, Academia Sinica, Beijing, China
12.	Feng Kang	Computing Center, Academia Sinica, Beijing, China
13.	Feng Xiaobing	Dept. of Math., Xi'an Jiaotong Univ., Xi'an, Shaanxi, China
14.	Friedman, A.	Dept. of Math., Northwestern University, Evanston, Illinois 60201, USA
15.	Guan Keying (Guan Ke-Ying)	Dept. of Math., Beijing Institute of Aeronautics and Astronautics, Beijing, China
16.	Han Houde	Dept. of Math., Peking Univ., Beijing, China
17.	Hoffmann, K. H.	Universitat Augsburg, Memminger Str. 6, 8900 Augsburg, FRG
18.	Huang Guangyuan	Dept. of Math., Shandong Univ., Jinan, Shandong, China
19.	Jiang Lishang	Dept. of Math., Peking Univ., Beijing, China
20.	Liang Min	Dept. of Math., Fuzhou Univ., Fuzhou, Fujian, China
21.	Li Daqian (Li Ta-tsien)	Dept. of Math., Fudan Univ., Shanghai, China
22.	Li Kaitai	Dept. of Math., Xi'an Jiaotong Univ., Xi'an, Shaanxi, China
23.	Li Likang	Dept. of Math., Fudan Univ., Shanghai, China
24.	Li Yishen (Li Yi-shen)	Dept. of Math., Univ. of Science and Technology of China, Hefei, Anhui, China
25.	Lin Longwei (Long-Wei Lin)	Dept. of Math., Hua Chiao Univ., Quanzhou, Fujian, China
26.	Mawhin, J.	Université de Louvain, Institut Mathématique B-1348 Louvain-La-Neuve, Belgium

27.	Mingarelli, A.	Dept. of Math., University of Ottawa, Ottawa, Ontario, Canada, K1N 6N5
28.	Mu Mu	Dept. of Math., Fudan Univ., Shanghai, China
29.	Pu Fuquan	Dept. of Appl. Math., Tsinghua Univ., Beijing, China
30.	Raphael, L.	Dept. of Math., Howard University, Washington D.C. 20059, USA
31.	Rautmann, R.	Dept. of Math., University of Paderborn, Warburger Str. 100, D-4790, Paderborn, FRG
32.	Schaeffer, D.G.	Dept. of Math., Duke Univ., Durham, NC27706 USA
33.	Shen Weixi	Dept. of Math., Fudan Univ., Shanghai, China
34.	Su Ning	Dept. of Appl. Math., Tsinghua Univ., Beijing, China
35.	Sun Nezheng (Ne-Zheng Sun)	Dept. of Math., Shandong Univ., Jinan, Shandong, China
36.	Tan Yongji	Dept. of Math., Fudan Univ., Shanghai, China
37.	Wang Jinghua	Inst. of Systems Science, Academia Sinica, Beijing, China
38.	Wang Yuanming	Dept. of Math., Nanjing Univ. of Science and Tech., Nanjing, Jiangsu, China
39.	Wang Zaigong	Dept. of Appl. Math., Northern Jiao-Tong Univ., Beijing, China
40.	Wang Zhaoling (Wang Zhao-lin)	Dept. of Engineering Mechanics, Tsinghua Univ., Beijing, China
41.	Wu Shengchang	Inst. of Appl. Math., Academia Sinica, Beijing, China
42.	Xiao Shutie	Dept. of Appl. Math., Tsinghua Univ., Beijing, China

43.	Xue Weimin (Wei-Min Xue)	Dept. of Engineering Mechanics, Tsinghua Univ., Beijing, China
44.	Ye Qixiao	Dept. of Math., Peking Univ., Beijing, China
45.	Ying Longan (Ying Lung-an)	Dept. of Math., Peking Univ., Beijing, China
46.	Zhang Kewei	Dept. of Math., Inner Mongolia Univ., Huhehot, Inner Mongolia Autonomous Region, China
47.	Zhang Tong (Chang Tung)	Inst. of Math., Academia Sinica, Beijing, China
48.	Zhang Xianglian (Zhang Xiang-Ling)	Dept. of Astronomy, Beijing Normal University, Beijing, China
49.	Zheng Dazhong (Da-Zhong Zheng)	Dept. of Automation, Tsinghua Univ., Beijing, China
50.	Zhou Baoxi	Dept. of Math., Shanghai Normal Univ., Shanghai, China
51.	Zhou Yulin (Zhou Yu-Lin)	Inst. of Appl. Phy. and Computational Math., Post-Office Box 8009, Beijing, China